V. Damić · J. Montgomery

Mechatronics by Bond Graphs

Springer

Berlin
Heidelberg
New York
Hong Kong
London
Milan
Paris
Tokyo

Engineering ONLINE LIBRARY

http://www.springer.de/engine/

Vjekoslav Damić · John Montgomery

Mechatronics by Bond Graphs

An Object-Oriented Approach
to Modelling and Simulation

With 406 figures and 39 tables

 Springer

Professor Dr. Vjekoslav Damić
The Polytechnic of Dubrovnik
Collegium Ragusinum
Ćira Carića 4
20000 Dubrovnik
Croatia
e-mail: vdamic@vdu.hr

Dr. John Montgomery
formerly The Nottingham Trent University, U. K.

now August-Bebel-Str. 21
50259 Pulheim-Brauweiler
Germany
e-mail: HaedickeMontgome@compuserve.de

ISBN 3-540-42375-3 Springer-Verlag Berlin Heidelberg New York

Library of Congress Cataloging-in-Publication-Data

Damic, Vjekoslav, 1941-
Mechatronics by bond graphs : an object-oriented approach to modelling and simulation /
Vjekoslav Damic, John Montgomery. p. cm. -- (Engineering online library)
Includes bibliographical references and index.
ISBN 3540423753 (alk. paper)
1. Mechatronics. 2. Object-oriented programming (Computer science)
I. Montgomery, John, 1936- II. Title. III. Series.
TJ163.12.D36 2003
621.3--dc21

Springer-Verlag Berlin Heidelberg New York
a member of BertelsmannSpringer Science+Business Media GmbH
http://www.springer.de

© Springer-Verlag Berlin Heidelberg 2003
Printed in Germany

Typesetting: print-data delivered by authors
Cover design: medio Technologies AG, Berlin
Printed on acid free paper   62/3020/M - 5 4 3 2 1 0

"I will have my bond."
The Merchant of Venice (Wm. Shakespeare)

To Mira, Renata, and Dražen   **VD**

To Helga, Lorna, and Stuart   **JM**

Preface

A short history of this book

This book had its origins in the authors' common interest in modelling and simulating dynamic engineering systems, especially those related to *mechatronics*. These interests date from the early 1970s.

We well remember—even somewhat nostalgically—our experiences with one of the first digital computer simulation tools that became available: IBM's Continuous System Simulation Program (CMSP) for IBM 1130 computers. C. W. Gear's famous DIFSUB code for solving stiff differential equations—the forerunner of modern differential-algebraic equations solvers—also appeared around the same time. Then, in 1975, Karnopp and Rosenberg's classic book, *System Dynamics: A Unified Approach*, was published. It introduced a system analysis methodology based on bond graphs. We loved it from the start for it laid a solid foundation for development of a systematic approach to modelling complex mechatronics systems. We were also aware of developments in field of electronic circuit modelling that led to the famous Berkley's SPICE program.

The difficulties posed by solving real-world design problems motivated the first author to begin development of a methodology for computer-aided modelling and simulation of engineering—particularly mechatronics—systems. It was targeted to developing a methodology that supports systematic model development by decomposition. Bond graphs where taken as modelling formalism, because they are well suited to modelling different physical processes taking place in a typical mechatronic system. It was expanded, however, by developing the concept of bond graph word models into complete component models. More attention was given to component ports as interfaces of the components. The ports are treated as objects in themselves that enable representation of the complex interconnections inside the components. This way, a model of a system can be built as a complex multilevel structure, in a form that mimics how a real system is built. The component can be reused as well to build the models.

Another departure from classical bond graphs and the Continuous System Simulation Language (CSSL) philosophy was made by putting aside the causality issues. Strict input-output relationships in the models are not supported. Thus, instead of mathematical models in state-space equation form, differential-algebraic equation models are used. This enables separation of modelling and model solving tasks. We believe that, taken together, this extends the applicability of methods to solving real engineering problems.

The first implementation of this methodology was made in the beginning of the 1980s with the release of *Simulex*. This program was implemented using

FORTRAN and run on Digital VAX-750 computers. *Simulex* models were described with SPICE-like scripts. The resulting equations were solved with a version of Gear's DIFSUB. *Simulex* was applied successfully to a range of practical problems in servo-systems and robotics.

The revolutionary appearance of PCs in the mid-1980s, followed by development of operating systems that supported user-friendly visual interfaces in the 1990s, spurred the next phase of development. This was also influenced by the paradigm shift in programming languages: The truly object-orient languages were replacing the procedural languages—such as FORTRAN and C—that we had all been using. Another important technological development became available around the same time—symbolic computational algebra.

In the beginning of the 1990s, the shift to object-oriented modelling paradigm was made. Class hierarchies were developed that enabled representing component models as objects. Also, computational algebra methods were developed that, as explained in the present book, simplified some important user-interface problems and the solution of model equations. Methods for solving differential-algebraic equations were further developed to support model solving during simulation. These all were implemented in a visual modelling and simulation program, *Bond-Sim*, the first version of which appeared in the mid-1990s. It fully automated many important operations. Thus, there was no need for the developer to use any traditional programming; rather, models were developed and solved simply by mouse clicks.

In 1995, the authors met at The Nottingham Trent University and started collaborative work on *Dynamic System Simulations Using Bond Graphs*, a project funded partially through an ALIS (Academic Links and Interchange Scheme) award (1995-1998), sponsored jointly by the Croatian Ministry of Science and Technology and the British Council. This co-operation continued *via* e-mail and reciprocal visits to Nottingham (England) and Dubrovnik (Croatia). One result of this fruitful joint work is this book that we here offer to the reader.

What is this book about?

The title suggests that this book is about mechatronics; this is, indeed, one of its central themes. It is not, however, another book on what mechatronics *is*; rather, it is about how mechatronic problems can be solved by a systematic approach employing bond graphs. *Why bond graphs?* Because they offer an efficient means of modelling interdisciplinary problems, such as those commonly found in mechatronics. (The book, by the way, assumes no previous experience with bond graphs, though it certainly would be useful.)

The book shows, in step-by-step fashion, how models are developed systematically, then simulated in a way that permits thorough analysis of the problem under study. Every chapter that deals with an engineering application starts with the exposition and solution of a simple problem relevant to that chapter. Then, the solution of related—though much more difficult—problems is explained.

The book is divided into two parts: *Fundamentals* and *Applications*.

Part 1, Fundamentals, consists of five chapters on Bond Graph modelling. It starts with an introduction to the subject, and then proceeds with describing a systematic object-oriented approach to modelling; implementation of object-oriented modelling in a visual environment; and the numerical and symbolic solution of the underlying model equations.

Part 2, Applications, consists of five chapters that apply bond graphs and component model techniques to Mechanical systems, Electrical systems, Control systems, Multibody Dynamics, and Continuous Systems. Great attention is given to modelling electrical components and systems, including semiconductors. The same holds for multibody systems, both rigid and deformable, such as found in various mechanisms and robots.

What readers can gain from the book?

There are several ways in which this book can be used, depending mainly upon the background and interests of the reader.

Researchers in mechatronics and micro-mechanics design, for example, can use it to find out how difficult problems in their disciplines can be solved using a combination of bond graphs and component model techniques.

For the reader interested in simulation technology, the book provides an introductory description of the object-oriented visual approach to modelling and simulation.

The reader whose background is in one of the applied disciplines covered herein can gain valuable insight into how bond graphs may be used to solve problems particular to his area of interest.

We also think that the book can be useful as a textbook, or as a supplementary text, in courses on physical modelling of engineering systems in general. We believe that it can help students learn the system way of solving a problem in electrical and mechanical engineering, as well as coupled problems that span disciplines.

Finally, it is our sincere wish that the text and software will aid the reader in his work. We invite, and will appreciate, all constructive feedback.

BondSim Research Pack

A special version of BondSim—*BondSim Research Pack* (beta version)—is bundled with this book. It provides a visual development environment for the modelling and simulation of engineering and mechatronics systems based on bond graphs. The problems presented in the book are solved using the *BondSim Research Pack*.

It runs on the Windows 2000 Professional operating system, but can be used on other Windows platforms, too. The reader can use this version of *BondSim* to analyse all of the problems presented in the book. (These are found in *BondSim*'s program library.) The projects that a reader might develop on his or her own are somewhat more restricted. The interested reader can order the complete version of *BondSim* from the first author. (*See* Appendix for details.)

Acknowledgements

A number of people have reviewed the initial outline (and the drafts) of this book. We are most grateful to them for their time and expertise.

Our special thanks go to the following people and institutions:

The Polytechnic of Dubrovnik (*Veleučilište u Dubrovniku – Collegium Ragusinum*) for facilities provided to both authors. We are especially grateful to the Rector, Professor Dr. Mateo Milković, for his encouragement and support.

Vlado Jaram, Mr. Sc., for initiating the whole publishing project and his help on getting the book published, as well as on his suggestions during writing the book

Professor Barry Hull of the Department of Mechanical Engineering of the Nottingham Trent University for his support.

The Croatian Ministry of Science and Technology and the British Council for the funds provided through the ALIS award.

Dr. Nick Staresinic, EcoMar Mariculture, for his careful reading of the manuscript and his helpful editing suggestions.

We would also like to thank Dr. Dieter Merkel of Springer-Verlag, Heidelberg, for his help, kindness, and patience during the preparation of the manuscript. We are also grateful to Ms. Petra Jantzen and Ms. Gaby Mass for their help.

And last, but in no way least, we wish to express our deep appreciation and love to our wives—Mira and Helga—for their love, support, patience, and sacrifice during the long period over which this book was produced.

Dubrovnik, June, 2002   Vjekoslav Damić
                                                         John Montgomery

Contents

PART 1

FUNDAMENTALS

Chapter 1 Basic Forms of Model Representation

1.1 Objectives

The solution of complex, real-world problems is based on *modelling*. A model simplifies the system of interest by abstracting some subset of its observable attributes. This focuses attention on those features of the system relevant to the problem of interest, and excludes others deemed not to be of direct relevance to the problem. The level of detail included in a model thus depends on the problem to be solved—as well as on the problem solver. Based on such an idealised picture, the system is described in a suitable form that is used as a basis for deriving a solution (Fig. 1.1).

After obtaining a solution, results are interpreted with respect to the real-world context of the original system. Thus, as well as being able to create a valid model of the system, and to solve it, it is of great importance to represent the solution in a form that can be understood readily and communicated.

The traditional modelling approach used in engineering is mathematical. That is, real-world physical processes are described by mathematical relationships that are solved using suitable analytical or numerical techniques. As real engineering systems are very complex, it is not an easy task to create a valid model and solve it. An added practical consideration is that problems must be solvable efficiently in terms of resources and time. Advances in computer technology have dramatically improved solvability; it is now possible to solve problems that formerly were intractable.

Solving problems with a computer means that a problem posed in one physical domain is solved in another physical domain, the computer domain. This naturally leads to the topic of *simulation modelling*.

Simulation models mimic the behaviour of engineering processes in their environment. By experimenting on models of equipment instead of on real equipment, the system's behaviour can be studied even before the hardware is built. Simulation models can be used at various stages of design, from the early stages of conceptual design to final prototype testing. There are many fields in which this technique has been applied profitably.

The fundamental question that naturally arises is: *How are such models best constructed?* A well-known adage suggests that modelling is more art than science. It is, in fact, a bit of both. On the "scientific" side, there are a number of approaches, methods, and tools that can be mastered, and then applied, to develop effective models; the "art", perhaps, is the insight that a modeller accumulates through practice and familiarity with the system being studied.

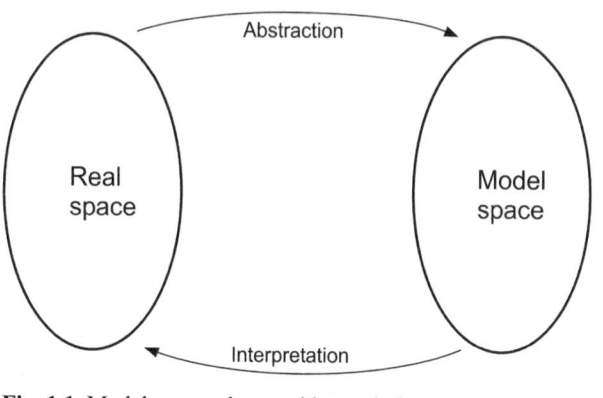

Fig. 1.1. Model approach to problem solution

This chapter reviews some of the more promising approaches and methods at the foundation of modelling. The perspective adopted is motivated mainly by problems in mechatronics. There are many definitions of mechatronics. [1] One that we choose to highlight states that mechatronics is a *synergistic combination of precision mechanics, electronic control, and system thinking.*

Perhaps the best-known examples of mechatronic design are found in robotics. There are also other, no-less-important applications of this design philosophy.

In today's highly competitive and demanding development environment, classical solutions without embedded microprocessors have little chance of success. Modelling and simulation play an even greater role in product design. To promote efficient solutions, the computer modelling and simulation environments should relieve designers of many routine, low-level tasks [1], as well as support collaborative work.

1.2 The General Modelling Approach

The concept of system plays a central role in model building. An efficient model need not embrace the entire universe to design just a part of it. This is not only an impossible task, but also an unnecessary one. We thus pay attention only to that part of the problem in which we are interested. This is termed the *system* for the given problem. Everything not included in the system constitutes its environment (Fig. 1.2).

The system might consist of the engineering equipment that is the subject of the problem, but it can include other parts, as well. In this system-centred approach it is tacitly assumed that the environment determines the behaviour of the system. Thus, the environment influences the system and can change its behaviour.

---

[1] In   http://www.engr.colostste.edu/~dga/mechatronics/definitions.html   definitions   of Mechatronics are collected from different resources.

It is, of course, also of interest to study the influence of the system on its environment, *e.g.* the current drawn by a system from an external source. In the case in which the system can change its environment in a way that there is a 'backwards' influence on its own behaviour, the system should, in most cases, be enlarged to include this part of its environment.

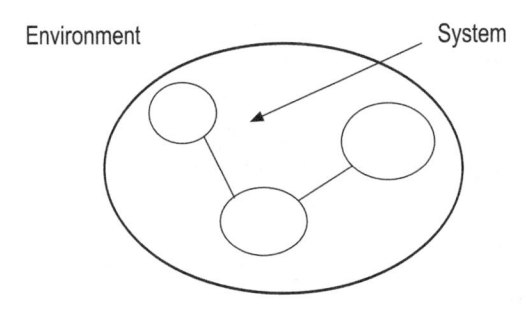

Fig. 1.2. System and its environment

It is often useful to decompose a system into components. For example, the simple actuator illustrated in Fig. 1.3 consists of an electric motor driven by a controller. The motor shaft is connected to a nut. The shaft rotation is transformed by the nut into a translation of an actuator shaft, which moves a load. The position information of the load is fed back *via* a sensor to the controller.

Every part of such a system, i.e. the electronic drive unit, motor, shaft, etc., may be modelled as a separate component. The complete model of the drive thus may be depicted as a system of interconnected components.

Decomposition of the complete system into its components generally simplifies the modelling task and gives a sharper insight into the system's structure. Such a representation is a great help in interpreting model behaviour in terms of the real engineering system.

System decomposition can proceed to ever-lower levels—essentially treating each component as another system that, in turn, consists of even simpler components. At some point a level of detail will be reached at which the components may be considered as elementary, i.e. not admitting any further useful decomposition. Such elementary components are modelled as *entities* and define the limit of detail of the model in question.

It should perhaps be pointed out again that a model is an abstraction of the real world: It is not necessary—or even possible—that the structure of the model represent the original physical system in all of its complexity. Model development, however, generally is an iterative process: Additional details may be added as the model matures, or expands to address additional problems.

Decomposing an engineering system into components also suggests a natural decomposition of tasks among members of a modelling team. Each group might be assigned development of one component. The overall model can then be built up by combining the separate sub-models.

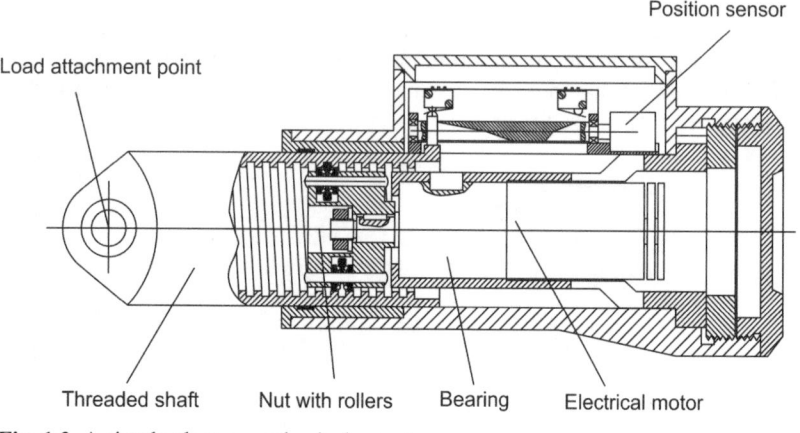

Fig. 1.3. A simple electro-mechanical actuator

Yet another advantage of this approach is that robust component models can be reused, e.g. components developed for one particular application might serve as building blocks for other, unrelated applications.

1.3 Physical Modelling, Analogies, and Bond Graphs

Both "top-down" decomposition and "bottom-up" composition are powerful modelling techniques. To use their full power requires describing those components treated as entities (elementary components), and modelling the interactions between them. In engineering, these considerations are based on physical reasoning derived from the various branches of physics. Such an approach is sometimes termed physical modelling [2].

Processes taking place in engineering systems thus may be classified generally as belonging to, for example, rigid and solid body mechanics, fluid mechanics, electricity and magnetism, semiconductor physics, thermodynamics, and so forth. Each of these branches has its particular methodology for solving problems. Thus, if the problem in question deals with a single physical domain, it is natural to apply the methodology of the field in question, including any specialised computational methods that may be available. This same approach can be applied even in multi-domain problems if the interactions between domains are weak; but this is rarely the case in engineering in general, and in the field of mechatronics in particular. We thus must cope with interacting, multi-domain physical processes.

One well-known approach designed to deal with multi-domain engineering problems is the bond graph method elucidated by Henry Paynter.[2] He presented

---

[2] Interested readers can review www page http://www.hankpaynter.com for more information.

this methodology for the first time in the lecture "Ports, Energy, and Thermodynamic Systems" delivered on April 24, 1959, at the Massachusetts Institute of Technology. This work later was published [3].

The application of Paynter's bond graph method began with the works of Karnopp, Rosenberg, Thoma, and others [4-13]. Over the last 40 years there have been numerous publications dealing with the theory and application of bond graphs in different branches of engineering.[3]

The method uses the effort-flow analogy to describe physical processes [7, 10]. These processes are represented graphically in the form of elementary components (bond graph elements) with one or more *ports* (Fig. 1.4). The ports represent places where interactions with other processes take place.

The process "seen" at a port is described by a pair of variables, *effort* and *flow*. These are termed *power variables*, and their product is *power*.

Through every port there is flow of power, either in or out of the component. The direction of power flow is depicted by a *half-arrow*.

In addition to the power variables, there also are internal variables that represent the accumulations of effort and flow over time. These variables are called *generalized momenta* and *generalized displacements*, respectively.

The typical association of variables in various domains is given in Table 1.1. It should be noted that thermal effort and flow variables, as defined in the last row of the table, are not power variables because their product is not power. Bond graphs corresponding to variables having this property are usually termed pseudo-bond graphs [10, 13].

Effort-flow pair

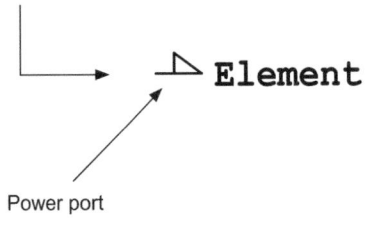

Power port

Fig. 1.4. Generic bond graph element

All physical processes are described using several elementary components, or *elements*:

- Sources of effort and of flow (denoted as SE and SF respectively),
- Accumulation of effort and of flow (I and C respectively),
- Dissipation of power (R),
- Transformers of power (Transformers and Gyrators) (TF and GY), and
- Branches of efforts and flows (denoted as e and f respectively)

---

[3] More information on the Bond graph method can be found in The Bond Graph Compendium held at http://www.ece.arizona.edu/~cellier/bg.html.

The processes that these components represent are described by constitutive relations expressed in terms of port and internal variables (generalized variables).

Table 1.1. Bond graph variables

Domain	Effort	Flow	Momentum	Displacement
Mechanical translation	Force	Velocity	Momentum	Displacement
Mechanical rotation	Torque	Angular velocity	Angular momentum	Angle
Electrical	Voltage	Current	Flux linkage	Charge
Hydraulic	Pressure	Volume flow rate	Pressure momentum	Volume
Thermal	Temperature	Heat flow	–	Heat energy

The ports of components are joined with a line. These lines are termed bond lines, or bonds, for short. They imply that the power variables at connected ports are equal. The graph that results is a bond graph model of the component.

A simple mechanical system consisting of a body, a spring, and a damper provides a useful illustration (Fig. 1.5a). The corresponding bond graph model is shown in Fig. 1.5b.

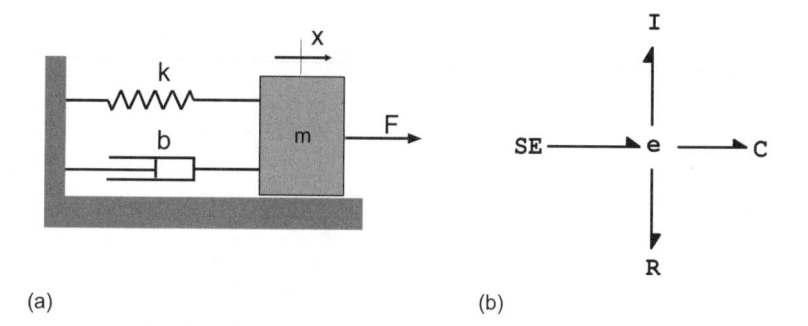

(a)   (b)

Fig.1.5. A simple mechanical system. (a) Scheme, (b) Bond graph model

The source effort SE represents an applied external force, component I represents the inertia of the body, component C describes the elasticity of the spring, and R represents friction in the damper. Branching element e denotes the summation of all forces acting on the body. This diagram, taken together with the corresponding component constitutive relations, completely defines the mathematical model of the system; it can thus be used as a basis for simulating the system. The bond graph also shows the structure of the model in a way that resembles the structure of the real system. This proves useful in efficiently communicating details of the model to interested parties outside of the modelling team.

There is also another analogy, introduced by Firestone [7,14] and based on *across-* and *through*-variables, that can be used. A variable defined at a point in space with respect to another point in space is termed an *across* variable. For example, velocity, voltage, pressure, and temperature all may be interpreted as across-variables. On the other hand, a variable defined at a single point without respect to any other point is termed a *through* variable. Examples of *through-variables* include force, current, and fluid flow. The *across-* and *through-variable* analogy naturally leads to representation of a model in terms of linear graphs.

There is no general agreement on which analogy is preferable. The effort-flow analogy corresponds to the force-voltage electromechanical analogy; and the across-through analogy corresponds to the force-current electromechanical analogy. We use the effort-flow analogy, as it perhaps better explains *efforts* as intensities and *flows* as extensities.

It should be stressed that system decomposition combined with the bond graph modelling method readily leads to a lumped-parameter model. For the case in which variables inside a component change continuously over some region of space, it is necessary to apply discretization; that is, to represent the model by a finite number of components. This can be done in various ways, such as using the well-known finite-element discretization method.

In spite of the attention this approach has attracted, the bond graph method has not received the widespread acceptance expected by its proponents. In the opinion of the authors, one of the drawbacks of the classical bond graph modelling technique is its "flat" structure. That is, the model is constrained to be represented as a single-level structure. This leads to quite complicated diagrams even for relatively simple systems, and these can be difficult to interpret. One remedy for this is to pay more attention to the modelling of components in general, as well as to their ports. This enables more systematic model development.

There is another important concept embedded in bond graph theory. This is the concept of causality. This refers to cause (input) and effect (output) relationships [10].

Thus, as part of the bond graph modelling process, a causality assignment is implicitly introduced. This leads to the description of bond graphs in the form of state-space equations. The problem that arises is that such a model is very restrictive. Furthermore, as pointed out in [15], there is no true notion of causality in physical laws. For example, there are no purely physical reasons to interpret a force on a body as the cause of its motion; nor to interpret a voltage on electrical terminals of a component as the cause of the current flowing through it. Thus causality will not be part of our focus. The point of view taken is that modelling can—and should—be treated as separate from the mathematical model developed and the solution derived thereof. Used in this way, and together with the general modelling technique mentioned in the immediately previous section, bond graphs can be used as a powerful *visual modelling language*.

1.4 Block Diagrams

Block diagrams are often used to denote input-output relations (Fig. 1.6). They have been used traditionally in control engineering, but also have found popular application in other fields, such as computer science, economics, and ecology.

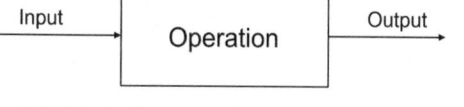

Fig. 1.6. Block diagram notion

Block diagrams depict operations on signals (information). The symbol inside the block represents a process applied to an input signal to generate an output. By connecting the output of one block to the input of another block, we can illustrate a procedure for the calculation of some quantity in which we are interested.

This approach can be used for modelling and simulating systems, too [2]. As an illustration, Fig. 1.7 shows a block diagram model of the simple mechanical system introduced earlier (Fig. 1.5a). The block diagram shows how to evaluate model variables given the time-history of the applied force and values of position and velocity of the body at the start of the motion.

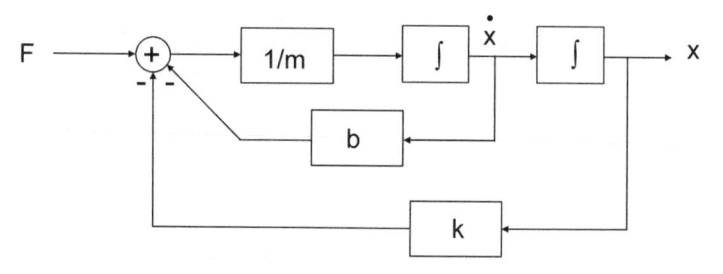

Fig. 1.7. Block diagram model of simple mechanical system

Many simulation programs are based on a modelling approach of this type (e.g. Matlab-SIMULINK). Unfortunately, many practical systems cannot be treated this way, except at some simplified conceptual level. In addition, modelling the system by block diagrams can be complicated and error prone, and also more difficult to interpret. This technique, however, can be useful as a complement to a more general modelling method, such as bond graphs, and is used at certain points of the present work, too.

Signals can be used for monitoring processes, describing control actions, and processing outputs. In the bond graph approach, signals are termed activated *bonds,* and there is no power associated with the transfer of signals (information). We allow bond graph components to have ports for the input or output of the signals. Such ports are depicted by a full-arrow and we call them *control ports.*

1.5 Symbolic Model Solving

Solution of a mathematical equation representing a system is usually accomplished by applying a suitable analytical procedure or, much more commonly, through some numerical routine. Developments in software engineering, however, have opened up the possibility of obtaining certain solutions symbolically. This approach relies on the tools of computational algebra.

Computational algebra software manipulates mathematical expressions symbolically [16]. There is currently a number of general-purpose symbol-manipulation applications available that can be applied to a wide range of practical mathematical problems. Some of the better-known packages are REDUCE, MAPLE, MATHEMATICA, and AXIOM. In the field of simulation, the best-known systems are MATHCAD and MATLAB.[4]

Solving problems symbolically requires processing power and memory resources usually well beyond the capability of most PCs and workstations. It thus becomes preferable to combine symbolical and numerical approaches to the practical problems that arise in engineering systems [1,17,18]. Thus, symbolic manipulation can be used to generate mathematical expressions that subsequently may be solved with numerical or analytical techniques. In this way, modelling of the system (the symbolic description) can be separated from its solution (the numeric result).

Once a model based solely on the physical considerations and requirements of the problem has been developed, the equations can be generated automatically in symbolic form. These equations can be simplified, either during the model's generation or after it. For example, constant expressions may be evaluated, trivial equations eliminated, or other actions taken to simplify the solution.

As the next step, the equation is prepared for numerical solution. This can entail a variety of operations, such as generation of functions that must be evaluated at run time, generation of the symbolic Jacobian matrix, and generation of any supplementary relationships necessary at start-up or for simulating across discontinuities. Symbolic manipulation also can be used for post-processing and the control of the complete problem-solving task.

This procedure can be accomplished without using the complete machinery of computational algebra. Simple arithmetical and logical operations plus symbolic differentiation is sufficient. What is really important is that the computational algebra should be seamlessly integrated with the other parts of the modelling and simulation environment.

It should also be pointed out that another technique for the evaluation of differentials has attracted much attention: *automatic differentiation* [19]. Automatic differentiation is a numerical technique for the evaluation of a differential based on an algorithmic approach using the rules of differentiation. This technique is often confused with symbol manipulation. It is usually argued that this technique is superior in terms of efficiency and memory usage; but symbolic manipulation is

---

[4] More information on symbol manipulations can be found on the Symbolic Mathematical Computation Information Center page at http://www.symbolicnet.mcs.kent.edu.

constantly being improved and can be applied to problems other than those that require the evaluation of a differential.

The approach taken here is based partly on the symbolic manipulation technique described in [20]. The constitutive relations for modelling the elements described in the last section are held in symbolic form, together with parameter expressions and model structure data. Before the start of a simulation, the model is assembled in the form of byte codes that can be evaluated efficiently during simulation. The implementation of the method is based on a combination of the interpretative and compiled approaches, and thus it is not necessary to recompile the model and link it. In this way, some efficiency in the solution is lost at the expense of flexibility.

1.6 The Object-oriented Approach

The object-oriented paradigm represents a major achievement in software engineering that facilitates modelling complex real-world problems [17, 21]. When properly applied, it yields robust models consisting of reusable, easy-to-maintain components. Only a very brief introduction is provided here. More information on object-oriented programming (OOP) can be found in [22, 23].

The focus of this approach is an *object*. An object is an abstraction of reality described by *attributes* and *methods* (Fig. 1.8).

Attributes define an object's state and can be represented as variables of fundamental types (such as integers, booleans, characters, or real numbers), other objects, or collections of various types.

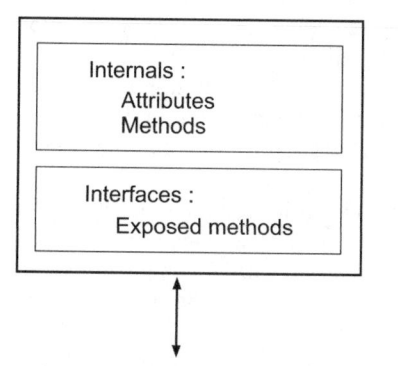

Fig. 1.8. Object attributes and methods

Methods describe an object's behaviour. From a related perspective, methods represent the services that an object provides.

Both attributes and methods are *members* of an object. Attributes sometimes are referred to as data-members, and methods as member-functions. An object is said to *encapsulate* its members.

One of the key concepts of the OO approach is *information hiding*. This means that an object exposes only those of its members that are required for use by other objects; all other attributes and methods are "hidden" within the object, inaccessible to the outside world.

To enable interaction with other objects in its environment, an object must provide an *interface*. An object interacts with its environment only *via* its interface. The interface exposes access methods.

Most real-world objects cannot be modelled adequately with only a single fundamental data type. The OO paradigm thus introduces the concept of a *class*. A class is a generalised data type that defines the attributes and methods shared by all objects of that class. The terms *object* and *class* are sometimes (improperly) used interchangeably; strictly, an object is an instance of a class. In practice, an object is created in the computer memory (constructed) at run time using the class definition as a template; and removed from memory (destructed) when no longer required.

Yet another fundamental principle of the object-oriented approach is *inheritance*. The inheritance mechanism permits classes to be organised in a logical hierarchy that describes their interrelationship. Thus, a class derived from an existing class inherits the members of its parent. The derived class may define additional attributes and functionality, as well as modify those inherited from its parent.

Relative to the derived (child) class, the parent class is a *generalization,* a *super class*, or, in the terminology of C++, a *base class*.

Viewed from the complementary perspective, the derived class is a *specialization* of the parent class; it is a *kind of* the base class. At the head of the hierarchy there is usually a class that does nothing more than define the interfaces of all lower level classes in a consistent way. Such a class is an *abstract base class*. We treat the elementary component of Fig. 1.4 as a kind of general component, which likewise is a kind of more abstract object class (Fig. 1.9).

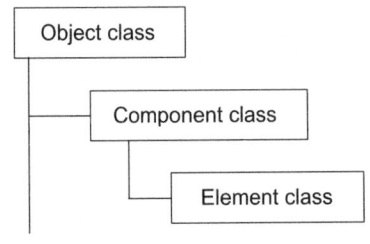

Fig. 1.9. Component class hierarchy

Another OO principle is *polymorphism*. Polymorphism enables methods with the same name to exhibit different functionalities. For example, methods in the lineage of a class hierarchy can have the same *function header*—that is, the same name, argument list, and return type—but different behaviour. Such methods are declared *virtual* in the base class. The particular method invoked by a function call depends on the object's class and is implemented through a mechanism termed *dynamic binding*. The decision on which method to call thus is made at run time

Another consequence of polymorphism is that two methods may share the same name, but have different argument (parameter) types and return types. The specific method executed depends on the type of the arguments passed by the calling function. This is an example of the so-called function *overloading*. In contrast to the previous example, these functions are not virtual: They have a unique *signature* (parameter list), so the decision on which version of an overloaded function to call is decided at compile time.

The general modelling approach set out in Section 1.2 cannot be supported using class hierarchies alone, as components generally contain other components; this is how real devices are built. To represent a model of a component we use two objects: a *component* and its accompanying *document*.

A component is represented by a *component object*. Such an object is compounded and contains *port objects* that serve as interfaces to other components (Fig. 1.10, component Platform). If the component is simple and hence doesn't contain other components, then it is just an elementary component. But, if it *does* contain other components, we use another object to define its internal structure. Such an object is not a kind of any other component, but a separate entity—a *document* (Fig. 1.10, bottom-right box).

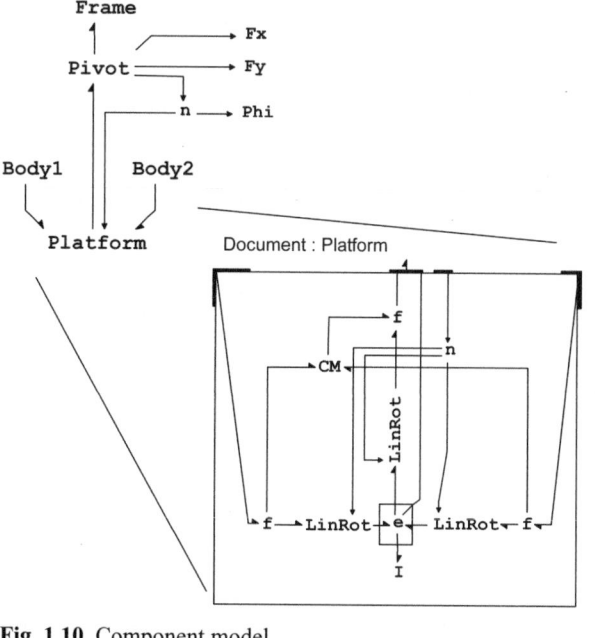

Fig. 1.10. Component model

The document object has *external* ports that correspond exactly to the component object ports and provide access to the component internals. In this way, the *component model* representation is based on two associated classes – the component class and the document class. These classes are designed in a way that supports access to the document through its accompanying component object, or by its ports.

This is very close to the way in which we deal with real components. That is, to see what is inside a component, we first have to find a component by its name or manufacturer's designation, or maybe how it is connected to other components; only then we can "open" it to look at what is inside. The component is usually contained in another component, this again in yet another component, etc. The set of all components constitutes the system. In this way a model of a system can be represented as a tree of components.

Elementary components are treated somewhat differently, as they do not contain other components. Ports of such components provide access to the component constitutive relations that describe the mathematical model of the particular elementary component (Sect. 1.3). Thus, the elementary components can be looked at as *leaves* of the model *tree*.

1.7 Computer Aided Modelling

One of the first computer programs developed for modelling and simulation of practical engineering systems was SPICE [24], which was developed at the University of California, Berkeley at the beginning of the 1970s for integrated electronic circuits. The program was accepted quickly by leading semiconductor manufacturers. Since that time, SPICE has undergone continuous development to follow technological advances in semiconductors. Today it is the *de facto* standard for electrical simulations. All of the main semiconductor manufacturers offer SPICE models of their components.

The success of SPICE—and of similar products that emerged shortly afterwards—was owed partly to the fact that, in electrical engineering, and particularly in electronics, systems and devices normally are modelled using electrical schemes. SPICE uses an input file that contains a description of electrical schemes using a simple textual language.

Another very successful modelling and simulation system in electrical engineering is SABER.[5] It uses the MAST modelling language. Recently, under the umbrella of the IEEE, a new modelling language has been developed: VHDL-AMS [25]. This offers a more uniform approach to modelling mixed analogue-digital systems. VHDL-AMS does not assume any causality, generates models in differential-algebraic form, and permits discontinuities.

From the start, bond graphs were looked at as a unified approach to the modelling of general engineering systems, in some ways similar to electrical schematics in electrical engineering. Many programs based on bond graphs have been developed over the last two decades. The first, ENPORT [26], uses a textual description of the bond graph model topology (similar to SPICE) as input. It supports macro

---

[5] The SABER is product of Analogy, Inc., Beaverton, USA, http://www.analogy.com.

capabilities to simplify bond graph model creation. The program was developed for workstations. There is also a PC version, but with somewhat reduced capabilities.

Historically, the next program to appear was TUTSIM, developed at Twente University, Netherlands. This application was originally developed for block diagram models, but was modified to accept causality-augmented bond graphs. The program translates models into state-space form and hence cannot treat models with dependent storage (masses, capacitors etc.) and algebraic loops.

The same research group later developed the CAMAS program [27], which is now known as 20-SIM [28,29]. This program uses bond graphs for model input and icons for component representation. The icons serve as placeholders for the component models. The models are expressed as equations using the SIDOPS language and are organised as model fragments. For real reusability of the fragments adequate caution must be exercised. Sub-model storage is not yet implemented as a complete database facility. System model processing, from model equations to assignment statements, is performed using computational causality analysis. The program has its own simulator based on routines from NETLIB.[6]

The MS1 program is designed for the modelling and simulation of dynamic systems with continuous elements.[7] Models are developed in the form of causality augmented bond graphs and solved using ACSL, Matlab-SIMULINK, and other programs. Similarly CAMP-G [30] was developed as a pre-processor for ACSL, Matlab-SIMULINK, and other programs, and is suitable for modelling systems that can be represented in state-space form only. There are other bond graph based programs that implement a similar philosophy.

Finally, the Dynamic Modelling Language (Dymola) [17] should be mentioned. Dymola was developed using another philosophy: the through/across physical analogy and the theory of graphs. It supports bond graphs, thus bond graph elements and structures can be coded directly. There is some incompatibility between these two approaches, but this can be circumvented. Dymola helps the user edit and compose programs, manipulate models from sub-models, and generate code for continuous system simulation languages such as ACSL, SIMULINK, and some others. It can also generate code in C or Fortran. This can be executed in a module for the simulation of time-continuous systems, which supports both ordinary differential equations (ODE) and differential-algebraic equations (DAE). Recently, under the auspices of EUROSIM, a new modelling language MODELICA has been developed [31].

This book describes an approach to the development of an integrated automated computer environment for visual model development and experimentation by simulation that is implemented in *BondSim,* a copy of which is provided with the book. To help the designer during the tedious task of model development, the modelling environment supports the fundamental modelling approach—systematic problem decomposition and model creation at every level of decomposition. The

---

[6] The NETLIB is web based public library repository that can be accessed att http://netlib.org.

[7] Lorenz Simulation SA, http://www.lorsim.be

number of levels of decomposition depends on the complexity of the engineering system being modelled and the depth of abstraction. For simple problems, a single level suffices, i.e. a flat model. But real systems, such as are usually found in mechatronics, consist of many components that are themselves constructed of many other components. Thus, two or more levels of model decomposition may be necessary.

The model can be created as a multi-level structure. This helps in understanding the model and in providing for its maintenance. Thus, the model for a problem can be treated as a tree with component models as its branches. The modelling environment also supports building the model from the "pieces", i.e. the component models. A combination of these two approaches is also possible. The environment also has facilities for the creation of suitable libraries in which the system and/or the component models are stored.

This new simulation tool, BondSim, supports the type of collaboration that has become essential in the development of complex models. Model development usually is teamwork, hence the distribution of modelling tasks and the integration of results is welcome, if not an absolute necessity. Also, the exchange of models with colleagues, or use of component models from manufacturers (similar to the SPICE models), is very useful. This aspect of the modelling environment also is implemented in BondSim.

BondSim's general concept of modelling and simulation is illustrated in Fig. 1.11. Development is done entirely visually—without coding—using the support of the Windows system [1]. In this *visual modelling* approach, the modelling process consists of creating objects in the computer memory that are depicted on the screen as bond graphs or block diagrams. It is also possible to represent them by common electrical and mechanical schemes. The model of a component discussed in the previous section plays a central role in this process.

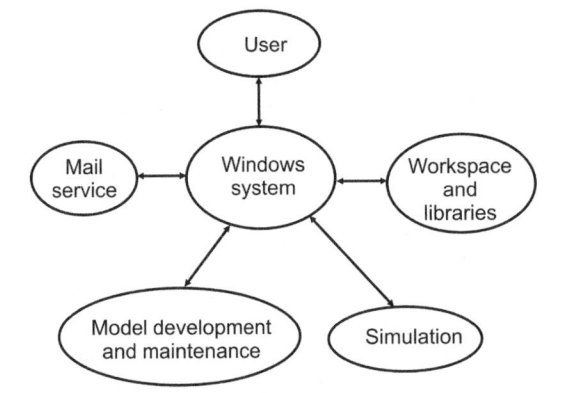

Fig. 1.11. Visual modelling and simulation environment

This approach should not to be confused with Microsoft's Component Object Model (COM) technology. Components used in BondSim are not placeholders for component models, but real objects that serve as interfaces to the document that contains the model. Thus, we can open a component to inspect it in more detail, edit it, and continue with multi-level model development. Such a component can be stored in a component library, or may be sent by e-mail to somebody else. Likewise, a component received by e-mail can be imported into the modelling environment.

The BondSim modelling environment automates many common utility operations, such as saving, loading, copying, deleting, inserting, importing, and exporting. The constitutive relations specifying the characteristics of the elementary components or block diagram operations are described as simple linear or nonlinear algebraic expressions.

After a bond graph model has been developed, its mathematical representation is built by examining the component tree. This results in a model in the form of differential-algebraic equations (DAEs), which then are solved by appropriate numerical routines. Symbolic processing of the model equations plays an important role in the solution.

BondSim also generates reports: Practically all of the bond graph diagrams contained in this book were created using BondSim's print-to-file support.[8] The application was also used to solve all of the mechatronics problems presented in this book.

1.8 The Book Summary

This book consists of two parts. The first part—*Fundamentals*—describes the basics of the design of the visual object oriented environment for the modelling and simulation of general engineering systems, with emphasis on mechatronics systems. This part contains five chapters.

Chapter 1 gives a short overview of the approaches and methods used for model representation. After a short discussion of the objectives, the general modelling approach is described which lies at the root of the modelling philosophy – model development by systematic decomposition. The concepts of system, environment and of component are discussed. Then physical modelling, analogies and bond graphs are described as well as an alternative approach to modelling based on the through/across variables approach. The block diagram approach is described next as a model representation method on its own or in combination with bond graphs. A short introduction to symbolic model solving is also given and it's potential for use in combination with numerical methods. The next important technique described is object-oriented programming with an emphasis on a component-based approach to the modelling of engineering systems. Finally computer aided modelling is described together with the concept of modelling and simulation in a visual environment.

---

[8] The BondSim supports, for documentation purposes, print of screen images into a file in emf (Enhanced Windows Metafile) format, which is supported by main word or graphic programs including MS Word, Corel Draw etc.

Chapter 2 gives an overview of the bond graph modelling technique. The perspective taken is an extension of conventional bond graphs to multilevel modelling. Starting from the concept of word models, ports, bonds, and power variables, the component model development approach is described. The elementary bond graph components used for modelling of the basic physical processes are defined. Also the components corresponding to basic block diagram operations are introduced. The systematic decomposition approach to bond graph model development is illustrated on examples of mechanical and electrical systems. Finally the notion of causality in bond graphs is discussed.

Chapter 3 deals with the systematic object-oriented approach to modelling. The basic idea is to do the modelling visually, i.e. without any coding, but solely by interacting with the modelling system in a suitably designed visual environment. The concept of the component model is introduced as the basic mechanism for systematic simulation models development. The special class hierarchies are designed to support creation of models for given problems as trees of linked objects in the computer memory. The models are represented visually as bond graphs and are stored as a set of linked files. Underlying physical processes are represented in terms of elementary bond graph components, which constitutive relations are described symbolically using a simple specially designed language. The design of a suitable modelling environment to support the modeller is outlined.

Chapter 4 describes an implementation of the object-oriented approach of Chap. 3. A program *BondSim* is described that offers a visual environment for the modelling and simulation of engineering and mechatronics systems. The program implements several services that are accessible to a user through a window system. The two basic services are Modelling and Simulation. The program also supports model database maintenance, library support, as well as collaborative support for model exchange using the e-mail service. The program supports Microsoft Windows (Windows 2000, NT, Windows 9x). A copy of the BondSim program is included with this book.

Chapter 5 closes the first part of the book and describes the methods used for automatically generation of the mathematical model equations and their solution. This is divided into two distinctive phases – the model building and the execution of simulation runs. During the first the mathematical model equations are generated based on the model object tree. The methods for generating the mathematical models implied by the component's structure are described. The model equations are machine generated in the form of differential-algebraic equations (DAEs). During the simulation phase such equations have to be solved. The well-known backward differentiation formula (BDF) is used. A special implementation of the method is described based on the variable coefficient formula. Also the problem of starting values and of discontinuities in the model equations are also discussed. The methods developed depend to a great extent on computational algebra support implemented in the program. This also adds to flexibility in the modelling.

The second part of the book deals with *Applications* to mechatronics of the bond graph modelling technique. It is divided into chapters 6 to 10.

Chapter 6 deals with simple mechanical problems. The intention is to familiarize the reader with using the *BondSim* program on relatively simple problems.

These problems are also of interest on their own right. Thus the well-known Body Mass Damper problem is used as the introduction to the bond graph modelling and simulation of mechanical systems. It is also shown that it is possible to visually represent mechanical components by mechanical schemes. After that the effects of dry friction are studied. Next another class of discontinuous problems is studied – impact. Using a simple model of impact the classical problem of a ball bouncing on a vibrating table is studied. The problem is better known for its chaotic behaviour. The chapter ends with a description of a see-saw problem. It is a pendulum, but it can be looked at as a multibody problem. It is shown that such problems – of interest in Mechatronics – can be readily solved by bond graphs.

Chapter 7 is devoted to the modelling of electrical systems. It is shown that the component models approach can be readily used for the modelling of electrical and electromechanical components and systems. This is important for this enables both the mechanical and electrical part of a system can be modelled and analysed on the same basis, i.e. from the bond graph point of view. Models of the most important electrical components are developed in terms of bond graphs such as resistors, inductors, capacitors etc., and also fundamental semiconductor components such as the diodes and the transistors, using SPICE like models. An important feature of the approach is the visual representation of bond graph electrical component models as electrical schemes. The chapter ends with analysing an electromagnetic system.

Chapter 8 describes modelling of control systems in terms of block diagrams. The approach is very popular in other fields as well. It shown how block diagram models can be developed and is illustrated on a simple control system problem. Some details of modelling that are specific to block diagram components are given. Then a short overview of the modelling approach to control systems is given concentrated mostly on the modelling of PID controllers in servo loops. Finally the modelling and simulation of a DC motor servo is given.

Chapter 9 is devoted to multibody systems. The models of planar rigid bodies and the basic joints are developed and applied to some practical multibody problems. The approach is then extended to multibody systems in space. The components developed are used for solving a problem of a hybrid control of a robot. This demonstrates that the component model approach developed in this book can be used for solving complex problems in mechatronics.

Chapter 10 is the last chapter of the book and deals with the modelling and simulation of continuous systems. Continuous systems are important in many engineering disciplines, mechatronics included. The approach used here is based on the *method of lines*. The system is discretized and represented as an assemblage of finite elements in the form of bond graph component models. The model equations are then generated and solved using the DAEs solver. The approach is applied first to a problem of the modelling of electric transmission lines. Then a bond graph component model of a beam element based on the classical Euler-Lagrange theory is developed and applied to the solution of two practical problems – a package vibration testing system and the analysis of a Coriolis mass flow meter.

References

1. V Damic and J Montgomery (1998) Bond Graph Based Automated Modelling Approach to Functional Design of Engineering Systems. In: GR Gentle and JB Hull (eds) Mechanics in Design International Conference, The Nottingham Trent University, Nottingham, pp 377-386
2. Lennart Ljung and Torkel Glad (1994) Modelling of Dynamic Systems. PTR Prentice Hall, Englewood Cliffs
3. Henry M Paynter (1961) Analysis and Design of Engineering Systems. MIT Press, Boston
4. AJ Blundell (1982) Bond Graphs For Modelling Engineering Systems. Ellis Horwood Limited, Chichester
5. Pieter C Breedveld (1984) Physical systems Theory in Terms of Bond Graphs, PhD thesis, Technische Hochschool Twente, Entschede
6. Peter Gawthrop and Lorcan Smith (1996) Metamodelling: Bond graphs and dynamic systems. Prentice Hall, Hemel
7. Peter MAL Hezemans and Leo CMM van Geffen (1991) Analogy Theory for a Systems Approach to Physical and Technical Systems. In: Paul A Fishwick, Paul A Luker (eds) Qualitative simulation modeling and analysis. Springer-Verlag, New York, pp170-216
8. Dean C Karnopp and Ronald C Rosenberg (1975) System Dynamics: A Unified Approach. John Wiley, New York
9. Dean C Karnopp, Donald L Margolis and Ronald C Rosenberg (1990) System Dynamics: A Unified Approach, 2nd edn. John Wiley, New York
10. Dean C Karnopp, Donal L Margolis and Ronald C Rosenberg (2000) System Dynamics: Modeling and Simulation of Mechatronic Systems, 3rd edn. John Wiley, New York
11. Jean U Thoma (1975) Introduction to Bond graphs and their Applications. Pergamon Press, Oxford
12. Jean U Thoma (1990) Simulation by Bondgraphs. Springer-Verlag, Berlin Heidelberg
13. J Thoma and B O Bousmsma (2000) Modelling and Simulation in Thermal and Chemical Engineering, A Bond Graph Approach. Springer/Verlag, Berlin Heidelberg
14. Firestone, FA (1933) A new analogy between mechanical and electrical systems. J. Accoustical Soc. 4:249-267
15. FE Celier, H Elmqvist and M Otter (1995) Modelling from Physical Principles. In: WS Levine (ed) the Control Handbook, CRC Press, Boca Raton, pp 99-108
16. Tan Kiat Shei and Willi-Hans Steeb (1998) SymbolicC++: An Introduction to Computer Algebra Using Object-Oriented Programming. Springer-Verlag, Singapore
17. FE Cellier (1996) Object-Oriented Modeling: Means for Dealing with System Complexity. In: Proc. 15th Benelux Meeting on Systems and Control, Mierlo, pp 53-64
18. EJ Kreuter (1994) Generation of Symbolic Equations of Motion of Multibody Systems. In: E. Kreuzer ed Computerized Symbolic Manipulation in Mechanics. Springer-Verlag Wien – New York, pp 1-66
19. Masao Iri (1991) History of Automatic Differentiation and Rounding Error Estimation. In: A Griewank, GF Corliss (eds) Automatic Differentiation of Algorithms: Theory, Implementation and Applications, Proc. of First SIAM Workshop on Automatic Differentiation, pp 3-16

20. Alain Reverchon and Marc Ducamp (1993) Mathematical Software Tools in C++. John Wiley, Chichester
21. W Schroeder, K Martin and B Lorensen (1998) The Visualization Toolkit 2nd edn. Prentice-Hall PTR, Upper Saddle River
22. J Rumbaugh, M Blacha, W Premerlani, F Eddy, W Lorenson (1991) Object-Oriented Modeling and Design. Prentice Hall
23. B Stroustrup (1998) C++ Programming Language, 3rd edn. Addison-Wesley, Reading
24. A Vladimirescu (1994) The Spice Book. John Wiley & Sons, New York
25. E Christen, K Bakalar, AM Dewey and E Moser (1999) Analog and Mixed Signal Modeling Using VHDL-AMS Language (tutorial). The 36th Design Automation Conference, New Orleans
26. RC Rosenberg (1974) A User's Guide to ENPORT-4. John Wiley & Sons, New York
27. JF Broenink (1990) Computer-Aided Physical Systems Modelling and Simulation: A Bond Graph Approach. PhD thesis, University of Twente, Enschede, Netherlands
28. APJ Breunese and JF Broenink (1997) Modeling Mechatronic Systems Using the SIDOPS+ Language. In FE Cellier and JJ Granda (eds) 1997 International Conference on Bond Graph Modeling and Simulation, Phoenix, Arizona, pp 301-306
29. JF Broenink and C Kleijn (1999) Computer-Aided Design of mechatronic Systems Using 20-SIM 3.0. In GN Roberts and CAJ Tubb (eds) Proceedings of 2nd Workshop on European Scientific and Industrial Cooperation, Newport, UK, pp 27-34
30. JJ Granda (1982) Computer-Aided Modelling Program (CAMP): A Bond graph Processor for Computer Aided Design and Simulation of Physical Systems Using Digital Simulation Languages. Ph.D. Dissertation, Dept. of Mech. Engineering, University of California, Davis
31. P Fritzon and V Engelson (1997) Modelica – A Unified Object-Oriented Language for System Modeling and Simulation. In: Modelica Home Page http://Dynasim.se/Modelica

Chapter 2 Bond Graph Modelling Overview

2.1 Introduction

The bond graph physical modelling analogy provides a powerful approach to modelling engineering systems in which the power exchange mechanism is important, as is the case in mechatronics. In this chapter we give an overview of the bond graph modelling technique. The intention is not to cover bond graph theory in detail, for there are many good references that do this well, e.g. [1, 2]. The purpose is to introduce the reader to the basic concepts and methods that will be used to develop a general, systematic, object-oriented modelling approach in Chapter 3.

2.2 Word Models

Many engineering systems consist of components, e.g. electric motors, gears, shafts, transistors etc. (Fig. 1.3). Simulation models of such components can be represented as objects in the computer memory and depicted on the screen by their name, i.e. any word description chosen to describe the component (Fig. 2.1).

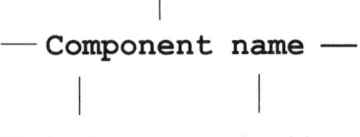

Fig. 2.1. Component word model

The component name is useful for reference to the model, but what is more important is to represent how the component is connected to other components. When we look at a component, the internals of its design are usually hidden (e.g. by its housing). What are seen instead are the locations at which it connects to other components. These places—such as electrical terminals, output shafts, fixing places, hydraulic ports, and boundary surfaces across which heat transfer takes place—are termed *ports*. In Fig. 2.1 the ports are shown by short lines.

When the component is connected and the system is energized from a suitable power source, there is a flow of power through these ports. Also, some ports serve to monitor or control the component. Thus, the ports serve as places where power or information exchange takes place. This is explained in Sect.2.3

A component represented by its word description (its "name") and its ports is taken as the most fundamental representation of a component model and is termed the *word model*. The word model is used as the starting point of component model development.

2.3 Ports, Bonds, and Power Variables

Ports, as explained in Sect. 2.2, are places where interactions between components take place. These interactions can be looked on as *power* or *information* transfer. Thus, two types of ports are defined.

Ports characterised by power flow into or out of a component are termed *power ports*. Such ports are depicted by a *half arrow* (Fig. 2.2). The half arrow pointing *to* the component describes *power inflow*. It is assumed that at such a port there is *positive* power transfer into the component. Similarly, a half arrow pointing away from the component depicts *power outflow* from the component and the corresponding power transfer is then taken as *negative*.

Another type of port is characterised by negligible power transfer, but high information content. These are termed *control ports* and are depicted by a *full arrow*. The arrow pointing to the component denotes transfer of information into the component (*control input*). Similarly the port arrow pointing away from the component denotes information extracted from the component (*control output*).

Control ports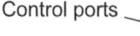

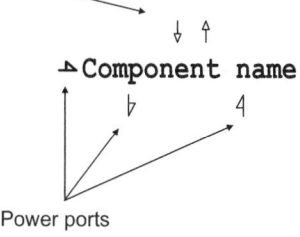

↓ ↑

➤Component name

Power ports

Fig. 2.2. Component word model with the ports defined

The word model, i.e. the component represented by the name and the ports, is taken as the lowest level of component abstraction (Fig. 2.2). Components interact with other components through their ports. These interactions are looked on as *power or information transfer* between components and are depicted by lines connecting corresponding component ports (Fig. 2.3a). The lines that connect *power* ports are termed *bond lines*, or *bonds* for short. A bond line joins a power *outflow* port of one component and a power *inflow* port of the other and clearly shows the assumed direction of power transfer between components. Similarly, lines connecting control output ports and control input ports are termed *active* or *control bonds*. These lines show the direction of information transfer between components. When a bond line is drawn, ports and connecting lines appear as a single line with a half or full arrow at one of its ends (Fig. 2.3b). In the bond graph literature emphasis is put on the bonds, with ports playing a minor role. In our approach just the opposite point of view is taken: Ports, the places where inter-component interactions take place, receive emphasis.

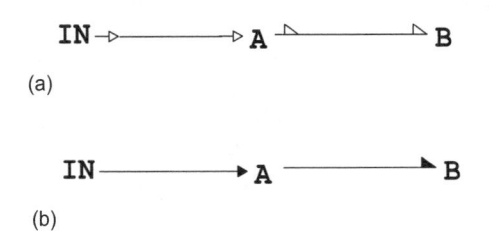

(a)

(b)

Fig. 2.3. Connecting components. (a) Connecting ports by bond lines, (b) Line and connected ports represented by bond line only

Power or information exchange between component ports can be quite complex. It generally depends on the processes taking place in the components. In the simplest case the process in the component as seen at a power port can be described by a pair of *power variables*, the *effort* and *flow variables*. Their product is the power through the port (Sect. 1.3). Connecting such ports by a bond simply implies that effort and flow variables of interconnected ports are equal. Similarly, information at component control ports can be described by a single *control variable* (signal). Connecting an output port of a component to an input port of the other just means that these two control variables are equal.

In general, the situation is not this simple. Thus, the revolute joint illustrated schematically in Fig. 2.4a may be used to connect robot links or, a door in a doorframe. The joint can be represented by a word model (Fig. 2.4b), with ports representing the parts of the joint provided for the connection.

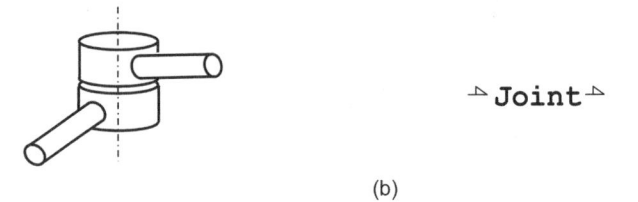

(a)   (b)

Fig. 2.4. Revolute joint. (a) Scheme of joint, (b) Word model representation

The function of the joint is to enable rotation of the connected bodies about the joint axis. To describe the interactions at the joint connection properly, pairs of effort and flow *vectors* are used. The effort vector can be represented by three rectangular components of the forces and torques, and likewise the flow vector by the rectangular components of linear and angular velocities. The meaning of these variables can be explained by defining a detailed model of the joint and the bodies in question. Hence the connection of a body to the joint can be represented by a bond, which denotes again that the efforts and flows of connected parts are equal. This time the power variables are not simple one-dimensional variables, but vector quantities.

Complex interactions at the ports can also be represented using multidimensional bond notation known as *multibonds* [3, 4]. We do not use this approach here, but instead treat the component ports as *compounded*. This means that the component ports are not simply objects, but define the structure of the mathematical quantities that describe the processes taking place inside the component. The bond lines simply define which port is connected to which, and hence which mathematical quantities should be equal. To define the structure of the ports the component model is developed in more detail.

2.4 Component Model Development

The detailed model of a component represented by the word model (Fig. 2.5, top-left) can be described in a document framed by a rectangle (Fig. 2.5, on the right). We call it a document because it will be represented on the computer screen in a document window and saved in a file (Chapt. 3). The document title uses the name of the component that it models. The document contains ports represented by short strips placed just outside of the frame rectangle. These *document ports* correspond to the ports of the component: Every component port has a document port. The component Comp A in Fig. 2.5 has three ports: a power-in port, a power-out port, and a control-out port. Thus, there are exactly three document ports of the same type. The document ports are depicted in the positions around frame rectangle that corresponds to the position of the component ports around the component text (name). This way it is easy to see which port corresponds to which.

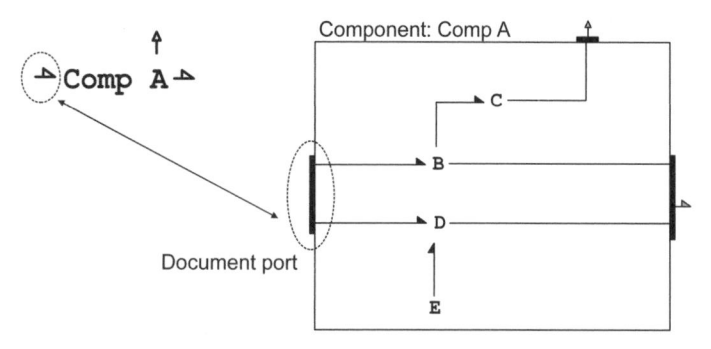

Fig. 2.5. Concept of component model

To develop the model, the component is analysed to identify the components of which it consists. Each is represented by its word model. Thus, in Fig. 2.5 there are four such components, named B, C, D, and E. Next we determine how these components are interconnected. Some are connected only internally, and this is represented by bonds connecting their respective ports, e.g. component B to component C. Some of the components are connected to the outside. In this case the respective component ports should be connected by the bonds to the document port strips, e.g. component B ports should be connected to the left and the right document ports.

Thus, document ports serve for the internal connection of the contained components. The document strip allows more than one bond to be connected to the port. We also assume that these bond connections are ordered, e.g. from top to bottom and from left to right. Hence, a component port (Fig. 2.5, left top) is represented by an array of internal component connections (Fig.2.5 on the right).

When all the word models of the contained components are connected, we obtain a diagrammatical representation of the component model structure termed the *bond graph model*. To complete the model it is necessary to continue developing the models of all contained components, which are represented by their word models, e.g. components B, C, etc. (Fig. 2.5).

The important question is how and when we end this process of systematic model decomposition. Formally, this happens when we get to a component that is fundamental, i.e. it doesn't contain simpler components. This is the problem of the level of abstraction we use when developing a model.

Normally we start model development from the system level (Sect. 1.2). At that level we define the model as a bond graph of the components. This is the lowest level of problem abstraction and component word models at this level usually correspond to the real-world components. In the next step we describe a model of the component by identifying the basic physical effects in the component, ignoring other, less important effects. Thus, the electrical resistor or mechanical spring can be described by an elementary model, such as Ohms law or the linear spring force-extension relationship. But we can also include the inductivity effect of the resistor, or inertial effects in the spring. Hence, even in such simple cases we can use either simple or compounded component models. In other more complex devices such as robot arms, we identify real components that constitute such an arm, e.g. links, joints, base, etc. But even then we reach a stage at which we decide on the level of detail to be included in the underlying model. Physical processes are usually distributed over the component space, not restricted to small regions only. In such cases distributed models are usually discretized and can be represented by a bond graph of components.

The bond graph modelling analogy enables the representation of models of basic physical processes taking place in engineering systems in the form of elementary components (Sect. 1.3). These components are described in more detail in the next Sec. 2.5. In addition, signal processing can also be described by several elementary operations (Sect. 2.6). Thus, starting from the system level, it is possible to develop the model gradually by applying the component decomposition technique. At every level of decomposition the components can be represented as elementary, or by a word model that is developed further. The resulting model thus can have one or more levels of decomposition. This depends on the system under study and the accepted level of abstraction of the problem being solved.

2.5 Modelling Basic Physical Processes

2.5.1 Elementary Components

The notion of elementary components has already been introduced in Sects. 1.2 and 1.3. These have a simple structure and serve as the building blocks of complex component models. In the bond graph method such components represent *basic physical processes*. Sometimes such components can be used as simplified representations of real components, such as bodies, springs, resistors, coils, or transformers.

There are, altogether, nine such components that represent underlying physical processes in a unique way. These are

- Inertial (I), Capacitive (C) and Resistive (R) components
- Sources of efforts (SE) and of flows (SF)
- Transformers (TF) and Gyrators (GY)
- Effort (e) and flow (f) junctions

The standard symbols used for the components are given in the parentheses.

In this way multi-domain physical processes, typical of mechatronics and other engineering systems, can be modelled in a unified and consistent way. A review of all the elementary components is given in Fig. 2.6. Components are described by their constitutive relations in terms of variables and physical properties.

The components can have one or more power ports. The processes seen at these ports are described by pairs of power variables – *effort* e and *flow* f. In addition, certain components have internal state variables. The next sub-sections give a detailed description of each component (Sects. 2.5.2—2.5.7). In Sect. 2.5.8 the *controlled* elementary components are described, i.e. common components with added control ports. At the control ports a control variable c is defined that is used for supplying information *to*, or for extracting information *from*, the component.

2.5.2 The Inertial Component

The *inertial* component is identified by the symbol I and has at least one power port (Fig. 2.6a). This component is used to describe the inertia of a body in translation or rotation, or the inductivity of an electrical coil.

The port variables are effort (e) and flow (f). In addition, there is an energy variable, *generalised momentum* p, defined by the relationship

$$e = \dot{p} \tag{2.1}$$

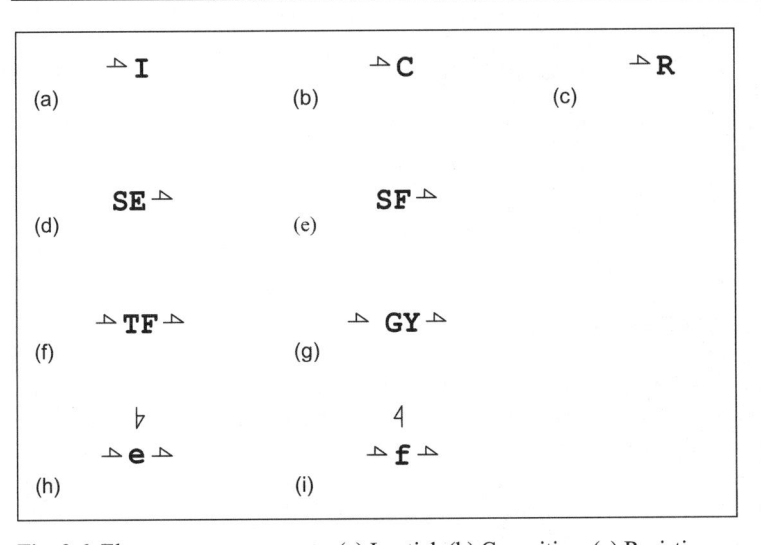

Fig. 2.6. Elementary components. (a) Inertial, (b) Capacitive, (c) Resistive,
(d) Source Effort, (e) Source Flow, (f) Transformer, (g) Gyrator,
(h) Effort Junction, (i) Flow Junction

The generalised momentum can be viewed as the accumulation of *effort* in the component,

$$p = p_0 + \int_0^t e \, dt \qquad (2.2)$$

The constitutive relation of the process reads

$$p = I \cdot f \qquad (2.3)$$

where I is a parameter. The constitutive relation also can be non-linear, of the form

$$p = \Phi(f, par) \qquad (2.4)$$

or, alternatively,

$$f = \Phi^{-1}(p, par) \qquad (2.5)$$

where Φ is a suitable non-linear function and par denotes parameters. If the component has n ports, the constitutive relation at the i-th port generally has the form

$$p_i = \Phi_i(f_j, par), \quad (i, j = 1, ..., n) \qquad (2.6)$$

or, alternatively,

$$f_i = \Phi_i^{-1}(p_j, par), \quad (i, j = 1, ..., n) \qquad (2.7)$$

where Φ_i is a suitable multivariate function.

A process represented by an inertial component is characterised by the accumulation of power flow into the component in form of energy

$$E = E_0 + \int_0^t efdt$$

Using Eq. (2.1) we get

$$E(p) = E(p_0) + \int_{p_0}^{p} fdp \tag{2.8}$$

2.5.3 The Capacitive Component

The *Capacitive* component is identified by the symbol C and has at least one power port (Fig. 2.6b). This component is used to model mechanical springs, electrical capacitors, and similar processes.

The port variables are effort (e) and flow (f). In addition, there is an energy variable, *generalised displacement* q, defined by relation

$$f = \dot{q} \tag{2.9}$$

Thus, generalised displacement can be viewed as the accumulation of the *flow* in the component,

$$q = q_0 + \int_0^t fdt \tag{2.10}$$

The constitutive relation of the process reads

$$q = C \cdot e \tag{2.11}$$

where C is a parameter. The constitutive relation also can be nonlinear, i.e.

$$q = \Phi(e, par) \tag{2.12}$$

or, alternatively,

$$e = \Phi^{-1}(q, par) \tag{2.13}$$

where Φ is a suitable non-linear function and par denotes parameters. If the component has n ports, the constitutive relation at the i-th port generally is of the form

$$q_i = \Phi_i(e_j, par), \quad (i, j = 1, ..., n) \tag{2.14}$$

or, alternatively,

$$e_i = \Phi_i^{-1}(q_j, par), \quad (i, j = 1, ..., n) \tag{2.15}$$

and Φ_i is a suitable multivariate function.

A process represented by a capacitive component is characterised by the accumulation of power flow into the component in the form of energy

$$E = E_0 + \int_0^t efdt$$

or by Eq. (2.9)

$$E(q) = E(q_0) + \int_{q_0}^{q} f \, dq \qquad (2.16)$$

2.5.4 The Resistive Component

The *resistive* component is identified by the symbol R and, like the inertial and capacitive components, has at least one port (Fig. 2.6c). This component models friction in mechanical systems, or electrical resistors.

The port variables are effort e and flow f. The component constitutive relation is given by

$$e = R \cdot f \qquad (2.17)$$

where R is a parameter. The constitutive relation can also be non-linear

$$e = \Phi(f, par) \qquad (2.18)$$

or, alternatively,

$$f = \Phi^{-1}(e, par) \qquad (2.19)$$

where Φ is a suitable non-linear function and par denotes parameters. If the component has n ports, the constitutive relation at the i-th port generally has the form

$$e_i = \Phi_i(f_j, par), \quad (i, j = 1, ..., n) \qquad (2.20)$$

or

$$f_i = \Phi_i^{-1}(e_j, par), \quad (i, j = 1, ..., n) \qquad (2.21)$$

and Φ_i is a suitable multivariate function.

2.5.5 Sources

Sources are components that represent power generation (or power sinks) such as voltage and current sources, certain types of force (e.g. gravity), volume flow sources (such as pumps) etc. In these sources efforts or flows are almost independent of the other power variable. It is possible to define two types of source components: *source efforts*, designated by SE; and *source flows*, designated by SF (Figs. 2.6d and 2.6e).

These are, basically, single port components. Denoting the port effort by e and port flow by f, the corresponding constitutive relations are given by the following relationships

Source Effort (SE)

$$e = E_0 \qquad (2.22)$$

or, more generally,

$$e = \Phi(t, par) \qquad (2.23)$$

Source Flow (SF)

$$f = F_0 \qquad (2.24)$$

or, more generally,

$$e = \Phi(t, par) \qquad (2.25)$$

In the relationships above, E_0, F_0, and par are suitable parameters, and Φ is a function of time t.

2.5.6 The Transformer and The Gyrator

The *transformer* TF and the *gyrator* GY are two important components that represent transformations of the power variables between their ports (Figs. 2.6f and 2.6g). Both have two ports; power is directed into the component at one port, and out of the component at the other. Thus, power is assumed to flow through the component.

An important characteristic of these elements is the conservation of power flow, i.e. power inflow is equal to power outflow. If we denote the power ports by labels 0 and 1, and the corresponding effort-flow variables by e_i and f_i ($i = 0,1$), this fact can be expressed by the relationship

$$e_0 f_0 = e_1 f_1 \qquad (2.26)$$

The Transformer

The transformer models levers, gears, electrical transformers, and similar devices. In robotics and multi-body mechanics, transformers are extensively used for the transformation of power variables between body frames.

In the transformer there is a linear relationship between the same types of port variables, i.e. efforts to efforts and flows to flows. Denoting the transformation ratio by m, we have

$$\left. \begin{array}{l} e_1 = m \cdot e_0 \\ f_0 = m \cdot f_1 \end{array} \right\} \qquad (2.27)$$

These relationships satisfy the power conservation relationship given by Eq. (2.26). It is sufficient to define one of these relationships; the other follows, owing to the power conservation requirement. There is some ambiguity in how to define the transformation ratio because the power conservation relation is also satisfied by the equations

$$\left. \begin{array}{l} e_0 = k \cdot e_1 \\ f_1 = k \cdot f_0 \end{array} \right\} \qquad (2.28)$$

The transformation ratio k in the last pair of equations is just the reciprocal of ratio m in the former equations, i.e. $k = 1/m$. The form to use is left to the discretion of the modeller.

The Gyrator

The gyrator is similar to the transformer, but relates to the other type of port variable, i.e. effort to flow. Denoting the gyrator ratios by m and k, the corresponding equations are

$$\left. \begin{array}{l} e_0 = m \cdot f_1 \\ e_1 = m \cdot f_0 \end{array} \right\} \qquad (2.29)$$

and alternatively,

$$\left. \begin{array}{l} f_0 = k \cdot e_1 \\ f_1 = k \cdot e_0 \end{array} \right\} \qquad (2.30)$$

Gyrators have their roots in the gyration effects well known from mechanics. Their use is essential in rigid-body dynamics. The gyrator is a more fundamental component than the transformer [1]. Two connected gyrators are equivalent to a transformer. A gyrator and an inertial component are equivalent to a capacitive element. Similarly, a source effort connected to a gyrator is equivalent to a source flow. Using such combinations makes it possible to reduce the set of elementary components necessary for physical modelling. We do not follow this approach here; there is little to be gained by using a smaller number of elementary components, as the resulting model would be more complicated and more abstract than necessary.

2.5.7 The Effort and Flow Junctions

Physical processes interact in such a way that there are restrictions on the possible values that efforts and flows can attain. Many physical laws express such constraints. In mechanics, forces and moments—including inertial effects—are governed by the momentum and the moment-of-momentum laws. In electricity, there is the Kirchhoff voltage law, and there are similar laws in other fields. Similar constraints on flows in rigid body mechanics are governed by the kinematical relative velocity laws, by the law of continuity of fluid flow in fluid mechanics, the Kirchhoff current law in electricity, etc. To satisfy such laws elementary components defined previously are connected to the junctions that impose constraints on efforts or flows. Such junctions are known as *effort* and *flow junctions* (Figs. 2.6h and 2.6i).

The Effort Junction

The *effort junction* is a multi-port component into which power flows in or out. We use the symbol e for this junction (instead of the more common 1 symbol). This junction also is called a *common flow junction* because the flows at all junction ports are the same, i.e.

$$f_0 = f_1 = ... = f_{n-1} \qquad (2.31)$$

where n is the number of ports at the junction. There is no power accumulation within the junction, thus the sum of the power inflows and power outflows equals zero

$$\pm e_0 f_0 \pm e_1 f_1 ... \pm e_{n-1} f_{n-1} = 0 \qquad (2.32)$$

In this equation the plus sign is used for the ports pointing towards the junction (positive power) and the minus sign for ports pointing away from the junction (negative power). Using Eq. (2.31) we get an equation of effort balance at the junction

$$\pm e_0 \pm e_1 ... \pm e_{n-1} = 0 \qquad (2.33)$$

The Flow Junction

The *flow junction* is similar to the effort junction, with the roles of efforts and flows exchanged. The flow junction is a multi-port component denoted by the symbol f (instead of the more common 0 symbol). This junction is also known as a *common effort* junction, as the efforts at all ports are the same, i.e.

$$e_0 = e_1 = ... = e_{n-1} \qquad (2.34)$$

There is also conservation of power flow through the junction (Eq. (2.32)). Thus, by Eq. (2.34) we get an equation of balance of flows at the junction

$$\pm f_0 \pm f_1 ... \pm f_{n-1} = 0 \qquad (2.35)$$

2.5.8 Controlled Elementary Components

The component constitutive relations introduced so far depend on port and internal variables only. In many instances it is also necessary to permit dependence on some external variable. This is the case when modelling controlled hydraulic restrictions in valves, variable resistors; capacitors, sources and other controlled components in electronics; and co-ordinate transformations in multi-body mechanics. For this purpose bond graphs use so called *modulated* components – modulated source efforts MSE and sources flows MSF, modulated transformers MTF and gyrators MGY. Some authors introduce other modulated components. We do not introduce such special components, but allow components to have control ports in addition to power ports. Fig. 2.7 shows some of the components with added control input or output ports.

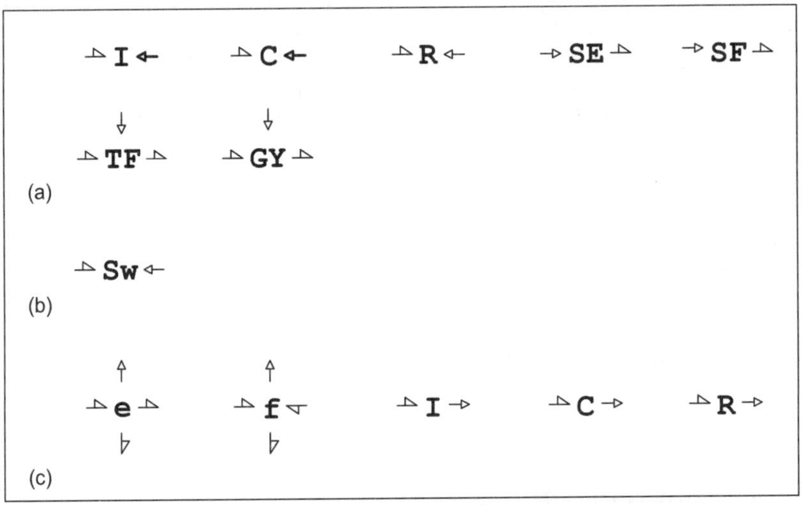

Fig. 2.7. Components with control ports. (a) Input, (b) Switch component, (c) Output

Components that have control input ports are called *controlled* (Fig. 2.7a). The constitutive relations for such components (see Sects. 2.5.1—2.5.6) can depend on the corresponding control variables. The transformers and gyrators must satisfy the power conservation requirement. This is satisfied not only by constant transformer and gyrator ratios, but also by the ratios dependent on a control variable c. Thus, the corresponding constitutive relations for controlled transformer and gyrators can have the same forms as given by Eqs. (2.27) to (2.30), but with variable transformation and gyration ratios, e.g.

$$\left.\begin{array}{l} e_1 = m(c) \cdot e_0 \\ f_0 = m(c) \cdot f_1 \end{array}\right\} \tag{2.36}$$

and

$$\left.\begin{array}{l} e_0 = m(s) \cdot f_1 \\ e_1 = m(s) \cdot f_0 \end{array}\right\} \tag{2.37}$$

respectively. The only components that cannot have control input ports are effort and flow junctions.

In addition to the already defined components that can be controlled, we define one specific component called the *switch*, denoted by Sw (Fig. 2.7b). This component has one power port and one control input port. The constitutive relation for the component is

$$e = 0, \ c > 0$$
$$f = 0, \ c \le 0$$
(2.38)

where e and f are the effort and flow variables of the power port, respectively, and c is the control variable. This component can be viewed as a controlled source that imposes the zero effort or zero flow condition, depending on the sign of the control variable. This component models hard stops and clearances in machines, switches and relays in electronics, and possibly of other discontinuous processes. The component can be generalised to allow effort or flow expressions, such as in sources (Eqs. (2.22) to (2.25)) and in systems with more complex switching logic than in Eq. (2.38).

Control output ports are used to access component variables that cannot be accessed any other way (Fig. 2.7c). Control output ports are commonly used for extraction of information on junction variables (efforts or flows). We also use such ports for access to the internal variables of inertial and capacitive components (momenta and displacements). They can be used for the extraction of information from other components, too.

2.6 Block Diagram Components

Finally, to complete the arsenal of components for modelling mechatronic systems, we define components that implement the input-output operations. These correspond to basic block diagram operations and can be used to define control laws of mechatronic devices, as well as for processing the results of simulations.

These are similar to other components, but can have only control input and control output ports (Fig. 2.8). We next give a short description of the elementary block diagram components.

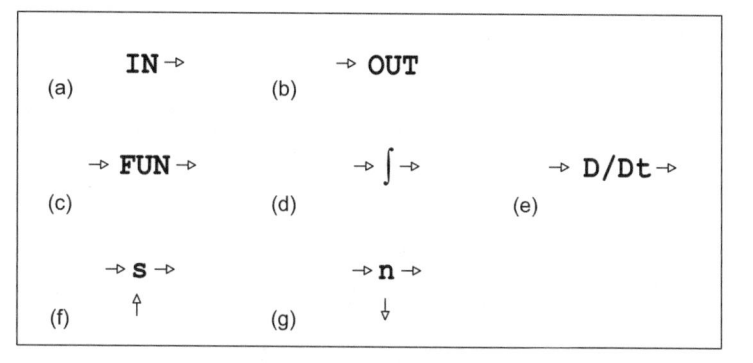

Fig. 2.8. Elementary block diagram components. (a) Input, (b) Output, (c) Function, (d) Integrator, (e) Differentiator, (f) Summator, (g) Node

2.6.1 The Input Component

The *input* component (Fig. 2.8a) generates control input action. This component can have only a single *output* port. Output from such a component could be of the form

$$c = \Phi(t, par) \tag{2.39}$$

where Φ is a suitable function of time and other parameters. This component typically is used to generate step input, sinusoidal input, pulse trains, and other input functions.

2.6.2 The Output Component

The *output* component (Fig. 2.8b) displays output signals. It can have only *input* ports. Typically such components are used for generating and displaying x-t and x-y plots, spectral diagrams, and the like.

2.6.3 The Function Component

The *function* component (Fig. 2.8c) generates output as a linear or non-linear function of its inputs. Thus the output is, in general,

$$c_{out} = \Phi(c_{in}, par) \tag{2.40}$$

The function can have one or more inputs but only a single output. Such a function can be used to represent linear gains, multiplications of inputs, or other non-linear operations on inputs.

2.6.4 The Integrator

As its name implies, this component gives the time integral of its input (Fig. 2.8d), i.e.

$$c_{out} = c_{out}(0) + \int_{0}^{t} c_{in}(t)dt \tag{2.41}$$

Obviously, this is a single input – single output component.

2.6.5 The Differentiator

The *differentiator* is similar to the integrator (Fig. 2.8e), but generates the time derivative of its input, i.e.

$$c_{out} = dc_{in} / dt \tag{2.42}$$

2.6.6 The Summator

The *summator* (Fig. 2.8f) gives the sum of its inputs, with optional positive or negative signs, i.e.

$$c_{out} = \pm c_1 \pm c_2 ... \pm c_n \qquad (2.43)$$

2.6.7 The Node

The node (Fig. 2.8g) serves for branching signals. This component has a single input and one or more outputs.

2.7 Modelling Simple Engineering Systems

The approach outlined in the previous sections can be used for the systematic computer-aided model development of engineering problems. We apply this approach to two simple, well-known problems, one from mechanical engineering and the other from electrical engineering. The technique is compared with the common bond graph modelling technique as given e.g. in [1]. We also consider a more complicated practical example from mechanical engineering (the See-saw problem).

2.7.1 Simple Body Spring Damper System

The first example models a single-degree-of-freedom mechanical vibration system (Fig. 2.9a). This consists of a body of mass m that translates along a floor, is connected to a wall by a spring of stiffness k, and has a damper with a linear friction velocity constant b. An external force F acts on the body. We neglect the Coulomb friction between the block and the floor for simplicity, as well as the weight of the body. In Fig. 2.9b the system is decomposed into its basic components. This is the free-body diagram well known from engineering mechanics. This decomposition clarifies the power flow direction assignment of component ports.

The bond graph model of the system is shown at the top of Fig. 2.10. The system consists of three components: spring, damper, and body. The spring and damper are represented by word models. There are two power ports, corresponding to the connection to the wall and to the body, respectively. The body is represented by a word model with three ports: one each for the spring and damper connections, and another for the external force. The wall (and floor) constitutes a component belonging to the system environment and is represented by a two-port word model. The source effort SE represents the external force applied on the system (body) by the environment. To complete the model, all the component ports are joined by bonds. Monitoring body motion is achieved by displaying the body position over time. Thus we add to the Body component a control-out port and connect it to a suitable Out(put) component used for displaying the body motion.

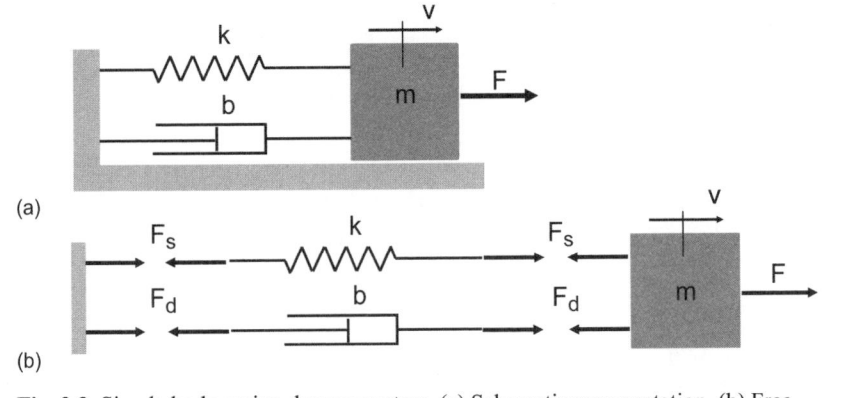

(a)

(b)

Fig. 2.9. Simple body spring damper system. (a) Schematic representation, (b) Free body diagram

The model at this level of abstraction has a structure that closely corresponds to the scheme of the system in Fig. 2.9a. The direction of power flow in the model is taken from the SE through the body, then through the spring and damper, and finally to the wall.

This corresponds to the physical situation in Fig. 2.9b. If the sense of the external force and the body velocity are as shown, the power at the external force port is positive; i.e., it is directed *into* the body. Assuming that the spring and damper resist the movement of the body—i.e., the sense of their forces is opposite to that of the body velocity—the powers at the corresponding ports are negative, and the power port arrows are directed *out* of the body. At the spring and damper ports, power again is positive, flowing into these components. Because of the direction of the body's force, according to *Newton's Third Law* their forces act in the opposite sense. A similar conclusion can be drawn regarding the wall side ports. Hence, by joining mechanical ports Newton's Third law is satisfied. Thus, to construct the bond graph model it is not necessary to draw a free-body diagram at all.

Next we develop the component models (Fig. 2.10). The force generated by the spring depends on the relative displacement (extension) of the spring. Thus, the model of the spring can be represented by a flow junction with three ports, two for connecting internally to the spring end ports and the third for the connection of the capacitive element that models the elasticity of the spring (Fig. 2.10 Spring). The junction variable is the force F_s in the spring; and the extension of the spring x_s is the generalized displacement of the capacitive element, with spring stiffness k taken as the parameter of the element. Thus, the constitutive relations for the capacitive element are (*see* Eqs. (2.9) and (2.11))

$$v_s = \dot{x}_s \tag{2.44}$$

and

$$F_s = k \cdot x_s \tag{2.45}$$

where v_s is the relative velocity of the spring ends.

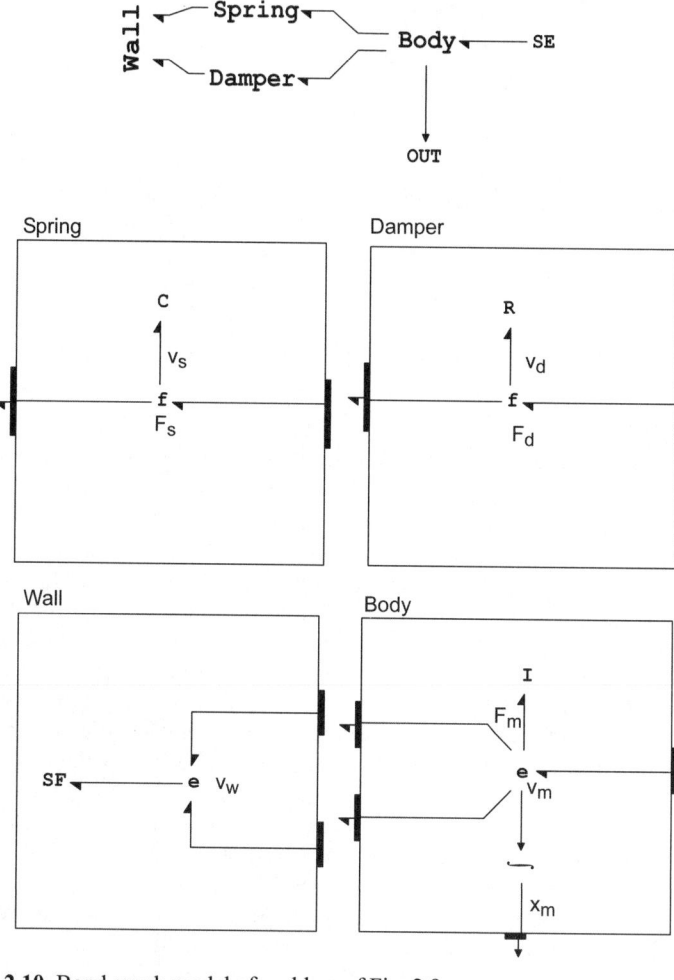

Fig. 2.10. Bond graph model of problem of Fig. 2.9

The damper has a similar model, with the resistive element used to model mechanical dissipation in the damper (Fig. 2.10 Damper). The junction variable F_d represents the force developed by the damper, v_d is the relative velocity of the damper ends and the velocity constant b is a parameter of the element. Assuming a linear constitutive relation for the resistive element we have (see Eq. (2.17))

$$F_d = b \cdot v_d \tag{2.46}$$

The third component of the system is the body of mass m (Fig. 2.10 Body). This component is represented by an effort junction that describes the balance of forces applied to the body including the inertial force of the block. This junction has four power ports: three for internal connections to the body ports and fourth for the connection of an inertial element. Denoting the body velocity taken as the junction variable by v_m (corresponding to velocity v of body in Fig. 2.9a) and the inertial force of the block with mass m taken as parameter by F_m, the constitutive relations of the inertial element are (see Eqs. (2.1) and (2.3))

$$F_m = \dot{p}_m \tag{2.47}$$

and

$$p_m = m \cdot v_m \tag{2.48}$$

To calculate the body position, a control output port is added to the junction, and the junction variable is fed to an integrator that outputs the body position x_m. The corresponding equation can be written as

$$\dot{x}_m = v_m \tag{2.49}$$

The next component is simply the source effort element SE, which generates the driving force on the system (body)

$$F = \Phi(t) \tag{2.50}$$

The spring and damper are connected to the fixed wall. The model of the wall is given in Fig. 2.10 Wall. The component consists of an effort junction with three ports. This describes the force balance at the wall. Two ports serve for the internal connection to the wall ports, and the third is for connecting the source flow, which imposes a zero wall velocity condition. The junction velocity is v_w. Thus, the relation for the source flow is

$$v_w = 0 \tag{2.51}$$

To complete the mathematical model of the system the equations of the effort and flow junctions are added. Corresponding variables can be found by following the bonds connected to junction ports until some elementary component is found that completes the bond. Thus, for the body effort junction the port effort variables are the spring force F_s, damper force F_d, inertia force F_m and driving force F, respectively. The equation of the effort balance at the junction thus reads

$$-F_s - F_d - F_m + F = 0 \tag{2.52}$$

If we denote by F_w the total force at the wall, the equation of effort balance at the wall junction reads

$$F_s + F_d - F_w = 0 \tag{2.53}$$

A similar equation can be written for the flow junctions. This time the summation is on the flows. Thus we have

$$-v_w - v_s + v_m = 0 \tag{2.54}$$

and

$$-v_w - v_d + v_m = 0 \qquad (2.55)$$

We therefore describe the motion of the system by eight equations of elements that describe the physical processes in the system, i.e. Eqs. (2.44) to (2.51), and four equations which involve the junctions Eqs. (2.52) to (2.55). There are, altogether, twelve differential and algebraic equations that have to be satisfied by twelve variables: F_s, v_s, x_s, F_d, v_d, F_m, v_m, p_m, F, F_w, v_w and x_m. Although we have arrived at a relatively large number of equations of motion for this simple problem, the equations are very simple, having on average only $27/12 = 2.25$ variables per equation.

The structure of the matrix of these equations is very sparse; this simplifies the solution process. We can simplify these equations further. Direct processing can be used to eliminate some, or all, of the algebraic variables (i.e., variables that are not differentiated). We can also simplify the bond graph first, then write the corresponding equations. We consider the second approach in more detail, as it leads to the sort of bond graphs usually found in the literature.

We can simplify the model by substituting every component at the top of Fig. 2.10 by its corresponding model, given at the bottom part of the same figure. The resulting bond graph is shown in Fig. 2.11a. The source flow on the left imposes zero wall velocity; thus, we can remove the effort junction and the source flow, as well as the two bonds connecting the junction to the flow junctions. We also remove the corresponding ports at the flow junctions. This yields a bond graph represented by Fig. 2.11b. We should also eliminate these flow junctions, for they are trivial, having only one power input port and one power output port. Thus, the C and R element ports can be connected directly to the right-hand effort junction ports. This results in the bond graph of Fig. 2.11c.

The model in Fig. 2.11c is much simpler than that in Fig. 2.10. The resulting equations now consist of

$$\left.\begin{array}{l} v_m = \dot{x}_s \\ F_s = k \cdot x_s \\ F_d = b \cdot v_m \\ F_m = \dot{p}_m \\ p_m = m \cdot v_m \\ F = \Phi(t) \\ -F_s - F_d - F_m + F = 0 \\ \dot{x}_m = v_m \end{array}\right\} \qquad (2.56)$$

We have reduced the system to eight equations with eight variables F_s, x_s, F_d, F_m, v_m, p_m, F, and x_m. This was achieved, however, by eliminating some variables that can be of interest, e.g. total force transmitted to the wall. This bond graph can be developed directly from Fig. 2.9a by the application of classical methods of bond graph modelling, as explained in [1].

Using this form of bond graph model, the equations of motion of the system can be developed in an even simpler form than that given above. It should be noted, however, that this is not true in general for engineering systems of practical interest. We address this matter in more detail in Sects. 2.9 and 2.10. The reduced model is, on the other hand, much more abstract. This makes it more difficult to understand and interpret: Unlike the component model of Fig. 2.10, there is no topological similarity to the system represented in Fig. 2.9a. A change in any part of the model affects the complete model. On the other hand, in the model of Fig. 2.10 we can change some of the components, leaving others unchanged. Such a model can be refined much more easily, thereby retaining the overall topological similarity to the physical model.

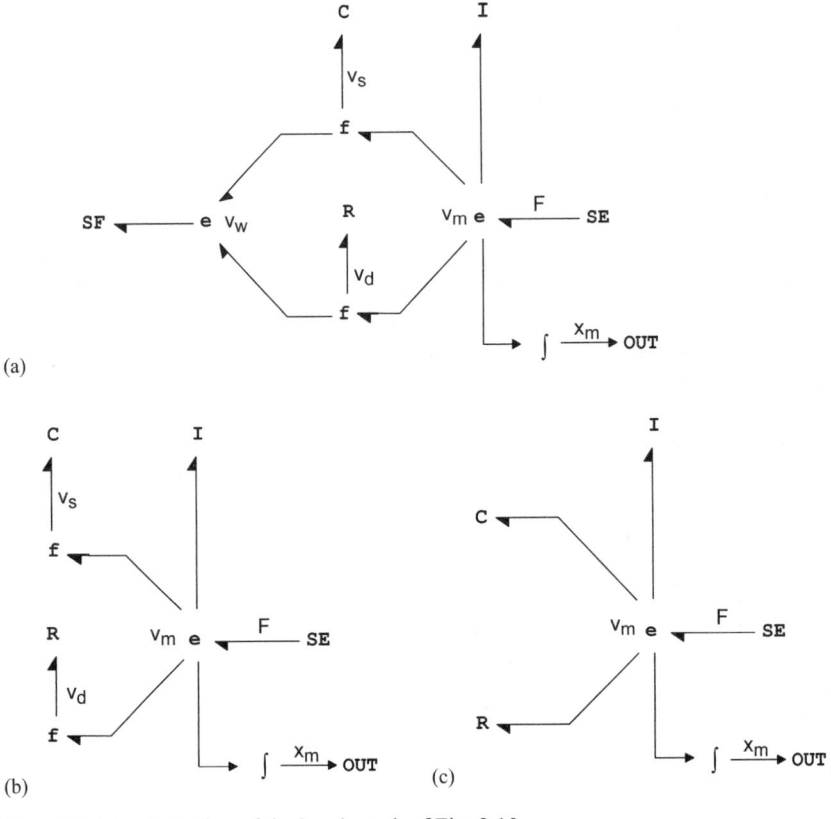

(a)

(b)

(c)

Fig. 2.11. Simplification of the bond graph of Fig. 2.10

2.7.2 The Simple Electrical Circuit

The second example considers the electrical Resistor Inductor Capacitor circuit (RLC circuit) shown in Fig. 2.12. The circuit consists of a series connection of a voltage source generating an electromotive force (e.m.f.) E, a resistor R, an inductor L, and a capacitor C. The polarities of the voltage drop across the electrical components are also shown, as well as the assumed direction of current flow. We can develop a bond graph model using an approach similar to the mechanical system analysed previously.

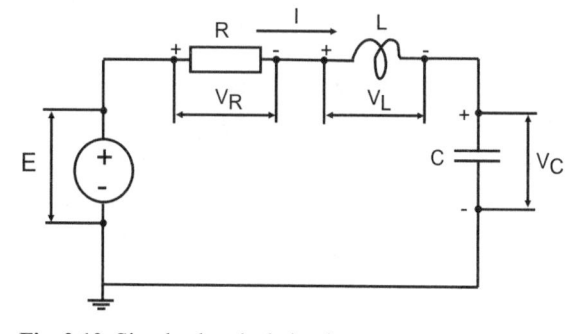

Fig. 2.12. Simple electrical circuit

We can represent the electrical components by suitable word models. But instead of the component names, we use the common electrical symbols. Standard component names—such as resistor, capacitor, and the like—are retained internally for compatibility with the usual word model representation. The resulting bond graph is shown at the top of Fig. 2.13.

The source voltage supplies electrical power to other parts of the circuit. The power port corresponding to the positive terminal is taken to be directed outward, and the other port inward. Power from the source flows through the resistor, inductor, and capacitor until the node component is reached where part of the power flow branches to the ground component. (Later it is shown that there is no power flow to the ground.) The other part returns back to the voltage source. The model has a very similar appearance to the electrical scheme in Fig. 2.12. What is different is the presence of power ports showing the assumed direction of power flow in the circuit. Thus, the correspondence of the bond graph model and the electrical scheme is really very close.

Component models for the voltage source, the resistor, the inductor, and the capacitor are shown in the lower part of the Fig. 2.13. Components have two ports used for connecting to other components. These are represented by source effort junctions with three ports. Two of these are used for internal connection to the component's ports, and the third is used for the connection of elementary components that describe the physical processes in the components. Power flow is chosen to flow into these elementary components. Thus, the effort of the elementary component port represents the voltage difference across the component. We model the components by idealised linear elements.

In the case of the resistor, the junction variable is the current i_R flowing through the component, and the voltage drop is v_R. Thus, the constitutive relation for the resistor is given by (Eq. (2.17))

$$v_R = R \cdot i_R \tag{2.57}$$

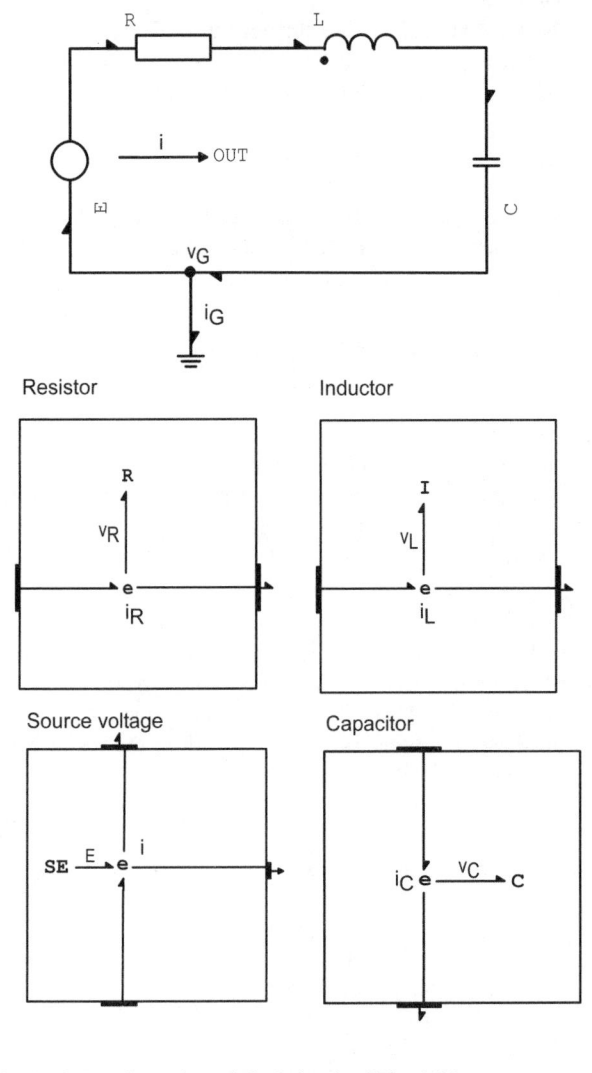

Fig. 2.13. Bond graph model of circuit of Fig. 2.12

with R the resistance parameter. Similarly for the inductor, the junction variable is the current i_L through the inductor and the voltage drop is v_L. If we denote the flux linkage of the coil by p_L, the constitutive relations read (see Eqs. (2.1) and (2.3))

$$v_L = \dot{p}_L \tag{2.58}$$

and

$$p_L = L \cdot i_L \tag{2.59}$$

where L is the inductance parameter of the inductor.

For the voltage source the joint variable is the current i through the source terminals (ports). The voltage E generated is described by the source effort element. The constitutive relation reads

$$E = \Phi(t) \tag{2.60}$$

Finally, for the capacitor, the junction variable is the current i_C through the component and the voltage drop is v_C. If we denote the capacitor charge by q_C, the constitutive relations can be written as (see Eqs. (2.9) and (2.11))

$$i_C = \dot{q}_C \tag{2.61}$$

and

$$v_s = q_C / C \tag{2.62}$$

where C is the capacitance.

The node component is simply another representation of the flow junction, and the ground is just the ground potential source effort. The node variable is the ground potential v_G. We take the ground potential as zero, hence the ground component constitutive relation reads

$$v_G = 0 \tag{2.63}$$

If we start from any of the component effort junctions and follow the bonds connected internally to the port, then out of the component to the next component port, and again into the component, we find that all effort junctions are interconnected. We thus can treat all these junctions as a single junction, the result being that all junction variables are, in essence, the same variable, i.e.

$$i \equiv i_R \equiv i_L \equiv i_C \tag{2.64}$$

Counting only ports connected to other components, the balance of efforts reads as follows

$$v_G + E - v_R - v_L - v_C - v_G = 0 \tag{2.65}$$

After cancellation of the ground potential v_G, we get

$$E - v_R - v_L - v_C = 0 \tag{2.66}$$

This is the Kirchhoff voltage law for the circuit.

Finally, if we denote the current drawn by the ground by i_G, and taking account of the identity represented in (2.64), the balance of flows at the node reads

$$-i - i_G + i = 0 \tag{2.67}$$

Again, owing to cancellation of current i, we obtain

$$-i_G = 0$$

or, after simplification

$$i_G = 0 \qquad (2.68)$$

This lengthy explanation shows that we arrive at the circuit equation by using the constitutive equations for elementary components and symbolically simplifying the junction equations. Thus, the mathematical model of the circuit consists of nine differential-algebraic equations (2.57) to (2.63), (2.66), and (2.68) with nine variables E, v_R, i, v_L, p_L, v_C, q_C, v_G and i_G. Eqs. (2.63) and (2.68) are trivial and could be eliminated from the system.

As in the previous example, we can simplify the bond graph instead of the equations. We first substitute the bond graph model of the components from the lower part of Fig. 2.13 into the system bond graph at the top (Fig. 2.14a).

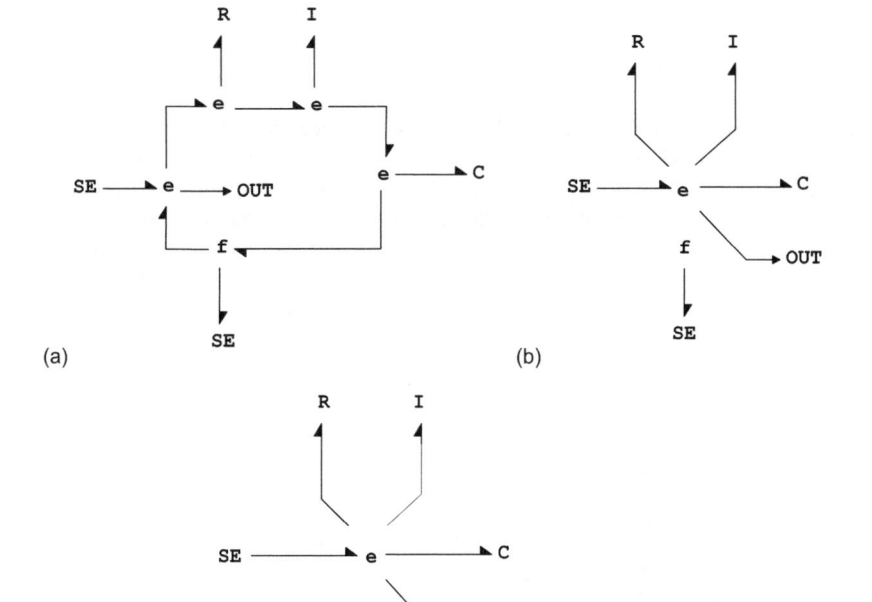

Fig. 2.14. Simplification of model of Fig. 2.13

We again see that the four effort junctions are interconnected and can be condensed into a single junction, retaining only the ports connected to other components. In addition, we see that this junction is connected to the same flow junction by two ports of opposite power-flow sense. Hence, such ports can be disconnected, and then removed. The corresponding ports of the flow junction must be removed, too. This yields the bond graph shown in Fig. 2.14b. We now have a flow junction connected only to a ground source effort. These can be removed, too. This results in the final simplified system bond graph (Fig. 2.14c). The last bond graph can be described by the same equations as before, but without Eqs. (2.63) and (2.68).

The above procedure shows that, instead of the simplification of junction Eqs. (2.65) and (2.67), we could directly arrive at Eqs. (2.66) and (2.68) by noting that interconnected effort junctions are connected to the *same* flow junction. Corresponding junction ports then could be treated as internal, and not taken into account when writing the junction equations.

Comparing bond graphs in Figs. 2.13 and 2.14c, we draw similar conclusions as in the previous example: The model in Fig. 2.13 is much easier to interpret and upgrade. Even people who are not too familiar with bond graphs could understand such a model. On the other hand, it retains the advantages that bond graphs enjoy over other modelling methods.

2.7.3 A See-saw Problem

As a third problem, we develop a model of a simple see-saw often found in children's playgrounds (Fig. 2.15). On a much larger scale, this same problem is known as the swing boat at the fair ground. The system consists of a platform that can rotate around a horizontal pin O, fixed in a frame, and having a body at each end, e.g. a boy and a girl sitting on the see-saw seats. If one of the bodies is pushed down, and then released, the system will begin to oscillate around its equilibrium position.

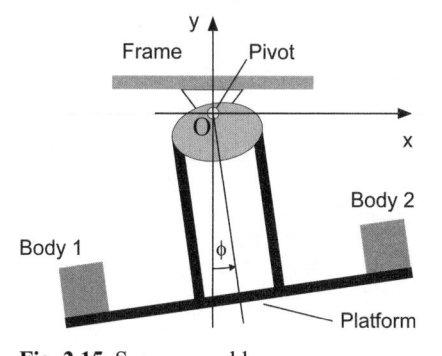

Fig. 2.15. See-saw problem

This problem is more complicated than those presented previously, as it consists of three interconnected bodies moving in the vertical plane. The model could be developed more easily by treating the system as a physical pendulum. Instead, we consider it as a multi-body system and demonstrate that the model can be developed systematically by decomposition.

We start by defining the overall model structure (Fig. 2.16). The word models **Body 1** and **Body 2** represent the bodies on the platform. They each have a port that corresponds to the location where the body acts on the platform. The component **Platform** has four ports, two of which correspond to the places where the bodies act, a third for connecting to the Pivot component, and the last for the input of information on the rotation angle. The Pivot permits only rotation of the platform about a horizontal pin fixed in the Frame.

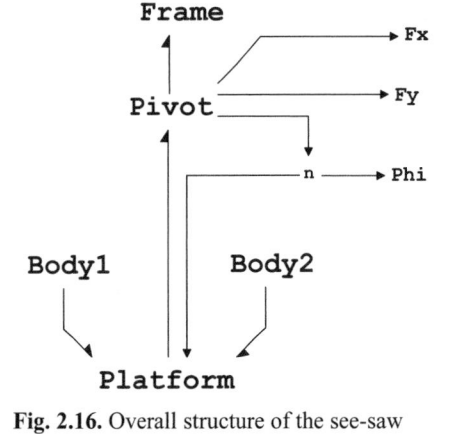

Fig. 2.16. Overall structure of the see-saw

We assume that the bodies are firmly placed on the platform and move with it. Hence, we join the ports of the bodies to the corresponding platform ports by bonds. We assume the power flow sense from Body 1 and Body 2 through the Platform and through the Pivot to the Frame. Further, information on the force at the pivot is of interest. Thus, we take the components of force at the Pivot and feed them to the displays Fx and Fy. Similarly, we extract information on the rotation angle and feed it to the node n. This information branches further to the Platform and to the display component Phi.

We proceed by developing models of the components. This requires defining more precisely the interactions taking place between them. Motion of the system is described in a global co-ordinate frame Oxy. The origin O is placed at the point of rotation of the platform in the vertical plane; the axis y is directed upward (Fig. 2.15).

Separate effort junctions are used for the summation of the x and y components of forces on the bodies (Fig. 2.17 Body 1 and Body 2). The junction variables are the x and y components of the velocities of the bodies. The junctions are connected internally to their respective ports. The order of connection is the x component first, then the y component, going from left to right. This order of connection is also used for the other ports. Hence, Body1 and Body2 port variables are pairs of effort flow vectors F_1/v_1 and F_2/v_2, respectively.

The inertial effects of the bodies in the x- and y-directions are represented by the inertial elements I connected to the corresponding effort junctions, with power flow directed into the inertial elements. The weights of the bodies, acting in the y direction only, are represented by source efforts connected to the y component junctions, with power flow directed into the junctions.

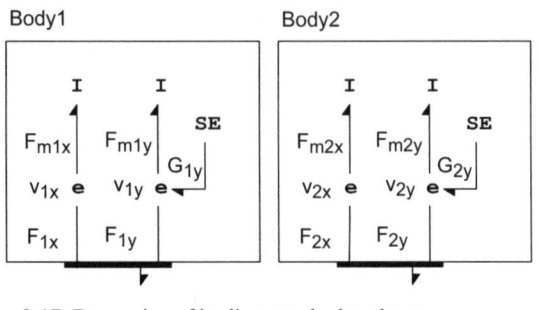

Fig. 2.17. Dynamics of bodies attached to the see-saw

The equations of motion of the bodies can be obtained directly from the bond graphs of Fig. 2.17. The relevant variables are also shown in the figure.

Body 1:

$$\left.\begin{aligned}
\dot{p}_{m1x} &= F_{m1x} \\
\dot{p}_{m1y} &= F_{m1y} \\
p_{m1x} &= m_1 v_{1x} \\
p_{m1y} &= m_1 v_{1y} \\
G_{1y} &= -m_1 g \\
-F_{1x} - F_{m1x} &= 0 \\
-F_{1y} - F_{m1y} + G_{1y} &= 0
\end{aligned}\right\} \tag{2.69}$$

Body 2:

$$\left.\begin{aligned}
\dot{p}_{m2x} &= F_{m2x} \\
\dot{p}_{m2y} &= F_{m2y} \\
p_{m2x} &= m_2 v_{2x} \\
p_{m2y} &= m_2 v_{2y} \\
G_{2y} &= -m_2 g \\
-F_{2x} - F_{m2x} &= 0 \\
-F_{2y} - F_{m2y} + G_{2y} &= 0
\end{aligned}\right\} \tag{2.70}$$

The masses of the bodies are m_1 and m_2, and g is the gravitational acceleration.

The Frame simply fixes the pivot, about which the platform rotates, against translation and rotation (Fig. 2.18). The equations are:

Frame :

$$\left.\begin{aligned}
v_{Px} &= 0 \\
v_{Py} &= 0 \\
F_{3x} &= F_{Fx} \\
F_{3y} &= F_{Fy} \\
\omega_P &= 0 \\
M_P &= M_F
\end{aligned}\right\} \tag{2.71}$$

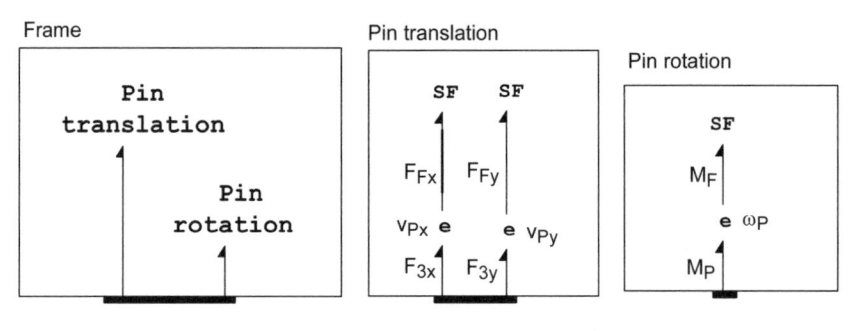

Fig. 2.18. Model of the see-saw frame

The Pivot similarly prevents translation of the platform with respect to the pin (Fig. 2.19). Two flow junctions are inserted to extract information about the x and y components of the force on the pin. Rotation is normally permitted, and it is assumed that this is frictionless. It is an easy matter to add friction, if required. For example, a resistive component R could be used instead of the source effort. The signal taken from the effort junction is integrated to get the platform rotation angle. The governing equations are again very simple:

Pivot :

$$
\left.
\begin{aligned}
v_{3x} - v_{Px} &= 0 \\
v_{3y} - v_{Py} &= 0 \\
\omega - \omega_s - \omega_P &= 0 \\
M_P - M_R &= 0 \\
M_R &= 0 \\
\dot{\phi} &= \omega_s
\end{aligned}
\right\}
\qquad (2.72)
$$

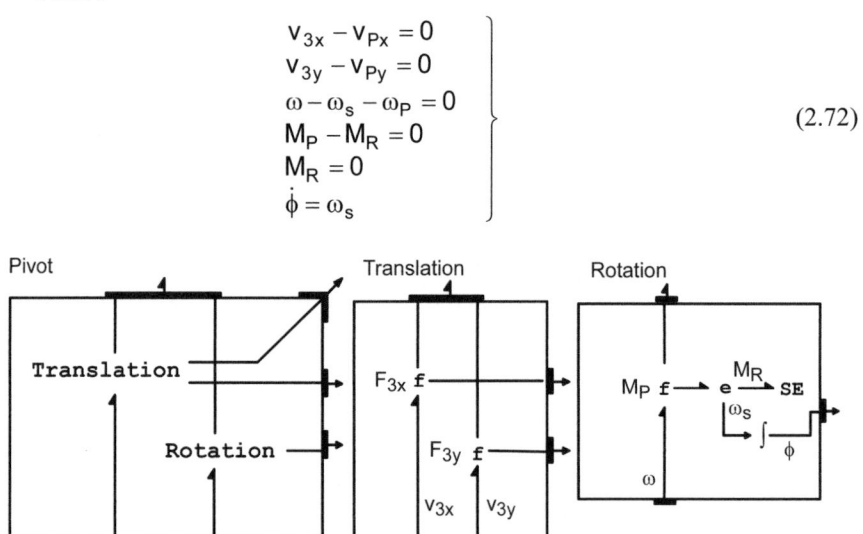

Fig. 2.19. Model of the see-saw pivot

The platform acts as a transformer of the velocities of the attached bodies. Simultaneously, transformation of the reaction forces of the bodies also takes place. To develop the bond graph model describing these interrelations we analyse the general plane motion of the platform in the global co-ordinate frame **Oxy** (Fig. 2.20).

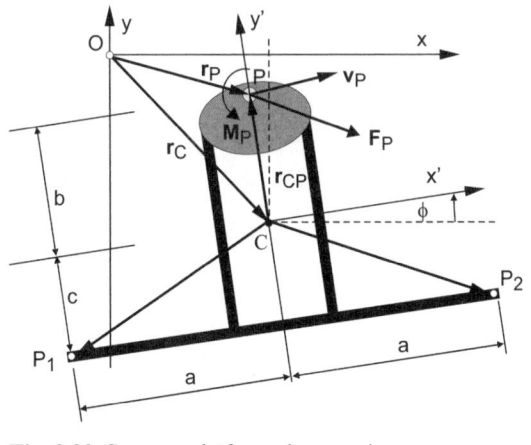

Fig. 2.20. See-saw platform plane motion

The position and orientation of the platform is described with reference to a body frame **Cx'y'** with the origin at its mass centre. The position vector of the origin **C** is described by a column vector of its global co-ordinates, i.e.

$$\mathbf{r_C} = \begin{pmatrix} x_C \\ y_C \end{pmatrix} \tag{2.73}$$

Orientation of the body is defined by the rotation matrix (see e.g. [5])

$$\mathbf{R} = \begin{pmatrix} \cos\varphi & -\sin\varphi \\ \sin\varphi & \cos\varphi \end{pmatrix} \tag{2.74}$$

The vector of the relative position of a point **P** in the body with respect to the origin of the body frame can be expressed in the body frame by a vector of its coordinates

$$\mathbf{r'_{CP}} = \begin{pmatrix} x'_{CP} \\ y'_{CP} \end{pmatrix} \tag{2.75}$$

The position of the same point **P** with respect to the global frame is defined by the vector of its global co-ordinates

$$\mathbf{r_P} = \begin{pmatrix} x_P \\ y_P \end{pmatrix} \tag{2.76}$$

The relationship between these vectors is given by

$$\mathbf{r_P} = \mathbf{r_C} + \mathbf{r_{CP}} \tag{2.77}$$

Note that vector $\mathbf{r_{CP}}$ is the relative vector expressed in the global frame, i.e.

$$\mathbf{r}_{CP} = \begin{pmatrix} x_{CP} \\ y_{CP} \end{pmatrix} \tag{2.78}$$

The relationship between the vectors of Eqs. (2.75) and (2.78) is given by the co-ordinate transformation

$$\mathbf{r}_{CP} = \mathbf{R}\mathbf{r}'_{CP} \tag{2.79}$$

Substituting the rotation matrix of Eq. (2.74) and evaluating the resulting expression yields

$$\begin{pmatrix} x_{CP} \\ y_{CP} \end{pmatrix} = \begin{pmatrix} x'_{CP}\cos\phi - y'_{CP}\sin\phi \\ x'_{CP}\sin\phi + y'_{CP}\cos\phi \end{pmatrix} \tag{2.80}$$

The velocity of a point P in the body can be found by taking the time derivative of Eq. (2.77), i.e.

$$\mathbf{v}_P = \mathbf{v}_C + \mathbf{v}_{CP} \tag{2.81}$$

which relates the velocity of the point P to the velocity of the origin C of the body frame and to the relative velocity of the point P with respect to the point C. These velocity vectors are expressed by their components in the global frame as

$$\mathbf{v}_P = \begin{pmatrix} v_{Px} \\ v_{Py} \end{pmatrix}, \quad \mathbf{v}_C = \begin{pmatrix} v_{Cx} \\ v_{Cy} \end{pmatrix}, \quad \mathbf{v}_{CP} = \begin{pmatrix} v_{CPx} \\ v_{CPy} \end{pmatrix} \tag{2.82}$$

and are defined by

$$\mathbf{v}_P = \frac{d\mathbf{r}_P}{dt}, \quad \mathbf{v}_C = \frac{d\mathbf{r}_C}{dt}, \quad \mathbf{v}_{CP} = \frac{d\mathbf{r}_{CP}}{dt} \tag{2.83}$$

Taking the time derivative of Eq.(2.79), we arrive at the expression for the relative velocity

$$\mathbf{v}_{CP} = \frac{d\mathbf{R}}{dt}\mathbf{r}'_{CP} \tag{2.84}$$

Note that the vector $\mathbf{r}'_{CP}$ is constant with time. The time derivative of the rotation matrix $\mathbf{R}$ yields

$$\frac{d\mathbf{R}}{dt} = \begin{pmatrix} -\sin\phi & -\cos\phi \\ \cos\phi & -\sin\phi \end{pmatrix} \cdot \frac{d\phi}{dt} \tag{2.85}$$

The time derivative of the body rotation angle is the body angular velocity

$$\omega = \frac{d\phi}{dt} \tag{2.86}$$

Thus, substitution of Eqs (2.85) and (2.86) into Eq. (2.84) yields

$$\mathbf{v}_{CP} = \mathbf{T}\omega \tag{2.87}$$

where $\mathbf{T}$ is the transformation matrix, given by

$$\mathbf{T} = \begin{pmatrix} -\dot{x}_{CP}\sin\phi - \dot{y}_{CP}\cos\phi \\ \dot{x}_{CP}\cos\phi - \dot{y}_{CP}\sin\phi \end{pmatrix} \tag{2.88}$$

Compared with Eq. (2.80), this matrix also can be expressed as

$$\mathbf{T} = \begin{pmatrix} -y_{CP} \\ x_{CP} \end{pmatrix} \tag{2.89}$$

Eq. (2.81) and Eqs. (2.86)—(2.88) are the basic relations describing the kinematics of rigid body motion in a plane. They apply to points P_1, P_2, and P_3, where the bodies and the pin act on the platform. Next, we consider the kinetic relationships of forces and moments applied to the platform.

A force $\mathbf{F}$ applied to the platform can be described by a vector of its rectangular components in the global frame, i.e.

$$\mathbf{F} = \begin{pmatrix} F_x \\ F_y \end{pmatrix} \tag{2.90}$$

The power delivered at point P is given by $\mathbf{v_P}^T\mathbf{F}$, where the superscript T denotes matrix transposition. From Eq. (2.81) and (2.87) we get

$$\mathbf{v_P}^T\mathbf{F} = \mathbf{v_C}^T\mathbf{F} + \mathbf{T}^T\mathbf{F}\omega \tag{2.91}$$

Evaluating the leading part of the second term on the right of Eqs. (2.91) yields

$$\mathbf{T}^T\mathbf{F} = -y_{CP}F_x + x_{CP}F_y \tag{2.92}$$

We recognise this as the moment M_C of the force at P about point C, thus

$$M_C = \mathbf{T}^T\mathbf{F} \tag{2.93}$$

By substituting in Eq. (2.91), we finally arrive at the equations of power transfer across the body

$$\mathbf{v_P}^T\mathbf{F} = \mathbf{v_C}^T\mathbf{F} + M_C\omega \tag{2.94}$$

This equation can also be read as a statement of *force equivalents*, well known from Engineering Mechanics (see e.g. [6]). That is, a force applied at a point P is equivalent to the same force applied at a different point C plus the moment of force about C. If at point P a torque also acts, its moment M_P should be added, too. Eqs. (2.93) and (2.94), jointly with Eqs. (2.81) and (2.86) to (2.88), constitute the fundamental equations of rigid-body motion in a plane. To complete the dynamical equations we need only add the inertias of translation and rotation. These equations clearly indicate how to represent the dynamics of the platform (Fig. 2.21 Platform).

At every point of application of the force (platform ports) we introduce a component f corresponding to the vector summation of the velocities, as given by Eq. (2.81). These components contain two flow junctions. The corresponding junction variables are the x and y components of the force at the port (Fig. 2.21 f). Next, we introduce an effort junction e corresponding to the angular velocity of the body; and the component CM, corresponding to the velocity of the mass centre

(Fig. 2.21 Platform and CM). We connect the vector flow junctions f to the angular velocity effort junction e by the components labelled LinRot, and to the mass centre velocity junction component CM.

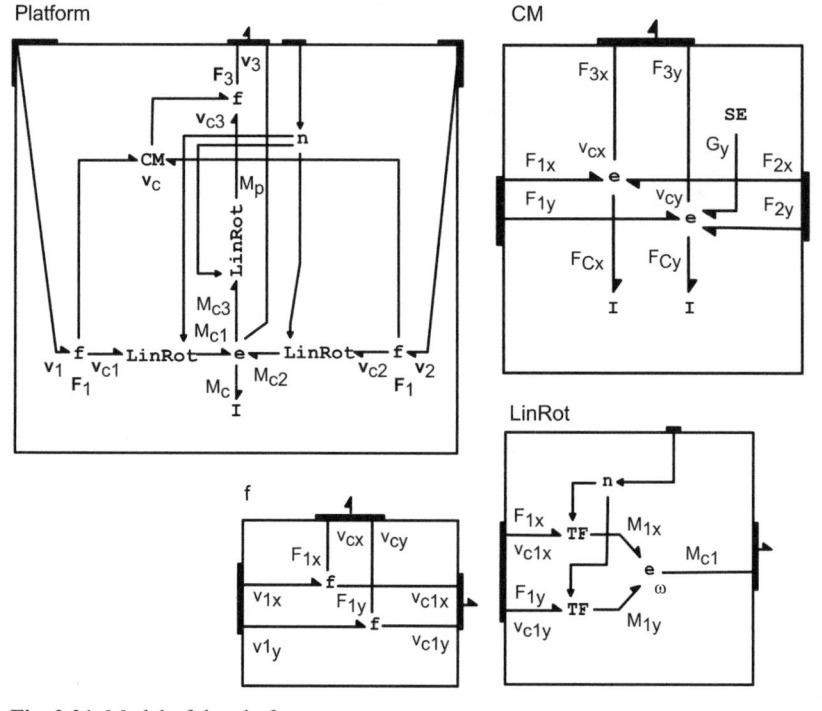

Fig. 2.21. Model of the platform

The LinRot components[1] represent the transformations given by Eqs. (2.87) and (2.93). The components consist of two transformers and a single effort junction that implements the transformation given by matrix (2.88) (Fig. 2.21 LinRot). The necessary information on the angle of rotation of the see-saw is taken from the input port. In addition to these force effects, any moment at a port is transmitted directly to the rotation effort junction e. An inertial element added to the junction represents the rotational inertia of the platform. The translation inertia is represented in component CM which consists of two effort junctions that add inertial elements corresponding to the x and y motion (Fig. 2.21 CM). The platform gravity is also added here.

The mathematical model of the platform can be written directly from the Fig. 2.21. Respective variables are given in the figure, and parameters a, b, and c are dimensions shown in Fig. 2.20; m is the platform mass, and I_C is its mass moment of inertia about its mass centre. The equations read:

---

[1] Because of space limitation, only one of the f and LinRot components is shown. The others have a similar structure.

Platform – left side:

$$
\left.\begin{aligned}
& v_{1x} - v_{C1x} - v_{Cx} = 0 \\
& v_{1y} - v_{C1y} - v_{Cy} = 0 \\
& v_{C1x} = (a \cdot \sin\phi + c \cdot \cos\phi) \cdot \omega \\
& v_{C1y} = (-a \cdot \cos\phi + c \cdot \sin\phi) \cdot \omega \\
& M_{1x} = (a \cdot \sin\phi + c \cdot \cos\phi) \cdot F_{1x} \\
& M_{1y} = (-a \cdot \cos\phi + c \cdot \sin\phi) \cdot F_{1y} \\
& -M_{C1} + M_{1x} + M_{1y} = 0
\end{aligned}\right\}
\qquad (2.95)
$$

Platform – right side:

$$
\left.\begin{aligned}
& v_{2x} - v_{C2x} - v_{Cx} = 0 \\
& v_{2y} - v_{C2y} - v_{Cy} = 0 \\
& v_{C2x} = (-a \cdot \sin\phi + c \cdot \cos\phi) \cdot \omega \\
& v_{C2y} = (a \cdot \cos\phi + c \cdot \sin\phi) \cdot \omega \\
& M_{2x} = (-a \cdot \sin + c \cdot \cos\phi) \cdot F_{2x} \\
& M_{2y} = (a \cdot \cos\phi + c \cdot \sin\phi) \cdot F_{2y} \\
& -M_{C2} + M_{2x} + M_{2y} = 0
\end{aligned}\right\}
\qquad (2.96)
$$

Platform – upper side:

$$
\left.\begin{aligned}
& -v_{3x} + v_{C3x} + v_{Cx} = 0 \\
& -v_{3y} + v_{C3y} + v_{Cy} = 0 \\
& v_{C3x} = -(b \cdot \cos\phi) \cdot \omega \\
& v_{C3y} = -(b \cdot \sin\phi) \cdot \omega \\
& M_{3x} = -(b \cdot \cos\phi) \cdot F_{3x} \\
& M_{3y} = -(b \cdot \sin\phi) \cdot F_{3y} \\
& M_{C3} - M_{3x} - M_{3y} = 0
\end{aligned}\right\}
\qquad (2.97)
$$

Platform – mass centre motion:

$$
\left.\begin{aligned}
& \dot{p}_x = F_{Cx} \\
& \dot{p}_y = F_{Cy} \\
& p_x = m \cdot v_{Cx} \\
& p_y = m \cdot v_{Cy} \\
& G_y = -mg \\
& F_{1x} + F_{2x} - F_{3x} - F_{Cx} = 0 \\
& F_{1y} + F_{2y} - F_{3y} - F_{Cy} + G_y = 0
\end{aligned}\right\}
\qquad (2.98)
$$

Platform – rotation:

$$
\left.\begin{aligned}
& \dot{K}_C = M_C \\
& K_C = I_C \cdot \omega \\
& M_{C1} + M_{C2} - M_{C3} - M_P - M_C = 0
\end{aligned}\right\}
\qquad (2.99)
$$

The model consists of fifty-seven very simple equations. No substitutions or other simplifications have been made, as we wished to develop the model strictly by describing every elementary component in terms of its variables and parameters.

This procedure, based on systematic decomposition, results in multi-level models. Such models can be developed and changed more easily, if necessary, than conventional "flat' models. Some of the components can also be reused in other models. For example, the Platform component can be used for problems dealing with the plane motion of rigid bodies. For comparison, a flat model corresponding to the model developed above is shown in Fig. 2.22. Such a model, however, is not easy to follow, particularly for people unfamiliar with bond graphs: There are many bonds, and it is not easy even to draw them correctly! It is thus more susceptible to errors and more difficult to change.

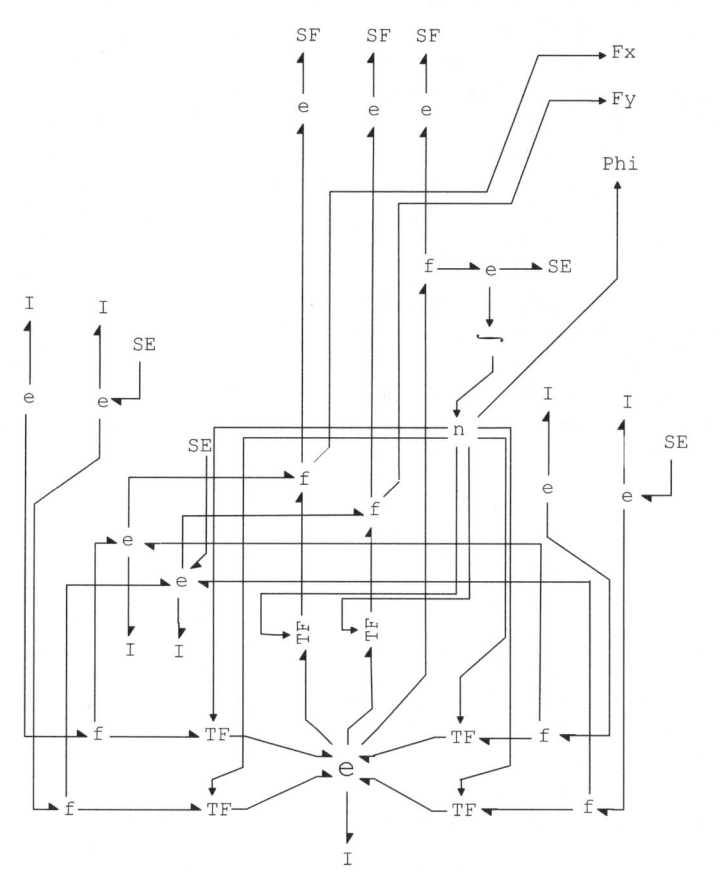

Fig. 2.22. Single level model of the see-saw

2.8 Causality of Bond Graphs

2.8.1 The Concept of Causality

The concept of *causality*, or *cause-effect relationships*, was introduced in the bond graph method to define the *computational* structure of the resulting mathematical equations at bond-graph level. Thus, the physical and computational structure of the model is defined in parallel during the modelling stage. It should be stressed that physical laws do not imply any causal preference: There is no physical reason to treat forces as the cause, and velocities of the body motions as effects; or voltages as the cause, and currents in circuits as effects. The assignment of causality can be looked on as a convenient—but not an essential—part of the modelling task. Further, it is arguable if it is convenient at all, in particular when using the object-oriented paradigm in simulation model building.

We nevertheless briefly describe causality and its consequences in bond graphs because of their close connection with bond graph theory (see e.g. [1,2]).

Causality means that, at every port of an elementary component, one of the power variables is the input (cause) and the other is the output (effect). Thus, power variables are treated as a pair of signals. Because bond lines in bond graphs connect ports, the same variable is the input variable at one port and the output variable at the other connected port. Causal relationships between connected port variables are depicted in the bond graph literature by *causal strokes*. These are short lines drawn at one bond end (port) perpendicular to the bond (Fig 2.23). This stroke denotes that the effort at the port is the input to the element and the flow variable is the output (Figs. 2.23a and b). At the other port just the opposite relation is valid; that is, the flow variable is the input and the effort variable is the output. Causal stroke assignment is independent of the power flow direction (Figs. 2.23c and d).

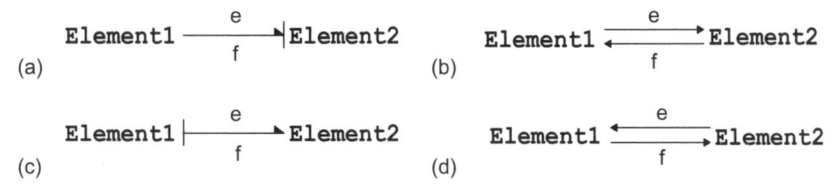

Fig. 2.23. Causality assignment denoted by strokes. (a) and (c) Possible stroke attachments, (b) and (d) Meaning of the attachments

2.8.2 Causalities of Elementary Components

The causality assignment defines the input-output relationship of the elementary component constitutive relations. Possible types of causalities of elementary component ports are summarised in Fig. 2.24.

Source effort ports (Fig. 2.24a) can have only one possible type of causality, i.e. the effort is always the output, because flow at the input port is not defined. Similarly, at source flow ports the output is the flow, because the effort is not defined (2.25b). Thus, sources have *fixed causalities*.

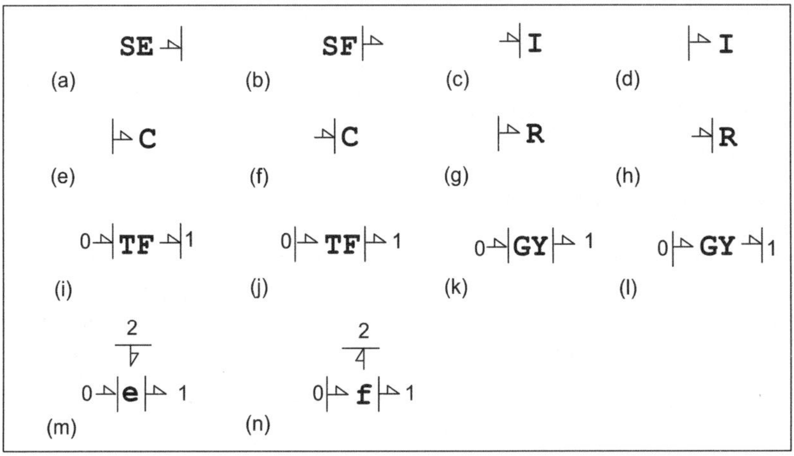

Fig. 2.24. Causalities of elementary components

The inertial component can have one of two possible causalities. If the effort at the port is the input and the flow is the output (Fig. 2.24c), then the constitutive relations are given by Eqs. (2.2) and (2.5)

$$p = p_0 + \int_0^t e\, dt \left.\begin{array}{c} \\ \\ \end{array}\right\} \qquad (2.100)$$
$$f = \Phi^{-1}(p, par)$$

Such causality is known as *integrating causality* because *integration* is used to calculate the output flow.

The other possibility is that the flow is the input and effort is the output (Fig. 2.24d). In this case evaluation proceeds by Eqs. (2.4) and (2.1), i.e.

$$p = \Phi(f, par) \left.\begin{array}{c} \\ \end{array}\right\} \qquad (2.101)$$
$$e = \dot{p}$$

This type of causality is known as *differentiation causality* because *differentiation* is used to calculate the output.

Analogous causal forms exist for capacitive ports. If we take the flow variable as input and the effort variable as output (Fig. 2.24e), by Eqs. (2.10) and (2.13) we have

$$q = q_0 + \int_0^t f dt \left.\vphantom{\int_0^t}\right\}$$

$$e = \Phi^{-1}(q, par) \qquad\qquad (2.102)$$

In this case we have integrating causality. On the other hand, if effort is the input and flow is the output (Fig. 2.24f), calculation proceeds by Eqs. (2.12) and (2.9), i.e.

$$\left.\begin{array}{l} q = \Phi(e, par) \\ f = \dot{q} \end{array}\right\} \qquad\qquad (2.103)$$

This yields differentiation causality.

Of these two possible causalities, integrating causality is *preferred* because integration is more easily implemented than differentiation. This is because integration works on the past values, whereas differentiation involves prediction.

For the resistor there are also two possible causalities. If the flow is the input and effort is the output (Fig. 2.24g), evaluation of the output is done using Eq. (2.18), i.e.

$$e = \Phi(f, par) \qquad\qquad (2.104)$$

On other hand, if the effort is input (Fig. 2.24h) and the flow is output, calculation is implemented by Eq. (2.19), i.e.

$$f = \Phi^{-1}(e, par) \qquad\qquad (2.105)$$

The first one is sometimes called *resistive causality*, and the other *conductive causality*. Preference of one over the other depends on which form is better defined, as some non-linear constitutive relationships are not invertible.

Transformers can also have two possible types of causality. If the effort at one port is the input at other port, then the effort has to be the output; the same applies to the flows. For causality as expressed in Fig. 2.24i, the constitutive relations are given by Eq. (2.27), i.e.

$$\left.\begin{array}{l} e_1 = m \cdot e_0 \\ f_0 = m \cdot f_1 \end{array}\right\} \qquad\qquad (2.106)$$

On other hand, if causality is as in Fig. 2.24j, the constitutive relations are given by Eq. (2.28), i.e.

$$\left.\begin{array}{l} e_0 = k \cdot e_1 \\ f_1 = k \cdot f_0 \end{array}\right\} \qquad\qquad (2.107)$$

Two possible causalities for gyrators are shown in Figs. 2.24k and 2.24l, respectively. Inputs at the gyrator ports can represent either the efforts or the flows. For the case in which inputs are *efforts*, output flows are given by Eq. (2.30), i.e.

$$\left.\begin{array}{l} f_0 = k \cdot e_1 \\ f_1 = k \cdot e_0 \end{array}\right\} \qquad\qquad (2.108)$$

Similarly, if the inputs are *flows*, the output efforts are given by Eq. (2.29), i.e.

$$e_0 = m \cdot f_1$$
$$e_1 = m \cdot f_0$$

$$(2.109)$$

Effort junctions represent the balance of efforts at junction ports. Hence, an effort can only be the output at one port; all others must be inputs. For the effort junction in Fig. 2.24m, the effort at port 2 is taken as the output and all others are inputs. Thus, output effort e_2 is given by

$$e_2 = -e_0 + e_1$$

A similar statement holds for the flow junctions: Flow can only be the output at one port; all other flows must be inputs. For the flow junction in Fig. 2.24n, the output flow f_1 is given by

$$f_1 = f_0 - f_2$$

In the expression for output effort or flow, the sign of all input efforts or flows should be positive if the sense of power flow is opposite to the sense of the output power flow. Otherwise, the sign is negative.

2.8.3 The Procedure for Assigning Causality

The causalities of junction, transformer, and gyrator ports are interrelated and thus imply constraints on the causalities of connected elements. The causalities of the complete bond graph can be assigned in a systematic way. The usual procedure is known generally as the *sequential causal assignment procedure* (SCAP) [1]. This procedure is summarised as follows:

1. Choose a source effort or source flow and assign causality to it. Extend the causality assignment, if possible, to the connected effort and flow junctions, the transformers, and the gyrators. Proceed in a like fashion until the causality of all sources has been assigned.
2. Choose an inertial or a capacitive element and assign to it the preferred (integrating) causality. Extend the causality assignment as in 1. Proceed until the causality of all such elements has been assigned. Otherwise, if the causality assignment of the all bonds is not achieved, go to the next step.
3. Assign causality to an unassigned resistor using any acceptable causality. Extend the assignment to the connected effort and flow junctions, transformers, and gyrators. Proceed until the causality of all resistors has been assigned. Otherwise, if the causality of all bonds is not already assigned, go to the next
4. Assign causality to any remaining bond. Extend the causality assignment to effort and flow junctions, transformers, and gyrators. Proceed until the causality of all bonds is assigned.

The bond graph to which causality has been assigned usually is termed a *causal* bond graph. Otherwise, it is termed an *acausal* bond graph.

The procedure can be illustrated on the simple body-spring-damper problem of Sect. 2.7.1. Other more complex examples can be found in, for example, [1]. Here we use the simplified bond graph of Fig. 2.11c, which is repeated as Fig. 2.25a.

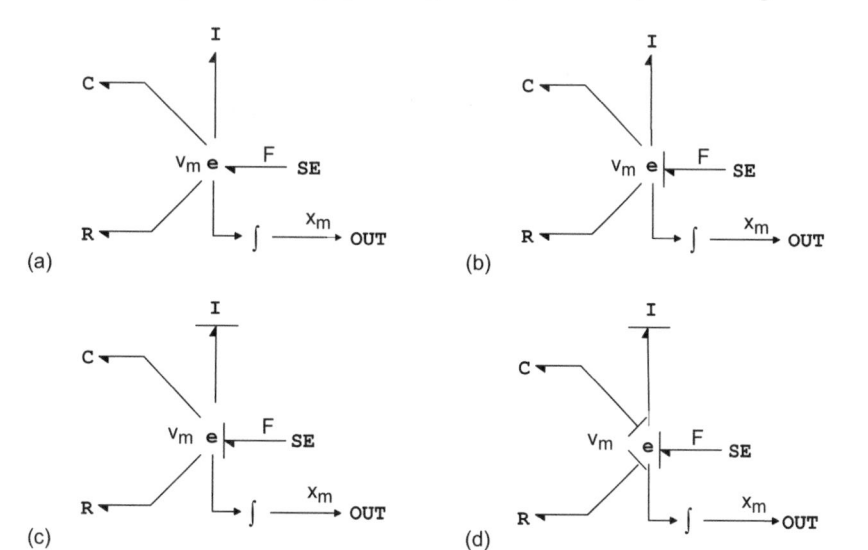

Fig. 2.25. Illustration of the causality assignment procedure

We start with the source effort SE (step 1 of SCAP) and assign its causality as shown in Fig. 2.25b. We cannot extend the causality assignment immediately to the effort junction, as the connected port is an effort input port. There are no more sources, thus we proceed with step 2 of SCAP. We can choose to assign causality to either the inertial or the capacitive element. Let us choose to assign integration causality to the inertial element I (Fig. 2.25c). We now can extend the causality assignment to the effort junction, because the port connected to the inertial port is an effort output port, and all other junction ports must be effort input ports (Fig. 2.25d). This completes the causality assignment of the bond graph.

We have obtained integration causality of the capacitive element C, as well. If we start at step 2 by choosing the capacitive element instead of the inertial element, we would have to assign the causality of the inertial element before we could proceed to the effort junction. The first procedure is somewhat shorter.

The causal assignment of Fig. 2.25 defines the order of evaluation of the equations. This is shown by the block diagram of Fig. 2.26. We start with the SE first. Next, we calculate the output flow of the inertial element. This is the input to the effort junction and the output of its all other ports. It is the input to the capacitor and the resistor used to calculate their outputs. These, together with the source effort output, are used to calculate the output of the effort junction, hence of the inertial element input. Independently, it is used as input to the integrator to calculate body position.

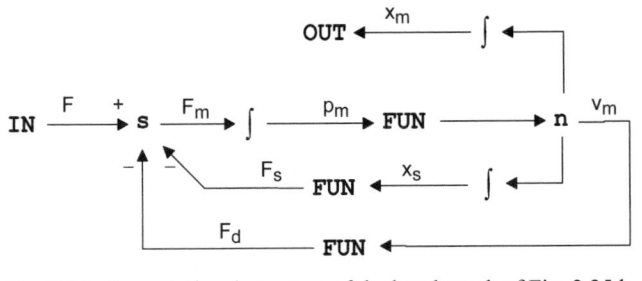

Fig. 2.26. Computational structure of the bond graph of Fig. 2.25d

2.9 The Formulation of the System Equations

The bond graph of a system completely defines its mathematical model. In Sect. 2.7 it was shown that the model could be generated directly from the bond graph by describing the elementary components, including junctions, in terms of their constitutive relations. This way of representing mathematical models is known as the *descriptor* form [7] and is widely used in electrical circuits. This is a non-minimum form because the equations are not expressed using the minimal number of variables. Some variables could be eliminated, e.g. by substituting into the equations of flow and effort junctions. This approach is in effect used in modified nodal analysis (MNA) of electrical circuits [8,9]. This also is the case with certain approaches used in multi-body dynamics [5]. In Sect. 2.7 it was shown that the matrix of the equations is typically very sparse, and this can be used to advantage in their solution.

The descriptor form of equation formulation leads to the model in the form of systems of differential-algebraic equations (DAEs). The success or failure of the descriptor formalism depends to a large extent on the possibility of solving equations in DAE form efficiently and reliably. Solving such equations has a relatively long history and started with the famous DIFSUB routine of CW Gear [10] for stiff systems. The work reported in this book also has its roots in software that solves DAEs in a way that is based on the DIFSUB routine. From that time significant advances have been achieved in the theory of DAEs and their application [11,12]. Today this is a viable approach to solving simulation models. We return to this again in Chap. 5.

Another common approach is to formulate the system in *state space* form. This technique uses a minimal set of independent variables to formulate the governing equations. It has its roots in the generalised coordinate methods of Analytical Mechanics [13], but it also is used widely, and is perhaps better known, from Control Theory. The theory of state-space equations has been a topic of research for a long time and is well understood. This approach is followed not only in bond graph theory, but is also used in many continuous system simulation languages (Sect. 1.7).

The usual approach in continuous system simulation languages is to create a system of sorted equations that is solved sequentially. Such systems can be solved relatively easily. Unfortunately, in many engineering problems of practical interest it is not easy to put the equations in such a form.

The sequential causal assignment procedure of Sect. 2.8.3 was really designed as an aid to the generation of mathematical model equations in sorted form. From that the equations can be reduced to the state space form. The bond graphs with completed causality assignment can be put in such a form if inertial and capacitive elements have integrating causality, and if there are no algebraic loops [1,14]. We illustrate this with the body-spring-damper system represented by the causal bond graph of Fig. 2.25d (or, equivalently, by the block diagram of Fig. 2.26). More elaborate examples can be found elsewhere [1].

We start with the source effort Eq. (2.50), following the order of causal assignment of Sect. 2.8,

$$F = \Phi(t) \tag{2.110}$$

The output of the inertial element I is given by (see Eqs. (2.100) and (2.48))

$$v_m = p_m / m \tag{2.111}$$

The variable v_m (by the effort junction) is used as the input to the capacitor C, resistor R, and the integrator. The order of evaluation of these elements is immaterial. From the first equation of Eq. (2.102), written in derivative form, or Eq. (2.44), we get

$$\dot{x}_s = v_m \tag{2.112}$$

Output of the capacitor is given by the equation (see Eqs. (2.102) and (2.45))

$$F_s = k \cdot x_s \tag{2.113}$$

Output of the resistor (see Eq. (2.104) or (2.46)) is

$$F_d = b \cdot v_m \tag{2.114}$$

Hence, all the inputs to the summator are found and we can calculate its output as

$$F_m = F - F_s - F_d \tag{2.115}$$

The output of the summator is the input to the inertial element. Thus, from the first equation of Eq. (2.100), written in differential form, or Eq. (2.47), we get

$$\dot{p}_m = F_m \tag{2.116}$$

To these equations we add the output of the integrator written as (Eq. (2.49))

$$\dot{x}_m = v_m \tag{2.117}$$

This completes the generation of the system of sorted equations.

The equations above consist of differential equations (2.112), (2.116) and (2.117), and algebraic equations (2.110), (2.111), 2.113), (2.114), and (2.115). Hence, it is a differential/algebraic system of equations (DAE), but of special structure. We classify all variables in these equations as being either differentiated or participating in algebraic operations only. The first are called *differentiated variables*, i.e. x_s, x_m and p_m. The others are *algebraic* variables; in the equations above these are F, v_m, F_s, F_d, F_m. All algebraic variables above can be expressed as functions of the differentiated variables and of time. Starting from the first Eq. (2.110) we see that the variables F, v_m, and F_s are already in this form (see Eqs. (2.110), (2.111) and (2.113)). From Eqs. (2.114) and (2.111) we get

$$F_d = \frac{b}{m} \cdot p_m \tag{2.118}$$

Finally, from Eqs. (2.115), (2.110), (2.113), and (2.118) we have

$$F_m = \Phi(t) - k \cdot x_m - \frac{b}{m} \cdot p_m \tag{2.119}$$

We now substitute these expressions in the differential equations (2.112), (2.116), and (2.117). We thus obtain

$$\dot{x}_s = p_m / m \tag{2.120}$$

$$\dot{p}_m = \Phi(t) - k \cdot x_m - \frac{b}{m} \cdot p_m \tag{2.121}$$

and

$$\dot{x}_m = p_m / m \tag{2.122}$$

Equations (2.120) - (2.122) represent the model in state-space form. Variables x_s, p_m and x_m constitute a minimal set of independent variables that completely defines the state of the system. Solving these equations with suitable initial conditions, all other variables can be found from equations (2.110), (2.111), (2.113)—(2.115).

In general, if all capacitive and inertial ports have integrating causality, then the corresponding differentiated variables, i.e. generalised moments and displacements of Sects. 2.5.2 and 2.5.3 can be looked upon as independent variables where accumulation of past histories of the efforts and flows take place. Such variables completely determine the future state of the system and usually are called *state variables*. All other variables can then be determined if the state of the system is known. If, in addition, there are no algebraic loops—that is, there are no implicit algebraic equations between variables—then all other variables can be eliminated from the governing equations. Thus, if all state variables are represented by a vector **p** and all external inputs (represented by the sources) by a vector **u**, then a change of system state can be described by a vector equation

$$\dot{\mathbf{p}} = \mathbf{\Phi}(\mathbf{p}, \mathbf{u}, t) \tag{2.123}$$

where Φ is a suitable vector-function of the state, inputs, and eventually time. This is an ordinary differential equation that can be solved given the *initial state* of the system. Such an equation is termed the *state-space equation* of the system.

2.10 Causality Conflicts and Their Resolution

The sequential causal assignment procedure (SCAP) of Sect. 2.8.3, in many cases of practical interest, leads to a causally augmented graph that cannot be described by equations in state space form [1,15–17]. We illustrate this using the examples of Sect. 2.7.

We first analyse the electrical circuit of Fig. 2.13, but with the resistor replaced by a diode (Fig. 2.27a).

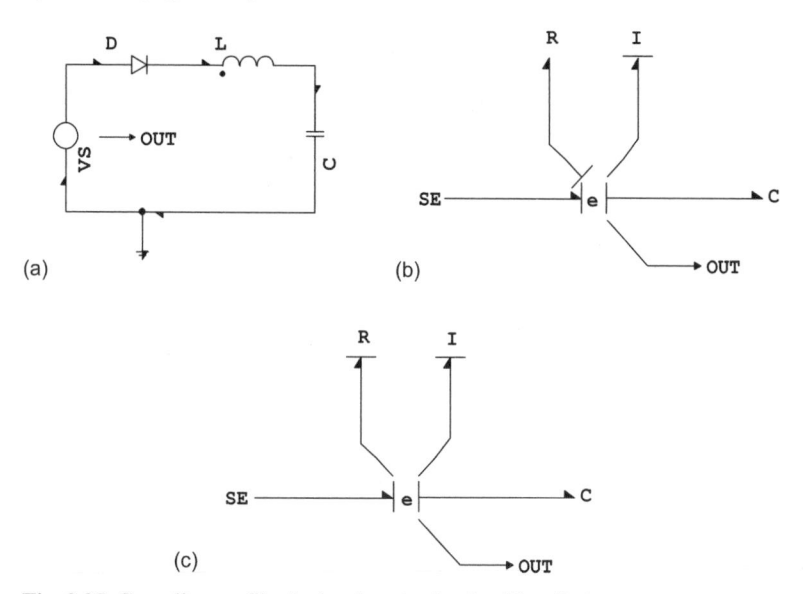

Fig. 2.27. Causality conflict in the electric circuit with a diode

If we model the diode as a non-linear resistor, we get the equivalent causally augmented bond graph shown in Figure 2.27b (see Figure 2.14c). The problem here is that the diode is a non-linear element normally described in conductive form, i.e. the diode current is a function of the voltage across the diode. It is thus in conflict with the assigned causality that implies resistive causality. In order to resolve conflicts caused by non-linear elements, the relaxed causal assignment procedure was proposed in [18] and its modification in [19]. This procedure requires that, at step 2 of the SCAP (Sect. 2.8.3), propagation of causalities over junctions may not violate the fixed causality of non-linear elements. Thus applying the causal assignment procedure again results in the augmented bond graph of Fig. 2.27c. The conflict caused by the fixed causality of the diode disappears, but a *causal conflict* appears at the effort junction because there is more than one output. Thus, the equation corresponding to the effort junction constitutes an algebraic constraint that the variables have to satisfy. The procedure permits casual conflicts at effort or flow junctions as an indication that the mathematical model is of the differential-algebraic equations (DAE) form, rather than of the state space form.

In the see-saw problem (Sect. 2.7.3) there also is a causality problem. The see-saw is a single-degree-of-freedom mechanical system. The motions of the bodies depend on the motion (rotation) of the platform, which is represented in Fig. 2.21 by the LinRot transformers. To show this we apply the SCAP to the bond graph of Fig. 2.22. The resulting causally augmented bond graph is shown in Fig. 2.28. There is only one inertial element with integrating causality. All others have differential causality.

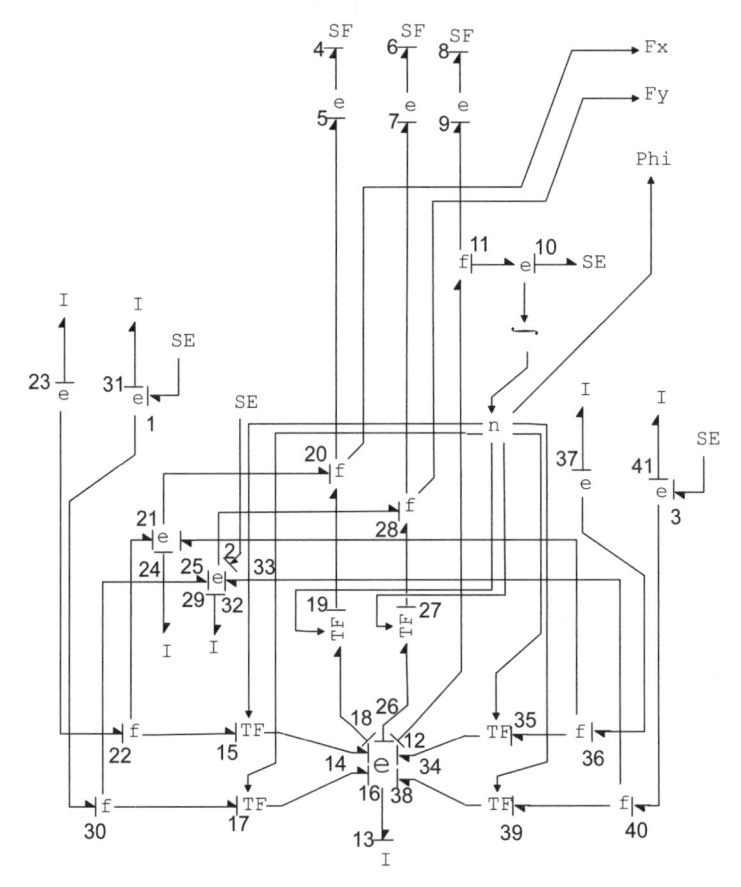

Fig. 2.28. Causal bond graph of Fig. 2.22

This bond graph is relatively complicated, so numbers are used to indicate the order of the causality assignments. Selection of the preferred (integrating) causality for one inertial element, e.g. the platform rotation, implies derivative causalities for all other inertial elements. Hence, there is only one state variable and all other generalized variables are non-state variables. The model again is a system of differential-algebraic equations.

There have been attempts to resolve causality conflicts by, for example, adding 'parasitic' compliances or inertias [20,21]. This is not an acceptable approach, however, because, in the first instance, it is not clear how to do this without adversely changing the model behaviour. On the other hand, such modified models are not much easier to solve numerically than the corresponding DAE models because they are very stiff.

The causality assignment defines the model's computational scheme based only on the model's structure. In cases in which the model changes sufficiently, such *a priori* schemes can lead easily to a loss of efficiency and even failure of the equation solving routines. This is the case with models that have discontinuities.

Discontinuities are present in engineering systems in various forms, e.g. switches in electrical circuits, hard stops, clearances, and dry friction. For example, the diode represented in the circuit of Fig. 2.29 is modelled as a switch in Fig. 2.29b.

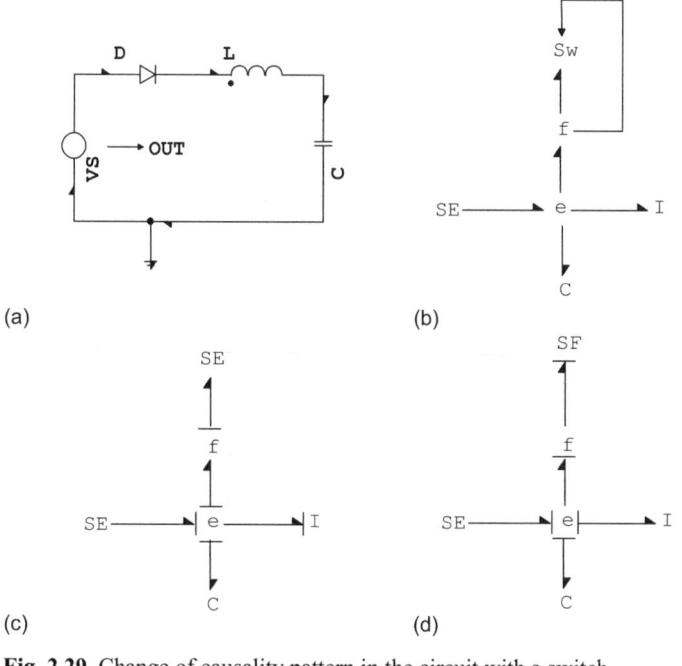

Fig. 2.29. Change of causality pattern in the circuit with a switch

If the diode is forward-biased (conducting), then the switch behaves as a source effort implying a zero voltage drop across the diode (Fig. 2.29c). When, on the other hand, the diode is reverse-biased, the switch behaves as a flow source of small reverse saturated current (Fig. 2.29d). The model structure and causalities are apparently different for these two states. In the conducting regime the system has two state variables, while in the non-conducting regime it has only one.

There have been various attempts to solve causality problems with switches [22–26]. Overall, these procedures are not completely satisfactory in the general case. This is particularly true if the discontinuities are not confined to switch elements, but appear in the element constitutive relations, too.

The concept of causality is generally not suitable for use in an automated object-oriented modelling environment. It is not only too restrictive with respect to the forms of models that can be used, but also puts restrictions on the design and usability of models libraries. A component that has one causality pattern in one system can have a quite different one when inserted in another system. We disregard causality issues when developing models of general engineering and mechatronic systems. Modelling is treated as a separate task from model simulation. The models will be generated in the form of DAE systems and solved as such.

References

1. Dean C Karnopp, Donald L Margolis and Ronald C Rosenberg (2000) System Dynamics: Modeling and Simulation of Mechatronic Systems, 3rd edn. John Wiley, New York
2. J Thoma and B O Bousmsma (2000) Modelling and Simulation in Thermal and Chemical Engineering, A Bond Graph Approach. Springer/Verlag, Berlin Heidelberg
3. PC Breedveld (1982) Proposition for an unambiguous vector bond graph notation. J. Dynamic Systems, Measurement, Control 104:267-270
4. EP Fahrenthold, JDWargo (1991) Vector and Tensor Based Bond Graphs for Physical Systems Modeling. J. of Franklin Institute 328:833-853
5. EJ Haug (1989) Computer-Aided Kinematics and Dynamics of Mechanical Systems, Vol. I: Basic Methods. Allyn and Bacon, Needham Heights, Massachusetts
6. FP Beer and ER Johnson (1990) Vector Mechanics for Engineers, 2nd SI Edn. McGraw-Hill Book Co., Singapore
7. RW Newcomb (1981) Semistate description of nonlinear and time variable circuits, IEEE Trans. Circuit Systems. CAS-26: 62-71
8. R Märtz and K Tischendorf (1997) Recent results in solving index-2 differential-algebraic equation in circuit simulation. SIAM J. Sci. Comput. 18:135-159
9. A Vladimirescu (1994) The Spice Book. John Wiley & Sons, New York
10. CW Gear (1971) Numerical initial-value problems in ordinary differential equations. Prentice Hall, Englewood Cliffs
11. KE Brenan, SL Cambell and LR Petzold (1996) Numerical Solution of Initial-Value Problems in Differential-Algebraic equations, Classics in Applied Mathematics. SIAM, Philadelphia
12. E Hairer and G Wanner (1996) Solving ordinary Differential Equations II, Stiff and Differential-Algebraic Problems, 2nd Revisited edn. Springer-Verlag, Berlin Heidelberg
13. H Goldstein (1981) Classical mechanics, 2nd edn. Addison-Wesley Publishing Co., Reading
14. RC Rosenberg (1971) State-space formulation of bond graph models of multiport systems. Trans. ASME J. of Dyn. Syst., Meas., and Control. 93:35-40
15. J Van Dijk and PC Breedveld (1991) Simulation of System Models Containing Zero-order Causal Paths – I. Classification of Zero-order Causal Paths, J. of The Franklin Institute 328:959-979

16. J Van Dijk and PC Breedveld (1991) Simulation of System Models Containing Zero-order Causal Paths – II. Numerical Implications of Class 1 Zero-order Causal Paths, J. of The Franklin Institute 328:981-1004

17. Peter Gawthrop and Lorcan Smith (1996) Metamodelling: Bond graphs and dynamic systems. Prentice Hall, Hemel

18. BJ Joseph and HR Martens (1974) The Method of Relaxed Causality in the Bond Graph Analysis of Nonlinear Systems, Trans. ASME J. of Dynamical Systems, Measurement and Control 96:95-99

19. J Van Dijk and PC Breedveld (1995) Relaxed Causality A Bond Graph Oriented Perspective on DAE-Modelling. In FE Cellier and JJ Granda (eds) 1995 International Conference on Bond Graph Modeling and Simulation, Las Vegas, Nevada, pp 225-231

20. D Margolis and D Karnopp (1979) Analysis and Simulation of Planar Mechanisms Using Bond Graphs. ASME J. Mechan. Des. 101:187-191

21. A Zeid and CH Chung (1992) Bond Graph Modelling of Multibody Systems: A Library of Three-Dimensional Joints. J. Franklin Inst. 329: 605-636

22. W Borutzky (1995) Representing Discontinuities by Sinks of Fixed Causality. In FE Cellier and JJ Granda (eds) 1995 International Conference on Bond Graph Modeling and Simulation, Las Vegas, Nevada, pp 65-72

23. FE Cellier, M Otter and H Elmqvist (1995) Bond Graph Modeling of Variable Structure Systems. In FE Cellier and JJ Granda (eds) 1995 International Conference on Bond Graph Modeling and Simulation, Las Vegas, Nevada, pp 49-55

24. IF Lorentz and H Haffaf (1995) Combination of Discontinuities in Bond Graphs. In FE Cellier and JJ Granda (eds) 1995 International Conference on Bond Graph Modeling and Simulation, Las Vegas, Nevada, pp 56-64

25. PJ Mosterman and G Biswas (1998) A Theory of Discontinuities in Physical Systems Models. J. Franklin Inst. 335B: 401-439

26. U Soderman and JE Stromberg (1995) Switched Bond Graphs: Towards Systematic Composition of Computational Models. In FE Cellier and JJ Granda (eds) 1995 International Conference on Bond Graph Modeling and Simulation, Las Vegas, Nevada, pp 73-79

Chapter 3 Object-oriented Approach to Modelling

3.1 Introduction

This Section describes the principles of object-oriented design based on the bond graph technique. The idea is to perform the modelling visually, without any coding; all interaction between the modeller and the model will be through a visual development environment.

We not develop a new modelling language, but instead design classes that are used to create a structured model as a tree of linked objects. This will be persisted as a set of linked files. It is not necessary to have the complete model in memory, as parts are loaded as required.

It is very important during the development phase to relieve the user of certain implementation details, such as how to create components, where and how to store data, etc. These operations are automated. The user communicate with the model in familiar terms, e.g. by the names of a project and of its components, how they are connected, and by defining component constitutive relations in terms of variables meaningful to the user. The developer thus concentrates on the problem. The modelling environment (framework) provides as much support as possible to fulfil this task.

3.2 The Component model

The concept of *component model* plays a central role in our approach to modelling [1].[1] The model of a component is created in two steps (Sect. 2.2). First, a *component* object is created as a *word model*. Next, by "opening" the component, the associated *document* object is created. The document describes the component model in terms of a bond graph. It contains lists of the word model components and interconnecting bonds that constitute a bond graph model of the component. The component and its associated document are intimately related and are used as the engine for systematic model development either by the top-down (decomposition) or down-up (composition) approach, or by a combination of both.

---

[1] This is not to be confused with Microsoft's Component Object Model (COM) [2].

3.2.1 The Component Class

CComponent is the base class used for the creation of the component word model (Fig. 3.1). It holds textual information, including its name, ports that interconnect to other components, as well as information that defines its position on the screen. Visually, the component is designed as a rectangle with a strip surrounding its name (Fig. 3.2). This strip is used to place the ports.

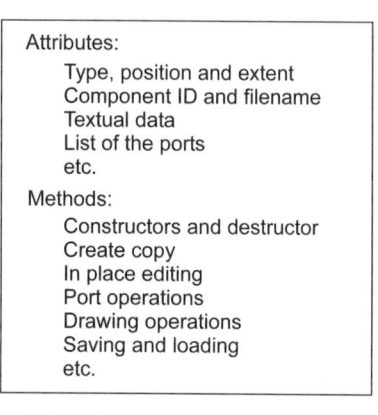

```
Attributes:
    Type, position and extent
    Component ID and filename
    Textual data
    List of the ports
    etc.
Methods:
    Constructors and destructor
    Create copy
    In place editing
    Port operations
    Drawing operations
    Saving and loading
    etc.
```

Fig. 3.1. The *CComponent* class

Creation of a word model component typically consists of

1. The selection of a position on the screen where the component is to appear.
2. The construction of the *CComponent* and initialising its position, type, and text font.
3. The creation of a unique identifier (*id*) and associated document file name, and storing them in the object.
4. Editing the name and storing it in the object.
5. Saving the bounding rectangle in the component object.
6. The creation of ports and adding them to the component object

Strip for the port placement   Component name box

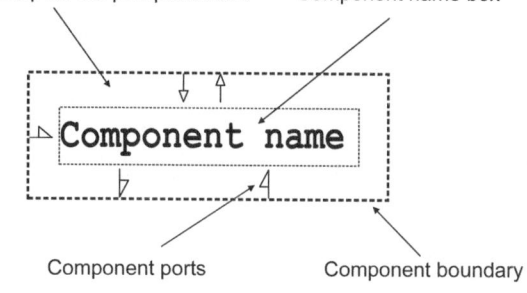

Component ports   Component boundary

Fig. 3.2. The component word model visual representation

The component word model is created within a suitable *document* object that manages the complete process of object creation (Sect. 3.2.2).

The component class supports simple text editing operations for the in-place editing of the object name (title). Typical of such operations are inserting or deleting of a character, new line creation, joining lines, jumping to the head or the end of the text. The component name is only for the user's convenience; the framework refers to a component solely by its *id*. Thus, two different components can have the same name and appearance, but be different objects with different *ids*. The component *id* and document filename are unique in the workspace in which all model files are stored.

The Component class implements several methods for port creation and deletion, and for moving the ports around the component periphery. All ports that are created are stored in the component object as a list.

The component is responsible for its own visual appearance. This means that the component class has methods for drawing itself on the screen and for printing to a printer or to a file. The component is drawn as part of the drawing operations of the document object, which contains the component (Sect. 3.2.2).

The component class supports creation of a copy of a word model component. In addition to the usual copy constructor, a virtual method is provided for creating a copy of component objects. Because this is a new component object, a unique component *id* and a unique document file name are created and stored in the object. All other necessary changes are also made, e.g. in ports (Sect. 3.4).

3.2.2. The Document class

The document class associated with *CComponent* is termed *CBondSimDoc* (Fig. 3.3). This is a container class used to create and store the bond graph model of a component. The model is held as a list of document ports, components, and connecting bond objects.

Visually, a document is designed as an area framed by a rectangle that is used for placement of the components and drawing interconnecting bond lines (Fig. 3.4). Document ports are created in the surrounding strip and correspond to the ports of its word model. The ports serve as outside connectors of the components contained in the document. Their positions correspond to the positions of the word model component ports. In this way it is visually clear which port correspond to which.

The document and its corresponding word model constitute the complete model of a component. Such an object typically is accessed through the corresponding word model. Hence, a word model serves as the interface to the document that contains its model. Important parts of this interface are the ports (Sect. 3.4). A new document object typically is created as follows:

1. Assuming that a document object (the current document) that contains a word model has already been created, a component for which the accompanying

document is created is selected and a message is sent to the document to create a new document object.

2. The document method creates a new document object. This method uses a pointer to the word model object, from which the necessary data are taken, such as the name of the component, the filename where the new document object is to be saved, and the component ports.

3. A new document is created and its corresponding attributes set. Document ports also are created. These correspond to the word model object ports (Sect. 3.4) and are added to the list of document ports. Otherwise, a new document is empty, i.e. without any component or bonds. The new document also creates links to the previous (lower level) document.

Attributes:

    Component name and document file name
    Links to previous document and the word model
    component
    List of document ports, components and bonds
    List of model parameters, etc.

Methods:

    Constructor and destructor
    Create new document
    Open saved document
    Closing and saving document
    Copy document
    Remove document
    Creation of document ports, components and bonds
    Creation of model parameters and their linking
    Support for mathematical model generation, etc.

Fig. 3.3. The *CBondSimDoc* class

A new document object is not created if the ports of the corresponding word model component are not already defined. Such a document object would make no sense.

Once a document object is created, new word model objects of the contained components can be created inside the document working area. The creation process is managed by the document as already described (Sect. 3.2.1). The components created are added to the list of components that the document contains. In the same vein, the document manages the creation of the bonds that are used to interconnect components contained in the document, or to connect them to the document ports (Sect. 3.4). Model development can continue by creating a new document object for every word model in the document, thus developing higher levels of abstraction of the component model.

The document class manages saving the document objects to the document file. Before any a new document object is created, the current object is saved. All objects that the document contains are saved to the document file by calling the corresponding method. During this operation a copy of its word model object also is saved as a header, as well as the filename of the previous document and other pertinent data.

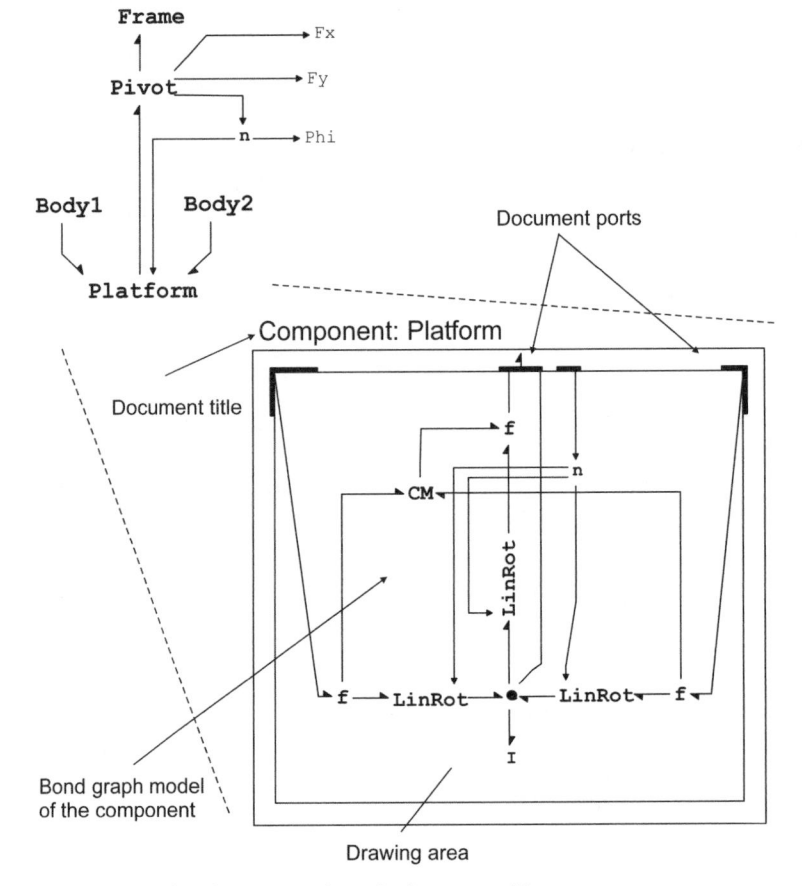

Fig. 3.4. The visual representation of a document object

This saves the model as a tree of interconnected document files. All the files that make up the model are double-linked, i.e. forward through the components contained in the document, and backward by the filenames of the previous document. Access to the component root document is normally made by the component word model contained in a document. We return to this problem again in Sect. 3.9.

Document objects are *persistent* [3] because they exist even after they are closed and deleted from memory. To remove a component completely from a document it is not enough to remove it from the list of the components that the document object maintains and to destruct it. The accompanying document file has to be removed, too. Removing the root document only creates dangling documents, i.e. documents of which nobody is aware. This means that it is necessary to open documents until the leaf documents are reached; then move backward by re-moving the document files until the component that is to be removed is reached. Only then can the component be disconnected from the container document and destructed.

The document class is complex and supports many operations of the automated modelling framework. The drawing operations are executed in a visual environment in which the modelling system is implemented, e.g. the Windows system. During its execution the objects contained in the document are called to draw themselves, e.g. the document ports, the components and their ports and the bonds. Similar actions are performed when printing a document to an external printer or to a graphic file.

One important and much used operation that the document class supports is copying. This creates a copy not only of the document, but also of all documents in the tree. This operation starts from the document root and proceeds toward the document tree leaves. In this process the *ids* and filenames of all contained components are changed. This causes a new document tree to be created that does not share any component or document with the original tree. Document copying is implemented in the following way:

1. Open the document object from a file and set the new filename and path where the copy is to be saved.
2. For components contained in the document:
   o   Get the original document filename
   o   Create a new component *id* and document filename
   o   Reset the component *id* and filename to the new values, saving the original filename to a temporary location. In the process, component *ids* are changed in the component ports, as well as in the bonds connected to these ports.
3. Save the document under the new filename
4. For every component contained in the document:
   o   If there is an existing document file repeat step 1
5. Close the document

This operation is called when a copy of a component model is needed. In this case a copy of the word model component is made first as explained in Sect. 3.2.1, and then a copy of its accompanying document is performed as described above. When this operation is called directly (Sect. 3.9), a copy of the word model component stored in the document header is made first, then a copy of the document is made, as explained above.

The document class also supports many other operations, such as creation of the mathematical model parameters defined at the level of a document object. These parameters are visible to all components contained in the document. Linkage of these parameters is designed so that a parameter defined in a document overrides (hides) a parameter of the same name that is defined in a lower-level document. The document class also has methods for mathematical model generation. This is addressed in Chap. 5.

3.3 The Component Class Hierarchy

The *Component* class is the base class from which more specialised component classes are derived. Fig. 3.5 illustrates the hierarchy of component classes.

One type of specialisation is visual appearance. The component can be depicted, for example, by graphical symbols, not only by its name. In electrical engineering such praxis is widely used and has been standardised (e.g. by ANSI, DIN). It is also used in mechanical engineering, but not to such an extent or with such versatility. Hence, to simplify modelling with bond graphs, several derived classes are defined that partly support such schemes.

In the first branch on the left of the hierarchy tree of Fig. 3.5 shows classes that represent some of the more common electrical components. These, of course, don't imply any specific model, for they are really word model classes represented differently. The model should be defined in the accompanying document object (see Fig. 2.13). Using these components and connecting them by bonds gives the bond graph an appearance very close to the usual electrical schemes. An important difference from the latter is the half-arrows used to indicate power flow directions in the circuit.

Mechanical engineering uses various schematics to depict, for example, systems in vibrations, hydraulics, and pneumatics. Here there are defined classes to represent simple mechanical components only, such as bodies in translation and rotation, springs, dampers, etc. (the second branch of the tree of Fig. 3.5).

Classes of fundamental importance involve derived classes that represent the elementary components discussed in Sect. 2.5. These are the classes in the third branch of the hierarchy tree of Fig. 3.4. In accordance with bond graph praxis, these components use predefined textual symbols that can readily be changed.

It should be noted that some simple electrical and mechanical components can be defined as a specialisation of certain elementary components, such as ground potential, and electrical or mechanical junctions.

Elementary components differ fundamentally from the base class component in that they do not have an accompanying document; they are entities in themselves. The question, then, is where to put the variables and constitutive relations of these components. The natural answer is: in their *ports*. Derived elementary classes support editing of the constitutive relations in the form given in Sect. 2.5. The elementary component classes also support creation of locally defined model parameters; that is, those valid only in the component.

Elementary component classes define various class-specific methods that override the base class methods. They also have an important role during mathematical model creation, and in simulations, too.

The last group of components deals with the block diagram operations of Sect. 2.6. Components in this group are shown as the last branch of the hierarchy in Fig. 3.5. They are similar to the elementary components discussed previously, but support only control ports. These components also use their ports to store the variables and input-output relations. An exception is the COutput class that serves for collecting the simulation output and displaying it on the screen.

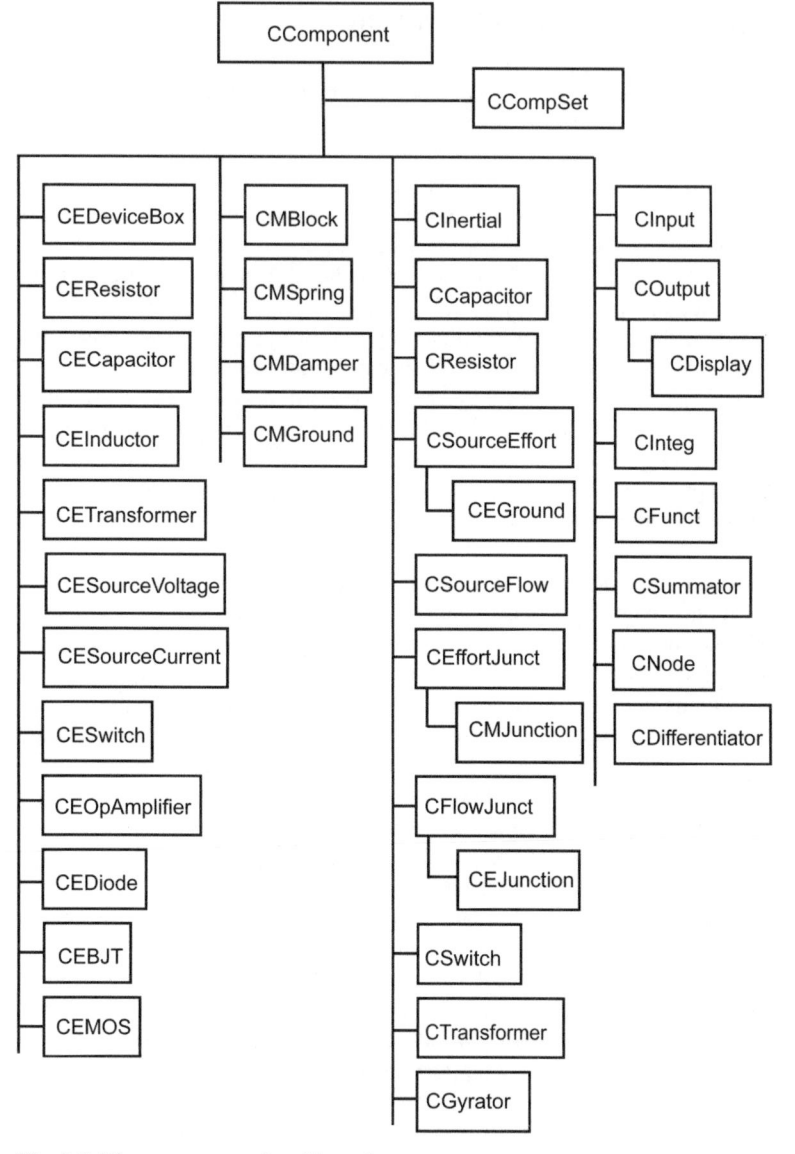

Fig. 3.5. The component class hierarchy

Finally, there is a derived class that differs from the classes discussed above. This is the *CCompSet* class, a container class for a group of component objects. It is desribed in Sect. 3.8.

The Document class, as previously described (Sect. 3.2.1), manages creation of component objects. Because there are no virtual constructors, the technique of the object factory is employed [4] to construct objects of the correct type. Similarly, every derived class overrides the method for creating a copy of a component. Thus insuring that a copy of the object of the correct type is created.

3.4 Port and Bond Classes

The component model introduced in the last two sections doesn't specify completely the interconnections between components. It is necessary to work out port interconnections, too. One approach is based on the notion of *multibonds* [5,6], a generalisation of the concept of bonds to the multidimensional case. We accept that another approach, based on the concept of *compound* ports, better fits the object-oriented philosophy. Thus, bonds are taken as simple objects that define only which port is connected to which; everything else is the responsibility of the ports. We define the necessary port classes, but first describe what we ask of them.

Looking at the component object and its accompanying document object (Fig. 3.6), we identify two types of ports: component ports and document ports.

The first type connects *external* components. Such a port "knows" that it belongs to a certain component and that it is, or it is not, connected by a bond. The port doesn't need to know what is on other side of the bond. This is the responsibility of the bond. The document port, on the other hand, belongs to the document object and serves for the internal connection of the document's components. This kind of port knows what, and how many, bonds are connected to it.

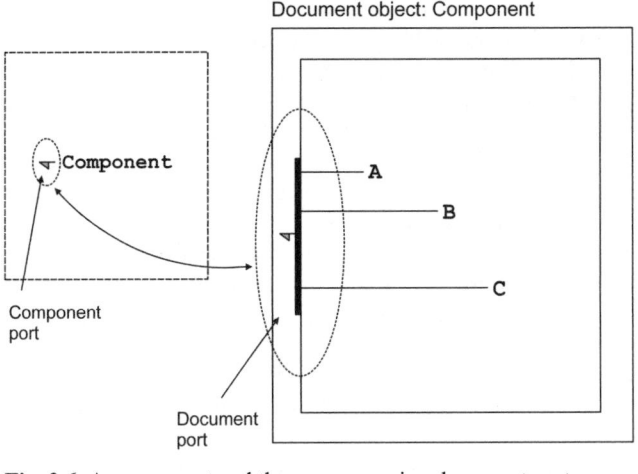

Fig. 3.6. A component and the accompanying document port

These two types of ports are used to describe connections looking from two sides of the same component, from outside and from inside. Looking from the outside—that is, at the component port side—it would be helpful to know how many bonds are connected inside. This we call the *dimension* of the port. On the other hand, looking from inside, it would be useful to know if the port is connected on the outside by a bond or not.

To simplify interchange of information between these two types of ports, we define the component port class *CPort* as the base class and the document port class *CDocPort* as a derived class (Fig. 3.7).

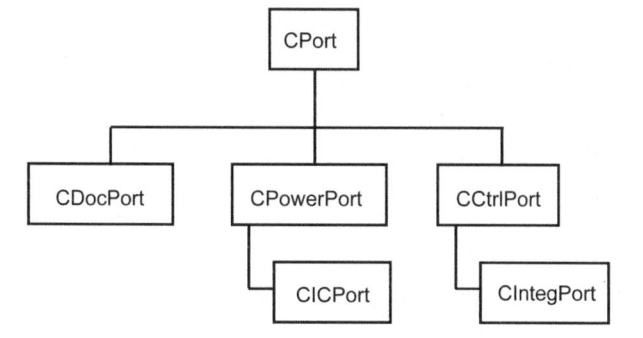

Fig. 3.7. The port classes hierarchy

Creation of these objects is the responsibility of their containers, *CComponent* and *CBondSimDoc*, respectively. A possible scenario is as described below:

1. When a user issues a command to the component object to insert a port, the component method responsible for port creation is called. It, in turn, calls the base class constructor, which supplies the port type requested, and its position on the component boundary. The port constructor creates a port object and assigns to it the component and the port *ids*, as well as the port type. The object initially is unconnected, both externally and internally (dimension 0). Depending on it's type and position on the component boundary, the necessary visual data are created that are used for drawing. The object subsequently is added to the list of the ports that the component maintains.
2. The document object is responsible for creation of document ports. These are created during document object creation, based on the component object ports (Sect. 3.2.2). The base class of the document port object is just a copy of the accompanying component port object translated to a position on the outside of the document drawing area (Fig. 3.6). The position is calculated such that it corresponds to the position of the accompanying component port in the strip surrounding the component name. Thus, they share the same information.
3. When a bond is connected to a component port, its *id* is sent to the port and stored there. The accompanying document port is updated at the same time. Similarly, when a bond is internally connected to the document port, the bond *id* is inserted in the list of bonds that the document port maintains. The position in the list corresponds to the position of the connected bond line in the connection rectangle. The dimension of the port is changed, as well as that of the cor-

responding component port. Thus, every change at one side of the interface affects the other side.

The ports belonging to elementary components are different because they have no document port counterparts. These ports serve mainly for storing data about the component model constitutive relations as "seen" at those ports. The necessary information is different for the power and control ports. Thus, two port classes, *CPowerPort* and *CCtrlPort*, are defined that are derived from the *CPort* base class (Fig. 3.7).

The power port class defines the port effort and flow variables, as well as the element constitutive relations. Similarly, the control port class defines the control port variables (signal). The constitutive relations are defined only at the output ports.

There are two other classes, *CICPort* and *CIntegPort*, derived from the *CPowerPort* and *CCtrlPort* classes, respectively (Fig. 3.7). These define differentiated (state) variables that the Inertial, Capacitor, and Integrator components need (Sects. 2.5 and 2.6).

Port object creation is managed by the corresponding elementary component objects in a similar way to the creation of ports of the base components, as described above. The variables are defined with default names and default constitutive relations. These depend on the type of the elementary component. Subsequent to creation, the names of the variables and the specific form of the constitutive relations can be changed. Details of the syntax used for description of the constitutive relations are given in the next section.

Port classes, like component classes, define methods for operations on the port objects, such as construction and deletion of the object, copying, saving, loading, and drawing.

The final object required to close the bond graph is the bond itself. Bonds are simple objects that indicate which port is connected to which. A *CBond* class is defined with attributes that hold the bond identifying label (*id*), the starting and the ending component, and the port *ids*, as well as data necessary for visual representation of the bonds. The bond class defines methods necessary for the creation and destruction of bond objects, copying, saving, loading, and drawing. Bonds, as objects, are contained in documents. The procedure for the creation of bonds is as follows (Fig. 3.8):

1. Select a component port or a document port. The starting position of the bond line is obtained from the port and corresponds to the position on the port boundary from which the bond will be drawn.
2. The document class is called to create a *CBond* object. The start-off position of the bond is set.
3. Select the next point in the document drawing area. Check if the point is within the other component port:
   - o   If it is not, add a point to the bond object and continue with step 3.
   - o   Otherwise, check if it is a port of another component and of the correct type—i.e. both are either power or control ports—but of different power or signal flow sense.

  ❑ If the answer is no, reject the point and continue with Step 3.
  ❑ Otherwise, continue with next Step 4.
4. Add the point to the bond object. Get the *ids* of the component and the port and set the ending component and port attributes of the bond object to these values. Create the bond object *id* and set the corresponding attributes of the bond object. Set also the bond attributes of the starting and ending ports to this value. Add the bond object to the list of bonds in the document object.

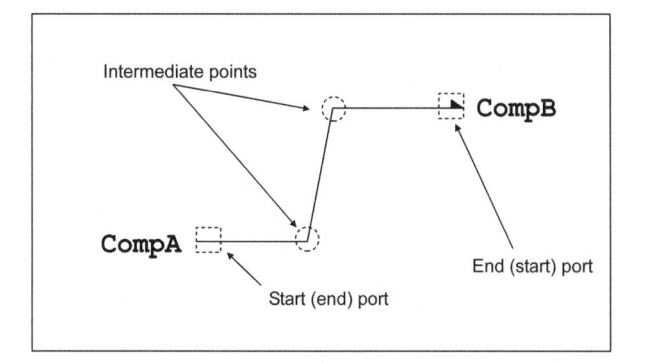

Fig. 3.8. The creation of a bond line

3.5 Description of the Element Constitutive Relations

The constitutive relations of elementary components are defined at the component ports in the form of symbolic expressions. These expressions are of the form

$$variable = algebraic\_expression$$

The algebraic expression is formed from variables, numerical constants, parameters, and operators.

  Acceptable variables in the symbolic expression are the port variables and time *t*, which is a reserved symbol, i.e. it cannot be used for a port variable or parameter name. Other port variables of corresponding component also can be used. There are also some restrictions on the implied form of the constitutive relations of specific components. These are described in Sects. 2.5 and 2.6.

  The standard forms of integer and floating-point constants are acceptable, e.g. 12, -1020, 2.7612, -1.36e-5. Integer constants are internally converted to floating-point constants. Constants can be defined symbolically in the form of parameter expressions formed from constants and other symbolically defined constants using the operators that are described later. Symbolically defined constants can be freely used as parameters in algebraic expressions. These can be defined at the level of the component or at some lower level, as has been described in Sect. 3.2.2.

The common arithmetic, relational and logical operators (Table 3.1) are permitted in algebraic and parametric expressions. The exponentiation operator ("^") can also be used, as well as function calls. To describe discontinuous relations, e.g. in Switch elements, the *if-else* type of control statement can be expressed by C/C++ like operators "?" and ":". These can be nested. Table 3.1 lists all the supported operators. The operators follow the common priority rules, as implemented in C/C++.

Table 3.1. Operators supported in an expressions

Operator	Meaning
()	Function call
!	Logical not
+ -	Unary plus and minus
^	Exponentiation
* / %	Product, division and mod
+ -	Addition and subtraction
< > <= >=	Relational operators
& |	Logical AND and OR
? :	Arithmetic if (conditional)

All common elementary mathematical functions are supported, such as *sin*, *cos*, *tan*, *log* (natural base). Functions sometimes are not known in analytical form, but as tabular data. For example, tables might result from finite element analysis or experiments. Such data can be interpolated by polynomials and used in the expressions. The BondSim program supports an interface that accepts one- and two-dimensional tabular functions. Functions are interpolated by B-splines [7] and are referred to in an expression by a user-assigned name, e.g. *flux(i,x)*.

3.6 Modelling Vector and Higher-dimensional Quantities

The power ports of elementary components define a pair of effort-flow variables. Similarly, a control port defines a single control variable. Using components—not necessarily elementary ones—it is possible to represent more complex variable structures. We describe these structures for the power ports, but the description holds for the control ports, too.

Thus, if three elementary components contained in a component are connected internally to a document port, the accompanying component port can be looked at as representing a pair of effort-flow *vectors* (Fig. 3.9a). This relationship can be seen more clearly if we substitute the component word model with its document, i.e. transform it from a multi-level to a single-level representation (Fig. 3.9b). The port of Component A (Fig. 3.9a) holds a list of bonds, each of which points to the elementary ports where the component effort-flow pairs are defined. Hence, such a port represents a pair of effort and flow vectors

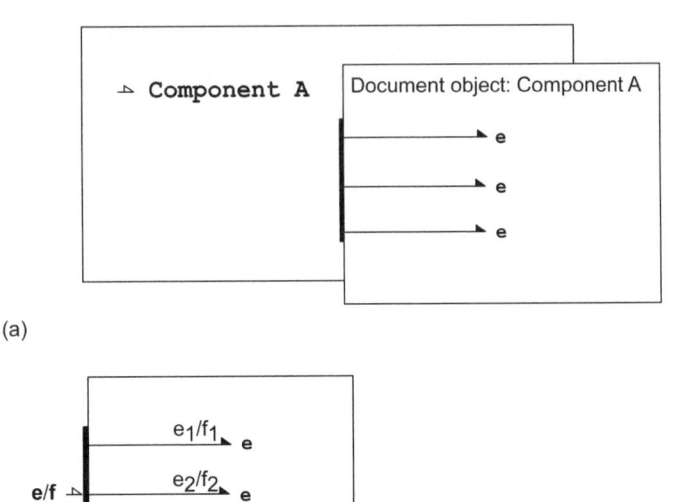

(a)

(b)

Fig. 3.9. The representation of a pair of effort/flow vectors. (a) Component representation,
(b) Single level representation

$$\mathbf{e} = (e_1, e_2, e_3)^T \tag{3.1}$$

and

$$\mathbf{f} = (f_1, f_2, f_3)^T \tag{3.2}$$

This approach can be used for 3D representation of a force applied at a body point
and the corresponding velocity (for 2D representation, see Fig. 2.17).

The dimension assigned to a port indicates the number of ports connected internally. In the case shown in Fig. 3.9, the port dimension is 3; but it could, of
course, be any number. The connected ports are the ports of contained components. These also can be of higher dimension than 1. Quite complex structures of
port effort-flow variables can thus be constructed.

As an example, Fig. 3.10a shows a component port to which two other ports are
connected internally. These are the ports of another component that contains six
elementary components. Two groups of three components out of these six elementary components are internally connected to the corresponding document ports.

The port of component A thus can be viewed as representing a pair of two 3D
vectors (Fig. 3.10b), i.e.

$$\mathbf{e} = \begin{pmatrix} \mathbf{e}_1 \\ \mathbf{e}_2 \end{pmatrix} \tag{3.3}$$

and

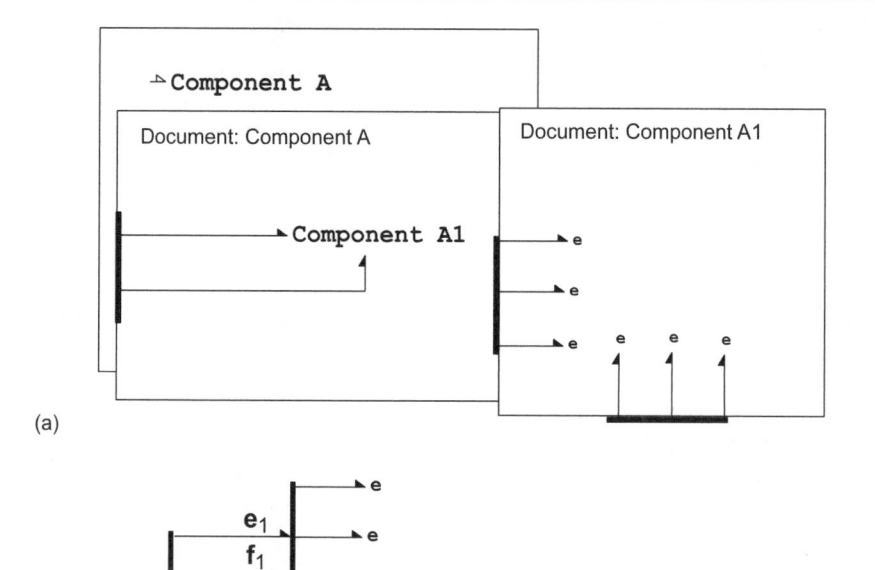

(a)

(b)

Fig. 3.10. The representation of complex effort/flow structures. (a) Component representation, (b) Single level representation

$$f = \begin{pmatrix} f_1 \\ f_2 \end{pmatrix}$$
(3.4)

Here the components of the effort vectors are

$$e_i = (e_{i1}, e_{i2}, e_{i3})^T, (i = 1,2)$$
(3.5)

and similarly for the flows,

$$f_i = (f_{i1}, f_{i2}, f_{i3})^T, (i = 1,2)$$
(3.6)

In general, the dimension of the connected ports can be different. Hence, a component port can be looked at more properly as a tree of connected component ports, with elementary ports as its leaves. A port can be interpreted as representing a pair of tree-like, higher dimensional effort-flow objects.

3.7 Port Connection Rules

An important question that must be answered is what component port connections are allowed. We have already discussed this in Sect. 2.3.

We impose additional restrictions to the permissible connection of elementary ports. This is motivated by physical reasons already discussed in Sect. 2.5.7. The ports of elementary components should be connected to junction ports only. For example, ports of inertial and capacitive components should not be directly interconnected, except to an effort or a flow junction.

The rule permits direct interconnection of two junctions of the same type, i.e. an effort junction to another effort junction, or of a flow junction to another flow junction. Such junctions can be interpreted as a single junction. The ports used for the interconnection of such components are treated as internal and are not counted in junction balance equations (Sect. 2.5.7). The connection of junctions of the same type is usually not permitted in the classical bond graph modelling approach. Nevertheless, we find this quite useful in modelling based on the component model approach. This can be easily seen from the example of Sect. 2.7.2.

We permit ports of word model components to be directly connected. It is necessary, however, to take into account the fact that such ports are, in general, compounded. That is, they correspond to a node of the tree of connected component ports. We require that the structure of connected ports is the same, which means that the port connection trees on each side of the bond line are symmetrical (Fig. 3.11a). In this case, it is easy to find out which port is connected to which. Ports are interconnected if they have the same position in their trees. Thus, in Fig. 3.11a, the port of component A1 is connected to the port of component B1; likewise, port "a" of component I is connected to port "b" of component e.

We further require that document ports be not interconnected directly; instead, we can simply bypass the component!

It is also possible to permit more flexibility in the structure of the ports connected by a bond line. Thus, in Fig. 3.11b, component A1 (Fig. 3.12a) was substituted with its document object, and the document ports were integrated without changing the bond connection order. This process does not change the interconnection of the elementary component ports of components A and B. Now, however, it is much more difficult to find out which elementary ports are connected to which, as there is no similarity in the structure of the connected component ports. It thus is necessary to find out the equivalent linear lists of elementary components ports using ordering of the bond connections in the corresponding document ports (Sect. 3.4). By comparing indexes of the ports in such lists, it is possible to discover which port is connected to which.

We do not allow the connection of ports with different structures, as it is more natural (not to mention also proper) to correlate ports of the same structure and, hence, the variables of the same type of structure. We also think such connections are more transparent and therefore easier to understand.

We next describe a procedure to find the port connected to a chosen component port (Fig. 3.12). This procedure treats only those ports that are branch ports—that is, ports where more than one bond is internally connected—and leaf ports. Ports where there is no branching bonds are taken as simple connector ports and are skipped. To describe this procedure, several variables are defined:

- o   nLevel   current port level
- o   nDim   current port dimension

o   nIndex   index of the current bond in the list of document port connections
o   Index   array for storing bond line indexes during port tree traversal
o   Dim   array for storing port dimensions during the tree traversal

Starting at level 0, the procedure searches outward, then inward.

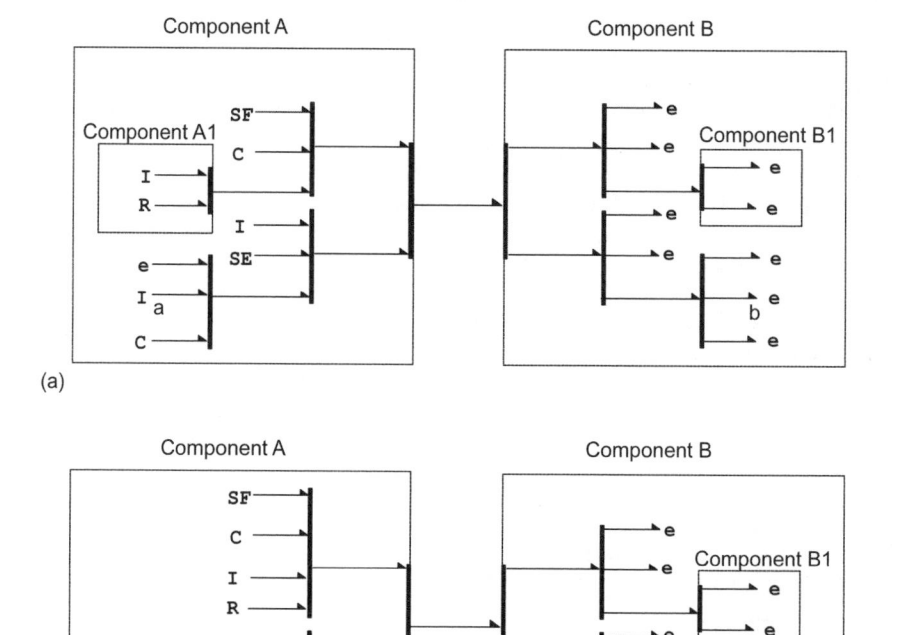

(a)

(b)

Fig. 3.11. The proper connection of the component ports. (a) Symmetrical structure,
(b) Unsymmetrical structure

At every step of the outward search we first check if the port has a connection.
If it does not, the search is stopped and a suitable error message is displayed. Otherwise, it continues along the bond line from the chosen port to the document port
to which it is connected. It then moves out of the component and on to the next
document port. The search continues in this way until a port is found that is either
unconnected or is not connected to a document port, but to a port of another component. At every step of the search we find the dimension of the document port
and the index of the bond connection. The level is decremented and these values

are stored in the corresponding index and dimension arrays. These actions are skipped if the port is a simple connection port (of dimension 1).

The inward search proceeds step-by-step, from the outside into a component, to find the connected port. At every step the dimension of the current port is found first. It then is compared with the dimension of the array corresponding to the current level. If it is OK, the index of the bond is taken from the index array corresponding to the current level and the level is incremented. This action is skipped if the port is a simple connector (dimension 1). If the dimensions don't match, the search and the complete procedure are aborted. To enter the component the accompanying document must be opened first. Next, the corresponding document port is obtained and then, using the index taken from the array, the corresponding bond is found. Finally, the port at the other side of the bond is found and the process is repeated. If the level reached at the end of the search is the same as that at which the search started, the port that was found indeed is the one for which we were searching, and the process ends.

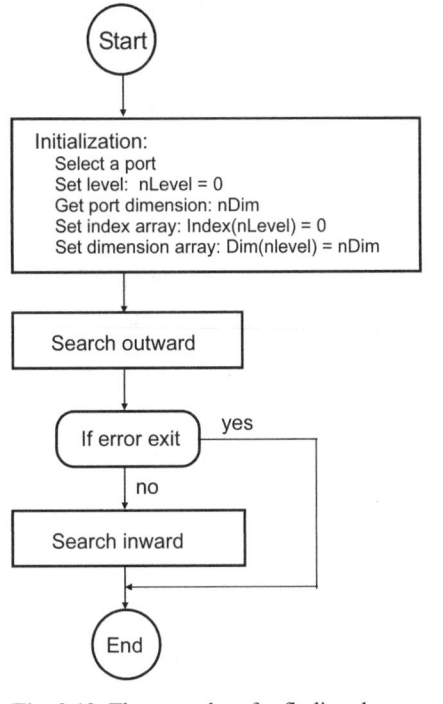

Fig. 3.12. The procedure for finding the port connected to a given port

The procedure can be used for checking the component's port connections. This can be implemented in a way that visually shows component port connections over various levels of the document. It can also be used during generation of the system's mathematical model.

3.8 The Component Set Classes

In addition to the component and document classes of Sect. 3.2, two other classes are useful for operations on a group of components. During model development, for example, it can be useful to copy, move, or delete not only single components, but also a set of components and their interconnecting bonds (Fig. 3.13). The set can even be unconnected. To support such operations we introduce a helper class *CSelSet* (Fig. 3.14).

CSelSet defines a list of components that makes up the set and a list that contains their internal bonds (i.e., the bonds that interconnect the components in the set.) There also is a list of external bonds to components not contained in the set. We use this class to create an object that contains a list of a document's components and their bonds. This is done in the following way:

1. Create a rectangle encompassing a group of components
2. Find all the components enclosed in the rectangle and add them to the list.
3. Check the bonds connecting the ports of selected components

If the bond connects the components from the set, add it to the list of internal bonds; otherwise, add it to the list of external bonds.

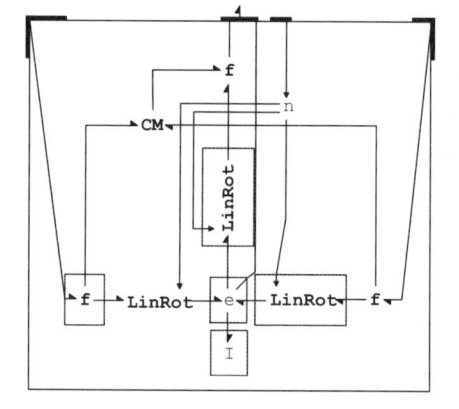

Fig. 3.13. A set of the components and the bonds

We also can add a component to the set and check its bonds. If the component already is in the set, it is then removed from the set. In this case, any bond of the component already contained in the list of internal bonds will be removed and stored in the list of external bonds.

The set of components is *free* if its list of external bonds is empty. Such a set can be moved or deleted. Otherwise, it is fixed and can only be copied. Moving the set means that every object in the set—i.e. the components and their bonds—is displaced by the same amount.

Delete and copy operations involve creation of an object of the other type that contains all information on the components and bonds in the set. *CSelSet* objects are lightweight components that contain only pointers to the components and bonds in a document. They are not persistent objects, but can be used to create persistent objects. When an object is destructed, the lists only are destructed, not the component and bond objects.

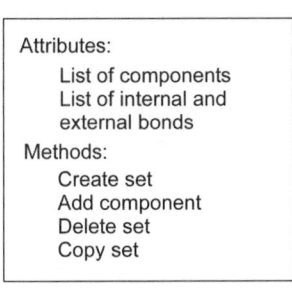

Fig. 3.14. The helper class *CSelSet*

We define a class that is used for creation of a persistent object using information from *CSelSet*. This is the *CCompSet* class we met when discussing the component class hierarchy in Sect. 3.3 (Figs. 3.5 and 3.15).

Attributes:
   List of components
   List of bonds

Methods:
   Constructors and destructor
   Copy operations
   Create from
   Saving and loading

Fig. 3.15. The component group class *CCompSet*

CCompSet is a kind of *CComponent*, but it differs from the latter in that it does not have a separate document to hold its model. It is a container for a set of components and their interconnecting bonds. It doesn't have a separate visual representation, as other components, because it is not used for model building.

CCompSet has one thing in common with *CComponent*: It can be saved and loaded like other component objects. The main use of this class is to create objects that contain a set of components and their bonds that are deleted or copied from a document. Such objects can be held in buffers, from which they can be reused (Sect. 3.9). In performing these operations, there is close co-operation between *CSelSet* and *CCompSet*.

The components contained in a *CCompSet* object can be inserted into a document using *Create from* method (Fig. 3.15), which creates a copy of all components, bonds, and associated documents, attaches them to the document where they should be inserted, and creates a corresponding *CSelSet* object.

3.9 Systematic Top/down Model Development

The previous Sections described the concept of the component model, the concept that lies at the core of the systematic top/down modelling philosophy developed for complex engineering and mechatronics systems. This Section describes an object-oriented environment that implements this modelling philosophy.

Physical model development normally begins at the *system level* (Fig. 1.2), a step that identifies the system and defines its interaction with its environment. The model at this level represents a world-view of the problem under study and is represented by a document object without the ports. That is, this is the system model root document.

We use the term *project* to mean the model development. We introduce a separate project class *CProject* (Fig. 3.18) to serve as the starting point of model development. Every project has a project object with a unique identification label (*id*) and filename that indicates where the project's root document is to be saved. This class is, in essence, a reduced version of the component class (Fig. 3.5).

```
Attributes:
      Project ID
      Document filename

Methods:
      Constructors and destructor
      Data access functions
```

Fig. 3.16. The project class *CProject*

Another class, *CBondSimApp*, manages the complete operation of the bond graph based project model development and simulation. It works in close cooperation with a suitable visual environment (Chapter 4). This is a quite complex class. The part that deals with the most important operations of model development is shown in Fig. 3.17.

The application class contains an index of the projects of interest. These generally are projects under development or currently being studied. Other projects are held in a separate storage (Sect. 3.10). Project indexes can be stored as a map (dictionary) to permit rapid access using the project name as the key.

At the start an application object is constructed and initialised. As a part of the initialisation the index of current projects is loaded. A new project may be started as follows:

1. Send a message to the application object to create a new object.
2. Enter the title of the new project. (The name should be unique.)

3. After the name is supplied and checked in the index of projects, a unique project identifier (*id*) is created, along with a file name under which the project's document will be saved. A new project object is constructed using the *id* and the file name as parameters; it then is added to the project index using the new project's name as the key.
4. A new document object is created.

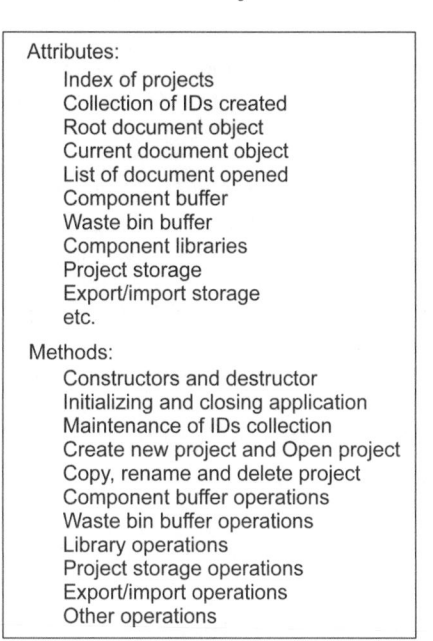

```
Attributes:
    Index of projects
    Collection of IDs created
    Root document object
    Current document object
    List of document opened
    Component buffer
    Waste bin buffer
    Component libraries
    Project storage
    Export/import storage
    etc.

Methods:
    Constructors and destructor
    Initializing and closing application
    Maintenance of IDs collection
    Create new project and Open project
    Copy, rename and delete project
    Component buffer operations
    Waste bin buffer operations
    Library operations
    Project storage operations
    Export/import operations
    Other operations
```

Fig. 3.17. *CBondSimApp* class – modelling support part

The document object that is so created takes as its name the project name. This is the document from which model development starts; it thus represents the root of the project's document tree. Further model development proceeds as described in Sect. 3.2.2.

A project is opened (or reloaded) by selecting its name in the index of projects. The project object to which the index refers is found and the corresponding document is obtained. The project document then is loaded, as in step 4 above.

The application class supports other important operations on projects that parallel those for components. Thus, to rename a project it is enough to change the key entry. To make a copy of a project, the procedure is:

1. Get the project object that is to be copied from the index of projects. Get the file name of the original document. Define the name of the project copy.
2. Create a new *id* and filename for the copy. Construct a new project object using the new *id* and the new filename as parameters. Add the new object to the index of projects using the name of project copy as the key.

3. Make a copy of the project's document, as explained in Sect. 3.2.2. The new document is saved under the new file name created in step 2.

To simplify model development the application object maintains a *Component* buffer that serves a function similar to the Windows clipboard. It also supports operations similar to the cut, copy, and paste operations of the Windows environment. Here, these operations are component based. This means that a component contained in a document can be selected and copied, as explained in Sect. 3.2.2, then added to the buffer. Similarly, a component can be cut—disconnected from the document—and added to the buffer. Components from the buffer likewise can be copied back and inserted in the same, or some other, document. These actions can be performed on a selected set of the components, as explained in Sect. 3.8.

Another useful service of the application object is the *Waste Bin* buffer. This buffer functions in a similar manner as the Windows Recycle bin. The project can be deleted from the project index using methods from the application class. Instead of removing all of the underlying document files, however, only the project entry is removed from the index; the corresponding project object is sent to the waste bin buffer. Similarly, other objects—e.g., a component or group of components and bonds—that are deleted from a document are disconnected, can be added to the buffer.

Projects or components in the Waste Bin buffer can be either restored or removed from the application. When restoring a project, the information object is moved from the buffer and inserted back into the project index. Similarly, when restoring a component, it again is moved from the Waste Bin buffer back to the Component buffer. From there, the component can be inserted into any document. On the other hand, removing a component or project from the buffer means removing the object and all underlying documents. This can be achieved as discussed in Sect. 3.2.2.

When closing the application the computer frees computer memory by destroying all objects and saving these data to files. The index of projects, Component and Waste bin buffers are saved too. When restarting the application, the buffers are restored. Hence, the models in the Component – and Waste bin buffers are available across modelling sessions. It is advisable, of course, to empty these buffers from time to time to conserve memory.

Finally, we return to the mathematical model parameters already mentioned in several places. These can be defined at the document level or at the component level. In the former case, they are visible in all contained components of this level and higher. Parameters defined in a component port, however, are visible only in that component. In this way, common parameters—such as gravity, and some other physical and mathematical constant—can be defined in a lower-level document, or maybe at the project root document level. Parameters that are specific to a process should be defined in the corresponding elementary component. Such flexibility comes with some dangers. If, for example, we delete a parameter defined in a certain document, this could create a problem if the parameter also is used by some higher-level components. The corresponding mathematical relationship then would be incompletely defined, thereby resulting in an undefined parameter. It is quite difficult to monitor all such changes. One remedy is to post-pone final checking until the phase in which the complete mathematical model is built.

3.10 Component Libraries and Model Reuse

The idea of code reuse is very old and very appealing, but well known not to be an easy problem to solve. Code reuse significantly improves the efficiency and qual-ity of development and, hence, of problem solving. There are many approaches and techniques developed and in wide use today that are based on code reuse, such as COM and DCOM technology. OOP languages are developed with code reuse as a goal. The following four fundamental reuse problems can be identified [8,9]:

1. Finding components
2. Understanding components
3. Modifying components
4. Composing components

As used in this sense, a component is any output of the solution process, e.g. code components, on-line documentation, specifications, etc.

Here we are interested in reuse of component models and projects. The concept of component oriented modelling, as discussed at the beginning (Sects. 1.2 and 1.6), is introduced not only to enable systematic model development, but aslo to support reuse of components for building system models. Component reuse can significantly improves the quality and efficiency of model building as well. Simi-larly this applies to projects. Here we describe approaches and methods used to that goal.

Component and project files are held in a library in a separate section of the workspace. The library is divided into two segments

1. Project repository
2. Component libraries

The project repository serves as a convenient storage place of complete pro-jects. Any project can be moved into the project repository and removed from the main workspace section (models section). Such a project can be reused at any time. To simplify the search for a particular project, the repository can be hierar-chically organised according to application areas, or by some other criterion. This is not considered further here, for it is a separate problem. Currently, only the ba-sic mechanisms of project managment is provided.

Component libraries are divided into three sets:

1. Word model components
2. Electrical components
3. Mechanical components

The first serves for components represented by a word model and are for general use. The electrical library stores electrical components represented by electrical circuit symbols. This library can store models of, for example, electrical resistors, coils, semiconductors, and electrical motors. The third is a library of mechanical components represented by suitable graphical symbols, e.g. springs, bodies in translation or rotation, connectors, etc. These libraries can be further specialised.

It is relatively easy to organise components into libraries. Project and component files of interest are put from the model section into a library section of the workspace simply by copying. The application object maintains an index of project and component files put into libraries. In this way, storing important components and project models separately from the model workspace is more secure; there is less chance that they will be accidentally modified or removed.

Projects or components of interest can be found by searching library indexes. To reuse a particular project it needs simply to be copied back into the model workspace. After that it can be opened and used as any other project.

Components from the library are inserted into the model workspace. It can be done as follows:

1. Find component in the library index
2. Open the document and extract a copy of the component object held in its header
3. Select the position in target document working area where the component is to be inserted. Change the component visualisation data to reflect this selection. Create a new component *id* and document filename, then reset the corresponding attributes in the component object.
4. Add the component to the document component list and save the document.
5. Copy the library document files the to the model workspace

The other three reuse problems stated in the beginning of section are solved by the component modelling approach. Any component or project in the library can be opened to analyse its structure, constitutive laws and parameters, and how it may be used. As with other coding methods, understanding of a project or component model depends also on how the model is constructed. If developed logically and by strictly applying systematic decomposition, the models will be transparent and more easily understood.

After their insertion into a document they can be modified easily. They can be further connected to other components and the connection checked as discussed in Sect. 3.7.

The component model concept presents another possibility for component reuse: Component use is not confined to the application in which they are created; different applications can exchange components (Fig. 3.18). To exchange models they are put in a package. This consists of the index file that specifies projects and components and the corresponding document files. To simplify its use, projects and components can be organised in the package in the same way as in the library. Further, the package can be sent as an e-mail attachment to another user. Projects and components from the package can be imported into another application. In effect, the component or project is copied into an application. This resolves any possible conflict in *ids* or filenames. Of course, there is a price to pay: Every application must maintain its database. To make it as robust as possible, the information class implements some recovery methods. These methods can recreate a complete model database in the event of file corruption.

Fig. 3.18. The exchange of components between applications

The exchange of models is a very useful means of supporting collaboration between people engaged in solving similar problems. A complex modelling project can also be divided into separate development tasks. After components are developed, they can be integrated into an application. We do not wish to imply that this is a simple task. For components to work properly, they must have ports designed to comply with the requirements of Sect. 3.7. This is not much different from the situation that arises when dealing with real components. Every such component can function properly only in an environment specifically designed for it. The approach developed here can help in understanding this problem and aiding design of real engineering components and systems.

References

1. V Damic and J Montgomery (1998) Bond Graph Based Automated Modelling Approach to Functional Design of Engineering Systems. In: GR Gentle and JB Hull (eds) Mechanics in Design International Conference, The Nottingham Trent University, Nottingham, pp 377-386
2. S Williams and C Kindel (1994), The Component Object Model: A Technical Overview. In: http://msdn.microsoft.com/library/techart/msdn_comppr.htm
3. Booch, G (1991) Object-Oriented Design with Applications, Benjamin Cummings, New York
4. B Stroustrup (1998) C++ Programming Language, 3rd edn. Addison-Wesley, Reading
5. PC Breedveld (1982) Proposition for an unambiguous vector bond graph notation. J. Dynamic Systems, Measurement, Control 104:267-270
6. EP Fahrenthold, JDWargo (1991) Vector and Tensor Based Bond Graphs for Physical Systems Modeling. J. of Franklin Institute 328:833-853
7. Carl de Boor (1998), A Practical Guide to Splines, Springer-Verlag, New York
8. T Biggerstaff and C Richter (1989), Reusability framework, assessment, and directions. In: TJ Biggerstaff and AJ Perlis (eds) Software Reusability: Concept and Models, Volume 1, Addison-Wesley

9. RA Walpole and MM Burnett (1997), Supporting Reuse of Evolving Visual Code. In: Proceedings of 1997 IEEE Symposium on Visual Languages, Capri, Italy, pp 68-75

Chapter 4 Object Oriented Modelling in a Visual Environment

4.1 Introduction

This Section describes BondSim, a program which offers a visual environment for the modelling and simulation of engineering and mechatronics systems, particularly those based on bond graphs. The general concept of the program is shown in Fig. 4.1.

BondSim implements several services that are accessible to a user through a window system. The two basic services are Modelling and Simulation. The first supports model development tasks and represents implementation of the ideas and methods of Chapt. 3. The program also supports model database maintenance, library support, as well as collaborative support for model exchange using e-mail.

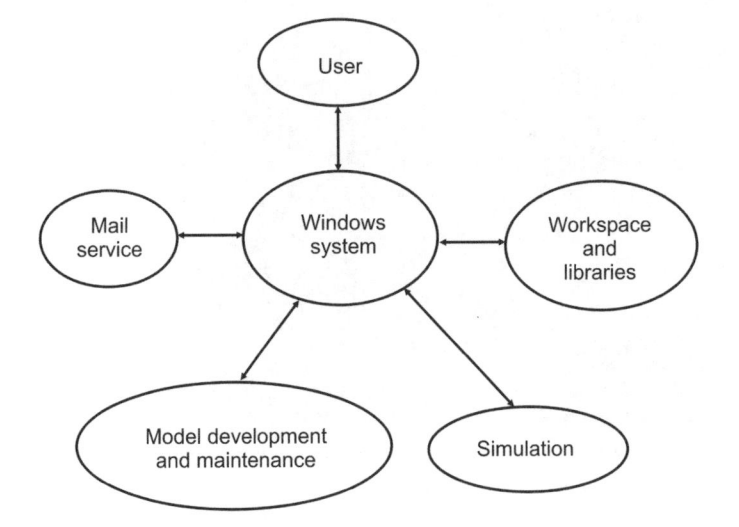

Fig. 4.1. The visual modelling and simulation environment

The simulation subsystem uses models developed by the application to study system behaviour. This Chapter focuses mostly on the modelling part of the program. The Simulation part is the topic of Chapt. 5.

The operating system for which the application has been developed is Microsoft Windows (Windows Professional, Windows NT, Windows 9x). The program has been developed using the Microsoft VisualC++ integrated development environment using the Microsoft Foundation Class (MFC) Library [1]. Use was also made of the ZLib library for data compression.[1] We greatly appreciate the permission of Zlib's authors for the free use of the library. BondSim is included with this book (see Appendix). Readers are encouraged to use it in conjunction with the text.

4.2 The Visual Environment

The main application class *CBondSimApp* was derived from the MFC class CWinApp and is used to construct a Windows application object, which in turn is used to implement modelling and simulation (Sect. 3.8). The application object is declared as global and is constructed at the start of *BondSim*. During its initialisation the main application window is created and appears on the screen as shown in Fig. 4.2.

Fig. 4.2. The BondSim main window

The main application window is based on the *CMainFrame* class. This class is derived from the MFC Library's *CMDIFrameWnd* class, which provides the functionality for a multi-document interface (Sect. 4.5). The class also defines some

---

[1] ZLib is a free data-compression library developed by Jean-Loup Gailly and Mark Adler, http://www.info-zip.org/pub/infozip/zlib/.

specific methods needed by the modelling and simulation environments. These belong to two groups. The first controls messages sent to the main window and gives information on the current status of operations. Such messages are displayed in the status bar at the bottom of the main window and include, e.g. *Ready*, *Create a new project*, etc. Attributes and methods are also provided for the creation and operation of a progress control bar located to the right of the status bar. This provides the developer with feedback on the percentage completion of some lengthy operations, such as are encountered during simulation runs. The other methods are used for the distribution of messages sent to the window during various phases of modelling and simulation, such as during the creation of bond graphs, operations on libraries, etc.

Methods accessible at the application level are organised in the *Project, Library*, and *Function* menus, in addition to the customary *View* and *Help menus*. There is also a row of toolbar buttons for the most important commands. Some of the most often used commands can also be accessed by keyboard shortcuts.

The *Project* menu contains commands for modelling operations implemented along guidelines given in Sect. 3.8. The first two are *New* and *Open*

The *New* command is used to create a new project (Sec.3.8). When this command is chosen, a dialogue window appears with an edit box into which the user inputs the name of the new project (Fig. 4.3).

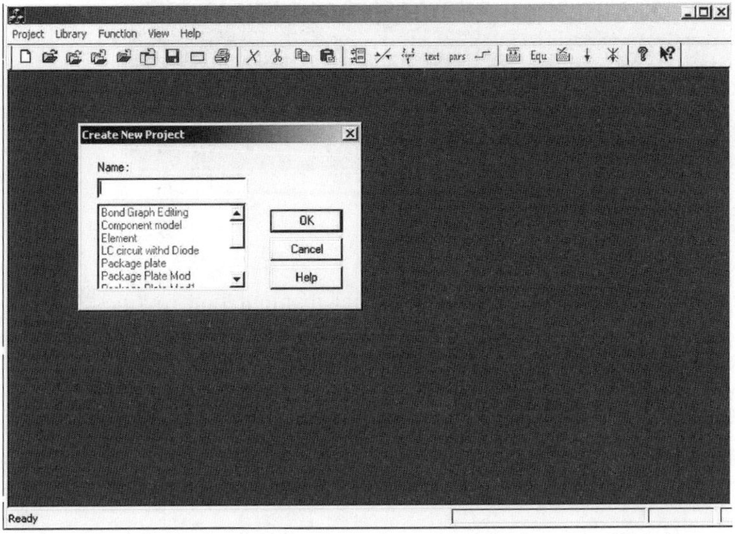

Fig. 4.3. The *New* modelling project dialogue window

The dialogue window also contains a list box with the names of projects that already exist. This information is held in the project index maintained by the application. The name of the new project is accepted, provided it is unique. The new project document is then opened in a suitable window. This is the starting point of the model development process (Sect. 4.6).

The *Open* command is used in a similar way. This command opens a modelling project already in the project files. It uses a similar dialogue window as the *New* command (Fig. 4.4). The name of the project can be selected from the accompanying list box. It is possible to open a project in the *Edit* or *Read only* mode by selecting one or the other from the *Mode* box. The *Edit* mode is the normal mode for opening projects, in which models are created or modified. The *Read only* mode, on the other hand, can be used only for reviewing. In this mode, project documents cannot be changed, thus protecting a model's original form. The document project is read from the project root document file and displayed in the window.

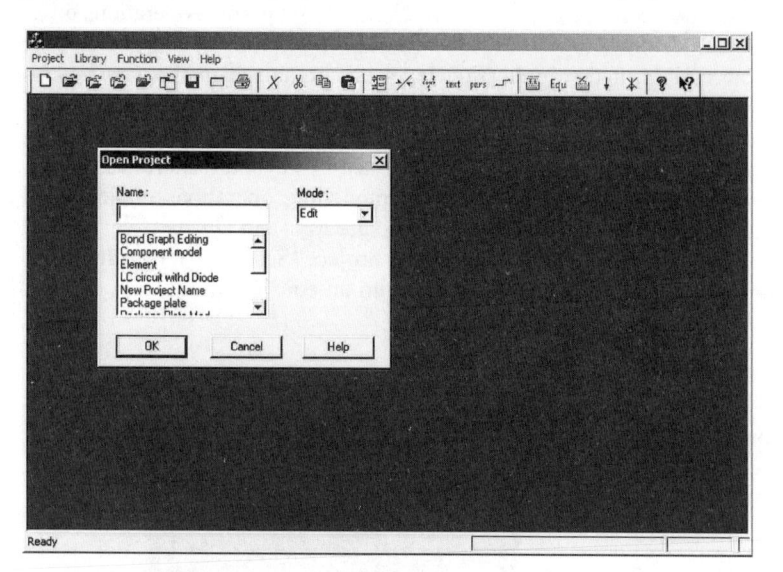

Fig. 4.4. The Open a new project dialogue

The next group of commands contains project manipulation routines, such as *Delete*, *Copy*, and *Rename*, each of which uses a similar dialogue. The *Delete* command removes the project from the project index and puts it into the *Waste Bin*. It is also possible to remove the project completely by bypassing the *Waste Bin*. The *Copy* command creates a copy of an existing project under a new name. This command invokes the copy operations of all document files that make up the project, as described in Sect. 3.8. The *Rename* command changes a project's name.

Two other useful commands are *Move To* and *Get From*. The first moves a copy of a project to the projects library; the second inserts a copy of a project from the projects library. Once copied to the library, the project can be removed from the project workspace. Any project in the library can be inserted back to the project workspace and subsequently opened and edited.

The *Repair Projects* command helps repair the model database in the case of corruption. The *Waste Bin* command provides access to the buffer that stores information on previously deleted projects or components. These can be completely removed from the application database, or recovered and inserted back into the workspace, or in the library.

The last group of commands in the Project menu uses e-mail to export models from, and to import models to, other BondSim applications. Their implementation and use is addressed in Sect. 4.9.

The *Library* and *Function* menus contain commands used for operations on the libraries and functions, respectively, that the application maintains (Sect. 3.9).

Thus, it is possible to *Open* a project or a component in the library, e.g. for reviewing. Similarly, we use the *Delete* command to remove a project or a component from the library completely or to move it to the *Waste Bin*. There is also a command to *Repair* the library database.

Commands in the *Function* menu support user-defined functions, which can be used in formulating the constitutive relations of the components (Sec. 3.5). Thus, it is possible to create a *New* function, *Open* an already defined function for editing, or *Preview* a function. It also is possible to create a *Copy* of a function, to *Rename* it, or to *Delete* it. Currently, only one- and two-dimensional functions defined in tabular form are supported.

The View menu contains the standard Windows commands. The Help menu contains *Help Topics* , *BondSim on the Web* as well as *About BondSim*. The first gives access to the underlying modelling and simulation help documentation. The next offers links to the BonSim related web pages, and the last gives some basic information on BondSim.

4.3 The Component Hierarchy

The component classes discussed in Sects. 3.2 and 3.3 are derived ultimately from *CObject*, MFC's base class. This provides basic support for dynamic creation and object persistence (serialization), as well as other services [1]. Dynamic creation is the MFC Library mechanism for the creation of an object of a given type at run-time. Of great importance is serialization. Objects derived from the *CObject* class are responsible for their saving to, and their reloading from, a persistent medium..

The hierarchy tree for the word model classes of Sect. 3.3 is given in Fig. 4.5. In addition to the functionality already discussed in Sect. 3.2.1, the base class, *CComponent*, also defines an internal state attribute and methods.

The component object supports four states that are distinguished visually in the development environment. These states are:

1. Normal
2. Selected
3. Text
4. Opened

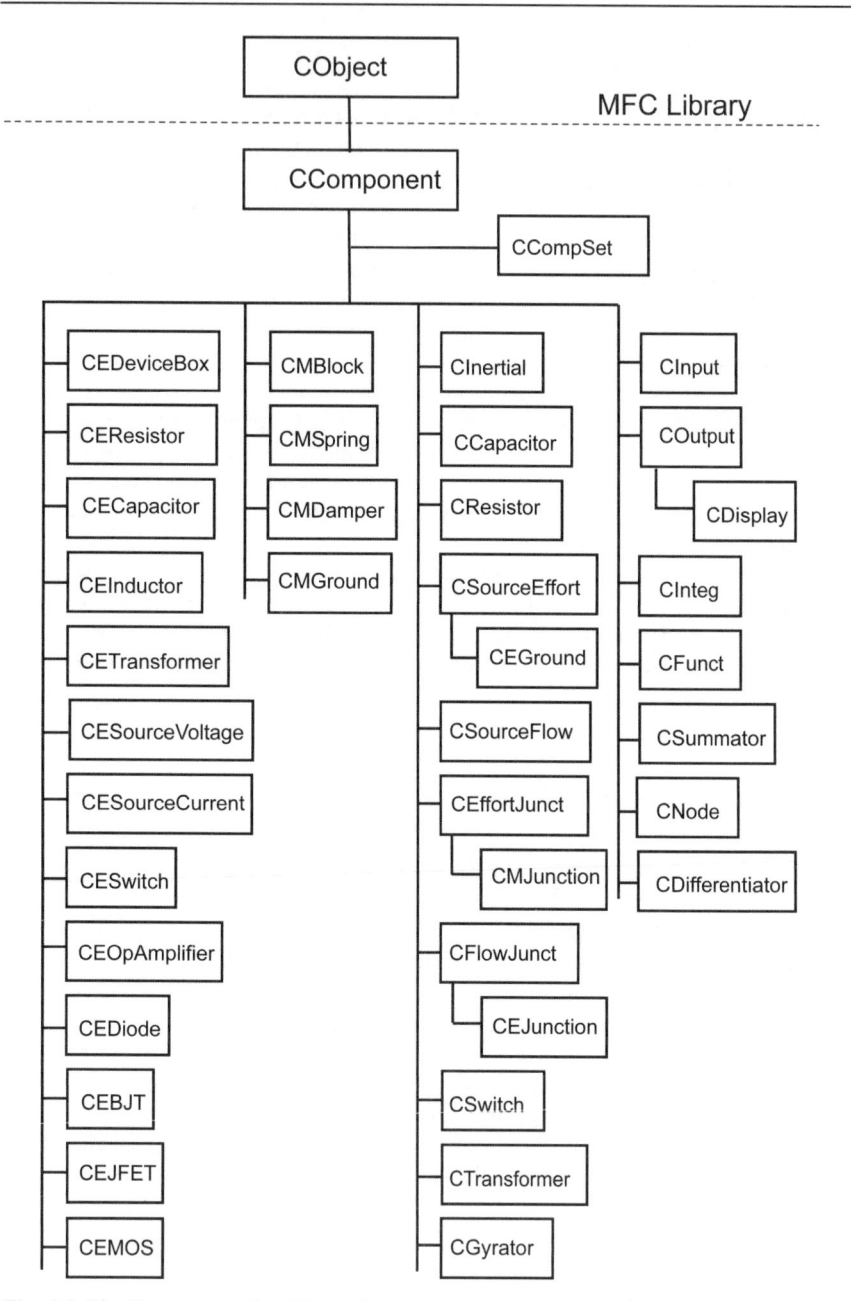

Fig. 4.5. The Component class hierarchy

The *Normal* state of a component is the default state. The component appears on the screen as the text (name) and the ports, without any visual adornment. To perform most of operations—such as copying, deleting, and moving—the component first must be selected. The *Selected* state is indicated by a bounding rectangle drawn around the component in a specific colour, e.g. in the red. When the object is deselected, it returns to the *Normal* state and the bounding rectangle disappears.

The component name can be edited in the *Text* mode. In this mode only the component text appears and it is drawn in a specific colour, e.g. in red. The object is responsible for redrawing any editing change. When redrawn, the component returns to the *Selected* mode.

The last state is the *Opened* mode. This indicates that the component is opened. When the component is closed, it returns to the *Normal* state.

4.4 The Port and Bond Classes Hierarchy

Similar to the component classes, the port and bond classes of Sec.3.4 are also derived from the *CObject* class. These also use *CObject*'s serialization support. The complete class hierarchy is given in Fig. 4.6.

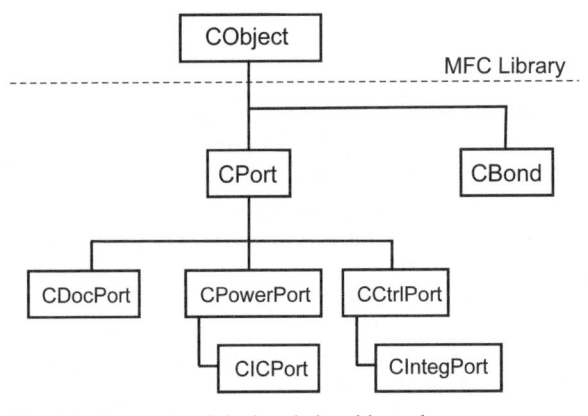

Fig. 4.6. The port and the bond class hierarchy

The port base class, *CPort*, in addition to the functionality already discussed in Sect. 3.4, defines an internal state attribute and methods used to change the state of a port object to enable certain operations, such as deleting, moving, etc. This is shown by changing the port visual appearance. The port states can be:

1. *Normal*
2. *Selected*
3. *Transparent*
4. *Connected-out*
5. *Connected-in*

The *Normal* state means that the port object is not in a specific mode and appears on the screen in the normal colour (black). To apply an operation to the port, such as moving it around the component object periphery, deleting it, or connecting it by a bond, the port object first must be *Selected*. To indicate this selection the port is drawn in red. When the object is deselected, it returns to the *Normal* state.

The *Transparent* mode is similar to the *Selected* mode, but only the port outline is coloured. The port is set to this mode when, for example, if it is moved. The cursor is constrained to move in the component, around its boundary. When a new position is chosen, the cursor returns to its normal appearance and the port reappears in the new position with the state changed to the *Selected*.

The last two states show ports that are interconnected. Thus, if a port is selected and we wish to find the other port to which it is connected, the operations described in Sec. 3.7 are applied. During the outward search all the ports found are changed to the *Connected-out* mode; similarly, the ports found during the inward search are set to the *Connect-in* mode. In these two modes the ports are shown in colours different from other modes. These colours are full blue and the blue outline, respectively. When deselected, the port returns to the *Normal* mode.

Similar to the port, the bond object can also be in one of the following modes:

1. *Normal*
2. *Selected*
3. *Connected*

In the *Normal* mode the bond appears as a line segment drawn in the normal colour (black). The *Selected* mode indicates the bond state to which certain operations can be applied, e.g. deleting the bond, or changing its shape by dragging it across the document drawing area. The bond object is set to this mode during creation; that is, when drawing a line between ports. Again, when deselected, it returns to its normal appearance.

The *Connected* mode is used to show the bond line segments that connect ports found during the outward and the inward searches, as explained above for the ports. In this mode, bond segments appear in a colour similar to that of the ports, e.g. in blue. When deselected, the bond returns to the *Normal* mode.

4.5 The Document Architecture

MFC supports the creation of document windows based on three associated classes: *CDocument*, *CView*, and *CFrameWnd*, or the classes derived from them [1]. The process is coordinated by a class derived from the *CDocTemplate*. For example, the *CMultiDocTemplate* supports multiple document interfaces.

The frame class, used to display documents, has a title bar and a border within which the view class displays the contents of the document. The view class can also handle events generated by keyboard input and mouse action. Typically, the document template object is constructed in the global application object at initialization (Sect. 4.2). When a command to create a new document is issued, the document template object dynamically creates document and frame objects. The latter creates a view object that displays an empty document. When opening a document that exists in a file, the procedure is similar, but this time the document is read from the file and displayed inside a window frame. Reading, as well as saving, a document is done by an archive object that supports serialization. Details of this and of related processes can be found in [1].

The document class described in Sect. 3.2.2 differs from the MFC's *CDocument* class in several aspects. The most important of which are:

1. Windows documents, as implemented in the MFC Library, are single-level documents, i.e. they are created at the application level. On other hand, the visual modelling system needs only the project root document to be opened at the application level. All others are opened within the previous document using word model objects as interfaces. In addition, all documents are linked.

2. To create a new (empty) document, Windows needs nothing. To open an existing document, it asks only for a filename. It uses the filename to read the file, but the title is also displayed in the frame title bar. In the modelling system described here, even a new document is not completely empty. It contains document ports! Thus, to create a document, the system needs data from its word model component object, preferably in the form of the object pointer or by a reference, not a copy. The component name can then be used for the title, and the component ports for the creation of the corresponding document ports (Sect.3.4). The file name is used for saving the document. A document already existing in the storage medium can be recreated by loading it from the file.

3. Some operations do not need the complete document-view-frame architecture. A document object contains a description of the model of a component. Thus, for operations on models, the document object can be treated simply as a C++ object. This is also the case when objects at both sides of a component interface—i.e., a word model object and its document object—must be updated. Such operations typically run in the background, hence it is inefficient to use windows resources and the corresponding processor time.

The approach adopted here is based on using the Windows document-view-frame architecture, but with changes that support the component model philosophy of Chapt. 3. Thus, common praxis when using the MFC Library is followed, and separate document, view, and frame classes are derived from their corresponding classes from the MFC Library (Fig. 4.7).

The document class *CBondSimDoc*, has already been discussed in some detail (Sect. 3.2.2). It overrides practically all virtual methods of the CDocument, and adds many other methods, too. The size of the document displayed on the screen is limited so as to fit to a page (A4) when printed. The page can be displayed vertically or horizontally. This really does not impose any restrictions on the size of model that can be developed because, using the component model technique, a large document can always be represented by smaller ones. We hope that such a restriction on the document size encourages model development to be done component-wise.

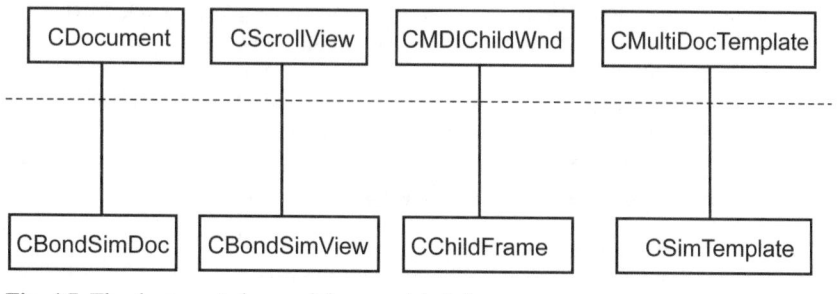

Fig. 4.7. The document class and the associated classes

The view class *CBondSimView* is derived from the MFC class that supports scrolling views. This class supports two modes: *Normal* and *Scale to Fit*. The model can be edited in the *Normal* mode. Scroll bars appear if the document extends beyond the size of the frame window. A new document is opened in this mode. This mode is also used when a project is opened from the file and the *Edit* mode is chosen (Sect. 4.4). Otherwise, the *Scale to Fit* mode is used. This mode displays no scroll bars and the document is fitted into the frame by scaling it up or down. No operations that change the model appearance are permitted in the *Scale to Fit* mode.

This class manages drawing the document in response to the paint commands from the Windows system. The corresponding drawing method forces all objects contained in the document to draw themselves. It also defines methods for handling commands forwarded to the view in response to the mouse or keyboard. The response depends on the *mode* chosen. Thus, the class implements an internal state attribute used to determine which action to execute.

The most important states are:

1. *Normal*
2. *Create component*
3. *Move component*
4. *Insert component*
5. *Insert library component*
6. *Edit text*
7. *Create port*
8. *Drag port*
9. *Move port*
10. *Create bond*
11. *Connect ports*
12. *Change bond*
13. *Size document*
14. *Size port*
15. *Set selection*

16. *Move selection*

Many of these states are not used alone, but in appropriate combinations. We explain here some of the actions that deal mostly with selecting and moving model objects. Those concerned with the creation and editing of model objects are explained in Sect. 4.6.

In the *Normal* mode the view is not set to any specific editing mode; it is ready to accept commands. A component can be selected simply by clicking it. [2] The component reacts by changing its visual appearance. The selected component can be opened using the command from the menu (Sect. 4.7). The component can also be opened simply by double-clicking it, or by the keyboard shortcut. Other commands can be executed on the selected component, as well. The component is deselected by clicking anywhere outside of the component rectangle and returns to the *Normal* mode.

We can drag a component around the document drawing area provided it is not connected to other objects by bonds. This is accomplished by pressing the left mouse button when the cursor is within the component name. This changes the editing state to *Move Component*. By dragging the mouse with the left button pressed, we can move a component rectangle around the document. When the mouse button is released, the component reappears in the new position and the mode changes to the *Normal* mode.

A component port can be selected by clicking on the port. In a similar way as with components, we can move a port around the component periphery, provided it is not joined by a bond. Simply put the cursor over a port, then press and hold the left mouse button. This changes the editing state to *Move Port*, and the cursor changes to the shape of the port. As we drag the cursor around the component periphery, the cursor shape changes to reflect the correct port shape. For example, a power-in port should point every time to the component. When the mouse button is released, the editing mode returns to *Normal* and the port reappears in the new position. Again, the port is deselected by clicking outside its boundary.

Similarly, we select a bond by clicking it, or deselect the bond by clicking outside of the bond. We can also move the bond by dragging it. This changes the bond shape in a manner similar to stretching a thin rubber band, the ends of which are fixed. If we press the left mouse button when the cursor is close to, or on, the bond, the editing mode changes to *Change Bond*. This creates a new intermediate point that is inserted into the array of bond points (Sect. 3.4). Thus, by dragging the cursor with the mouse button pressed, the coordinates of the point under the cursor change, as does this the shape of the bond line. By releasing the mouse button, the view again returns to the *Normal* state.

The operations described above are, in effect, single-object selection operations, i.e. we select or move or apply a menu command to a component, a port, or a bond. It also is possible to select a set of objects. Thus, if we put the mouse cursor outside of any object, then press and drag it, a rectangle appears. When we re-

---

[2] We will use terms like clicking or pressing mouse to mean using the left mouse button. Otherwise the button used will be stated.

lease the mouse, all components inside this are selected, as well as all of the bonds between these components (Fig. 3.6). This action creates a temporary CSelSet object that contains a list of pointers to the components contained within the rectangle, as well as of the bonds joining them (Sect. 3.8). The view object changes its state to *Set Selection*, and all components and bonds in the set change their state to selected.

We can also add a component to a selection by holding down the *Ctrl* key and clicking on the component outside the selection. Similarly, we deselect a selected component by clicking on it. If the components in the selected set are not connected to other parts of the bond graph, they can be moved jointly within the document area, as is the case for a single selected component. To do this, we put the cursor somewhere within the rectangle enclosing all the selected components, then press the left mouse button. This changes the view state to *Move Selection.* By dragging the mouse while the button is pressed, all selected objects move as a block. Other operations can be applied to a set of selected components (Sect. 4.7). We can remove a selection simply by clicking outside of the rectangle encompassing all selected components. This returns the states of the objects to *Normal*, the temporary CSelSet object is destructed, and the view object state also returns to *Normal*.

CChildFrame (Fig. 4.7) is derived from the MFC's *CMDIChildWnd* class, which supports multi-document frame windows. Finally, we come to the document template class, *CmultiDocTemplate*, from which we derived a document template class, *CSimTemplate*. This class inherits all functionality of its parent class. It is used for minor adjustments of menu items during normal document-view-frame creation. For example, it adds the *Last* command in the Windows control menu, which opens the previous document. It also defines a method for creating a document-only object, i.e. a document object without the associated frame and view objects. This is used when executing certain background tasks, as discussed at the beginning of this section.

4.6 Editing Bond Graphs

When a new project is created (Sect. 4.2), an empty document is created in a window with the name of the new project in the title bar (Fig. 4.8). The menu changes to reflect commands accessible at document level. We discuss these commands in Sect. 4.7. Here we explain how bond graphs are developed systematically.

4.6.1 The Bond Graph Palette

The bond graph model is created using an *Editing Palette* (Fig. 4.8). The palette is invoked by the command *Bond Graph Palette* on the *Tools* menu, or by pressing the corresponding toolbar button. The palette contains controls used to create bond graph diagram objects (Fig. 4.9). It is the top-level window that stays until closed either by the user or by the modelling system.

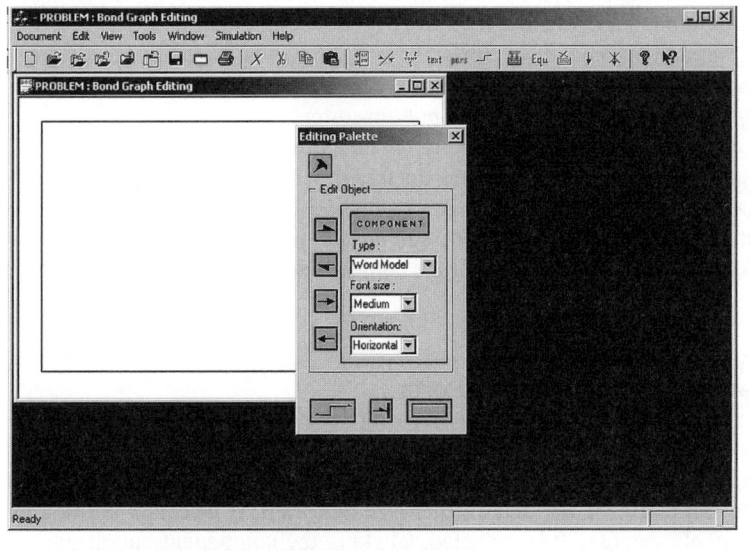

Fig. 4.8. The palette for the editing of a bond graph

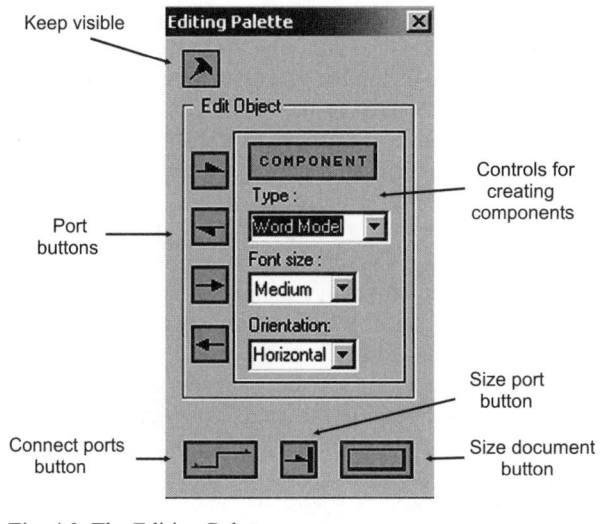

Fig. 4.9. The Editing Palette

The first button is used to fix the palette window. Otherwise, the palette is closed automatically and disappears from the screen when the first component is created. The middle of the palette contains control buttons for creating components. On the left are four buttons used to create component ports, i.e. power in, power out, control in, and control out ports. The sunken rectangle area on the right contains controls for the creation of a component in the document working area.

The *Component* button is used to create a component of the type appearing in the combo box immediately below it. A word model component or any elementary and block diagram components of Sects. 2.5 and 2.6 can be chosen. It is also possible to choose between one of three font sizes for the component name. The component text can be edited left-to-right or bottom-to-top.

Three buttons at the bottom are used to create a bond line connecting the ports, sizing the document ports, and sizing the document working area.

When the document is created, the corresponding view object is created in the *Normal* state. The *Ready* message appears in the status bar and bond graph editing can start. We normally start by creating the components, and then connect them by bonds.

4.6.2 Creating Components and Ports

To create a component, its type must be selected first. By default, this is the word model component, but specific elementary or block diagram components can be selected, too. Similarly, if the default font size and the text direction are not appropriate, suitable values can be selected. Clicking the component button sends a message to create the component object to the mainframe window. The message contains information on the component type, font size, and orientation of the component name. The message is distributed to the active view, which changes its state to the *Create Component* mode. A message also appears in the status bar advising the user to choose a position in the document where the component will be created. Thus, the mouse cursor must be moved in the document area framed by the rectangle. When the cursor is in this area, it changes to a cross. This indicates that a position can be picked where the user can start editing the component name.

When this position is chosen by clicking in the document working area, the view object calls its document method to create a component, supplying it with its type, position, font size, etc. (Sect. 3.2.1). The component is not created if the point selected is too close to the document bounding rectangle. It is better to create it somewhere well inside the document area, for it can easily be moved later. After the component is created, its state is changed to *Text* mode, and the state of the view object is changed to *Edit Text*. A text cursor (caret) appears and editing the component name can begin. A message appears in the status bar asking the user to edit the component name. The view object handles messages from the keyboard. These are translated into corresponding component object text editing actions, such as inserting a character, deleting character, opening a new line, joining the line with the last one, etc. The input is echoed back to the screen and the edited component name appears on the screen. Text editing is ended when the mouse is clicked outside of the text area. The view state returns to *Normal* and the component state to *Selected*. The *Ready* message appears at the status bar and the next action can be undertaken.

When creating an elementary or block diagram component, predefined names, such as C, I, R, and Fun, appear. These can be accepted or changed. The edited component name can also be changed. The component must be selected and suitable options in the palette changed, if necessary. We can then choose the *Edit Text* command on the *Edit* menu, or click the corresponding toolbar button. This changes the state to the text mode, and editing then may continue in the manner explained above.

After a component is created, we can add ports to complete the component (Sect. 3.4). This can be done using the port buttons. When the appropriate port button is clicked, the *Edit Palette* object creates a message to create the port object and send it to the mainframe window. This message, containing information on the port type, is distributed to the active view, which changes its state to *Create Port* mode. A message also appears in the main window status bar advising the user to insert the port inside a component boundary. As soon as the cursor is moved within the document rectangle, its shape changes to a cross. When the cursor is moved across a component, the component becomes temporarily selected and it is outlined with a rectangle. To insert the port, a position inside the boundary is chosen and the mouse button clicked (Fig. 3.2). The position selected must not be too close to existing ports. After the mouse button is released, the document creates the component port, as explained in Sect. 3.4, and the port appears at the component boundary in the selected state. Instead of picking and placing the port, it is possible to drag it with the mouse button pressed. This procedure is similar, but the cursor changes its shape to the half (power) or full (control) arrow, corresponding to the type of port chosen.

If the port has the wrong power direction, it can be changed easily to the opposite direction. This is accomplished by first clicking the port to select it, then selecting the *Change Port* command on the *View* menu. Alternatively, we can click on the corresponding button in the tool bar. The port power direction can be changed if the port is not connected by a bond line. If it *is* connected, the bonds first must be removed, as explained in Sect. 4.8), and the procedure applied.

4.6.3 Creating Bond Lines

Once the components are created, it is a simple matter to join their ports by bond lines. To create a bond, we click on the *Connect ports* button on the palette (Fig. 4.9), which sends a message to create a bond. The active view changes state to the *Create Bond* mode. Simultaneously, the status bar displays a message advising the developer to select a component port from which the bond line will be drawn. As soon as a start port is clicked, a line appears that moves as the cursor is moved. The view state changes to *Connect Ports*. It is possible to draw a line directly to another component or document port, and then click; but it sometimes is more convenient to click to some intermediate points first, e.g. to go around some other components (Sect. 3.4). When the end port is clicked, editing the bond object is completed and the view object returns to the *Create Bond* mode. The view object stays in this mode until the palette button is reset.

4.6.4 Editing Bond Graph Models

We proceed with editing the bond graph model by developing the models of word model components contained in the current document. To create the model of a component, we double-click the component (Sect. 4.7). A new empty document is created, as shown in Fig. 4.10. The document is displayed in a frame window that has as its title the name of the component and which shows the document ports that correspond to the ports of the component. The document drawing area is empty. We edit the bond graph model of the component as explained above.

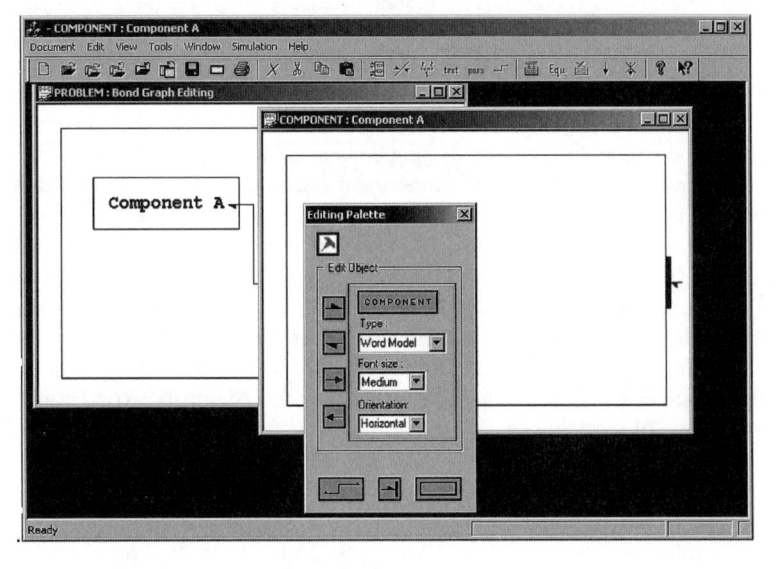

Fig. 4.10. Opening a new component document

During model development the drawing area can appear to be too small or too large. We can enlarge the document frame size by dragging one of the window borders, an operation familiar to Windows users. This does not change the document-drawing area framed by the rectangle. To change this, we use the *Size document* button at the bottom of the *Editing Palette* (Fig. 4.9).

When this button is clicked, the palette sends a message to the active view to change its state to *Size Document* mode. An appropriate message appears in the status bar. When the user moves the cursor over the right side or bottom side of the document rectangle, the mouse pointer changes its shape to the double arrow (different than when changing the size of the window). The mouse button now may be pressed to drag the object. As it is dragged, a rectangle appears that shows how the document area changes. When the mouse button is released, the view object changes to the *Normal* mode and the document is redrawn to show the changed working area. It is also possible to drag the right-bottom corner and, in that way, simultaneously change both the width and the depth of the document area.

There is both a minimum and maximum drawing area between which the document size can be set. The exact size depends on the orientation of the document, i.e. whether it is horizontal or vertical (Sect. 4.7). The maximum size is limited to the size of the page (A4), and the minimum is such that it can be the host of a single component. If the document cannot fit into the frame, scrollbars appear on the right side and/or bottom.

When a document is created, or when it changes its size, the document ports are created as well. The dimensions of the connecting strip depend on the size of the document area and the number of ports. If a number of bonds must be connected to the document ports, the size of the connecting strip can appear to be too small; in such a case, it is necessary to enlarge its width or height. Similarly, when the document strip size is too large, it would be nice to reduce it. We can use the *Size port* button on the *Editing palette* (Fig. 4.9) for these purposes.

This procedure is very similar to that involved in changing the document size, as described above. We first click the size document port button, which sends a message to the active view to change its state to the *Size Port* mode. An appropriate message also appears in the status bar. Next, the cursor is positioned over the side of the document port strip that we would like to drag. When the cursor changes to the double arrow, we press the mouse button and drag the selected strip side along the document working area. When a suitable size is achieved, the mouse button is released and the document port is redrawn to reflect the new size. The view and port states are then reset to *Normal*.

Starting at the project root document and working up, the bond graph model can be systematically edited as explained above. Great care should be taken when drawing the bonds. The rules discussed in detail in Sect. 3.7 should be followed closely; otherwise, the model can easily be incorrect. It is relatively easy to connect the ports of elementary components properly when these are contained in the same document. As a rule, however, systematic model development generally results in elementary components connected over several levels of documents. It is then of paramount importance to insure the symmetry of port connections, as shown in Sect. 3.7. The *Show Joined* command on the *View* menu aids in this important task by invoking the search out and search in procedures (Figs. 3.12).

Thus to find a port that is connected to some particular port, the port is selected, then the *Show Joined* command is chosen. The search displays all ports and connecting bond lines, starting with the selected port and ending at the other port. During the search the states of the ports and connecting bonds are set and coloured as explained in Sect. 4.4.

We illustrate the procedure and its result with the example of the *See-saw* model of Sect. 2.7.3. Thus, we open the Body 1 component of Fig. 2.16, then select the outgoing port of the left junction, i.e., the x-velocity junction (Fig. 2.17 Body1). We wish to find another port connected to this one. The results are shown in Fig. 4.11. Starting from the selected port, we first see the bond line connecting to the Body 1 document port, then out of it and into the Platform. The line proceeds further into component f, and finally ends at the x-component flow junction.

Thus, the connection is valid, i.e. the force component applied at the Body1 x-junction really is the x-component of the reaction force of the Platform. The port connections are apparently symmetrical.

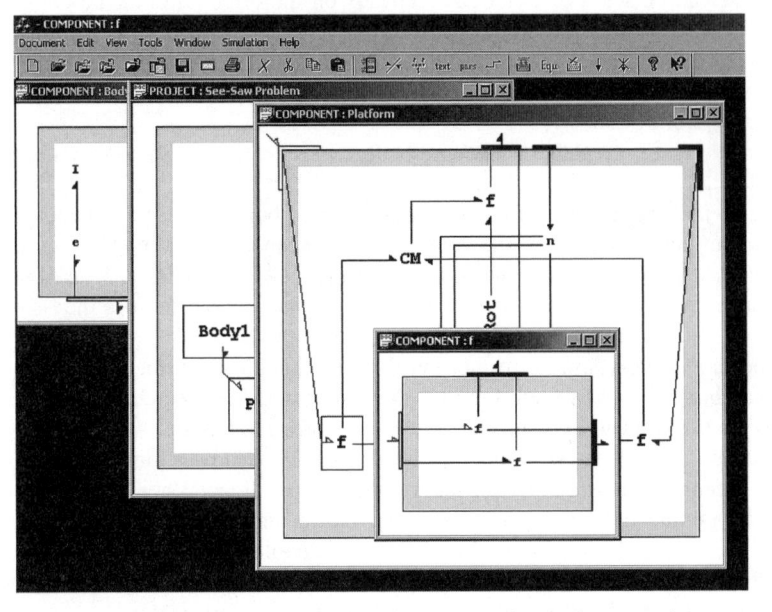

Fig. 4.11. Interconnection of ports found by the *Find Joined* command

4.6.5 Editing Electrical and Mechanical Schemas

When editing bond graph models of electrical devices, it is convenient to represent word model components using the familiar electrical circuit symbols (Sect. 3.3). This is achieved by using the *Use icon* command on the *Tools* menu, and then *Electrical*. The command opens the *Electrical Components* palette (Fig. 4.12) that contains a series of buttons for creating electrical word model components represented by their circuit symbols.

Creating a component uses the same mechanism as when using the *Editing Palette*. Thus, to create a component we click the corresponding button on the palette. Then, we position the cursor over the document where we wish to create the component and click the left mouse button. The component created in this way is a word model expressed in the form of the appropriate electrical circuit symbol. A difference, however, is that electrical components usually have a fixed number of ports. Thus, the ports are created as a part of the component object creation. The ports are just bond graph power or control ports. These cannot be moved around the component boundary, as is case with common word models; nor they can be deleted. As an illustration, in the document of Fig. 4.12 a Bipolar Junction Transistor (BJT) word model is created using the palette.

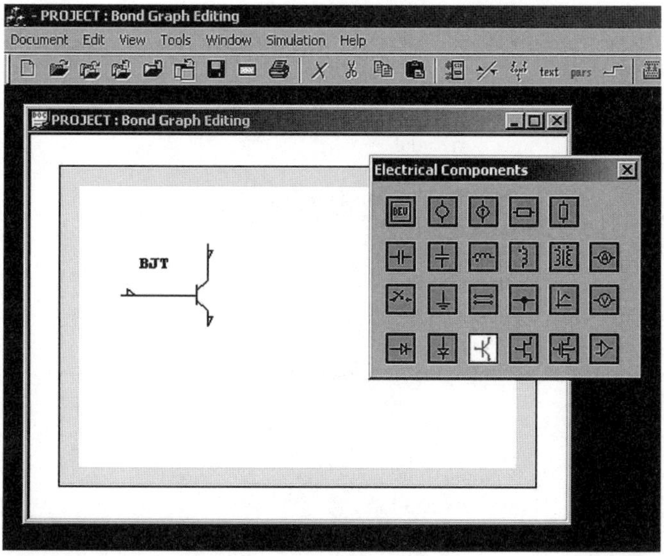

Fig. 4.12. Editing electrical components

A predefined text is also added to the component, but this is restricted in size, for it is used mainly for component designation. The component otherwise behaves just as any other word model component. It can, for example, be opened to define its model in terms of bond graphs. To interconnect such components, the *Editing Palette* again is used.

Only the most common electrical circuits symbols are currently implemented (Table 4.1). Some of these components—e.g. resistors, capacitors, coils, and diodes—can be placed either horizontally or vertically. It should be noted that the node is the only variable-number port component. Thus, ports to such components can be deleted or added using the *Editing Palette*.

Mechanical components are treated in a similar way. The *Mechanical Components* palette is opened using the *Use icons* on the *Tools* menu, and then the *Mechanical* command (Fig.4.13). Currently some of the more common mechanical component symbols are supported as given in Table 4.2.

Mechanical components are created in the same manner as described above for electrical components, using this time the *Mechanical Components* palette. The ports are created also during component creation and are fixed. Only node component ports can be changed, e.g. by deleting, or adding, a port using *Editing Palette*. Also there is a short text denoting the component. Fig. 4.13 shows a damper component created by selecting the *Damper horizontal* button in the palette and placing it into the document.

Table 4.1. List of electrical circuit symbols implemented

Component	Component
Device	Voltmeter
Voltage source	Ground
Current source	Transmission line
Resistor horizontal	Node
Resistor vertical	X-Y Plotter
Capacitor horizontal	Diode horizontal
Capacitor vertical	Diode vertical
Inductor horizontal	npn Bipolar Junction Transistor
Inductor vertical	n-channel JFET
Coupled inductors	n-channel MOSFET
Switch	Operational Amplifier
Ammeter	

Table 4.2 List of mechanical components symbols implemented

Components	Components
Spring horizontal	Node
Spring vertical	X-Y Plotter
Damper horizontal	Ground vertical left
Damper vertical	Ground vertical right
Body	Ground horizontal down
Dry friction	Ground horizontal up

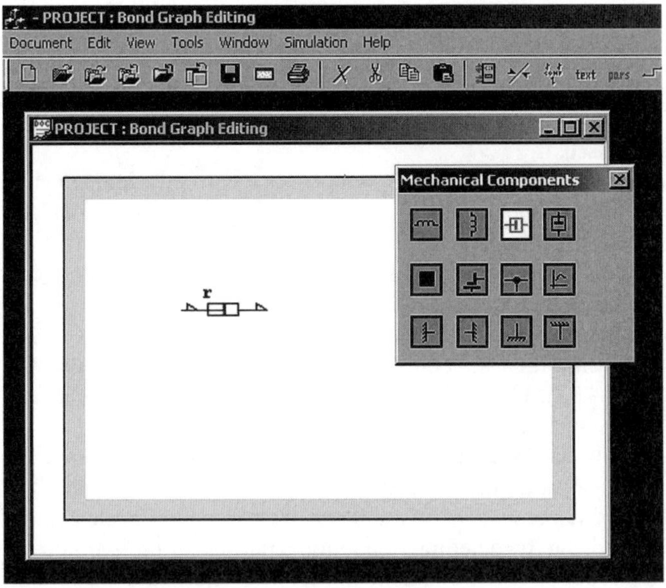

Fig. 4.13. Editing simple mechanical components

4.7 Important Operations at Document Level

This section reviews some of the most important document-level operations. These commands are collected under menus that appear when the first document—the project root document—is opened. These are: *Document*, *Edit*, *View*, *Tools*, *Window*, *Simulation*, and *Help*. Some of these commands can also be invoked by pressing the appropriate tool bar button, or using keyboard shortcuts.

We here deal only with commands intended for model development. These are found in the first four menus: *Document*, *Edit*, *View*, and *Tools*. They implement methods already discussed in Chapt. 3. The Windows menu contains standard Windows commands, such as *Cascade* and *Tile*, used for arranging windows on the screen. There is a list of all opened documents at the bottom. These are used to activate the corresponding frame windows. This menu also has one specific command: *Refresh*. This is used to update (redraw) the active window, e.g. during editing of the bond graph model. Commands in the Simulation menu deal with simulation tasks. We postpone discussion of these commands until Chapt. 5. The *Help* menu contains commands that link to online program documentation, as already described at the end of Sect. 4.2.

4.7.1 The Open, Close, and Save Commands

First on the Document menu is the *Open Next* command. It is used to open the next document level. We usually speak of opening a component because the component model document is what is opened. Elementary and block diagram components are treated differently (Sect. 4.8.). Thus to open a component it is selected first, and then the *Open Next* command is chosen. We can also open a document by double clicking a component.

The action executed next depends on whether the component model in question is new, or if it has already been persisted. In the first case, a dialogue window appears asking if the user wishes to create a new component functional description. If the answer is yes, a new document object is created. On the other hand, if the model document exists, it is loaded from disk using the file name stored in the component object. After the document is successfully opened, the component state is set to *Opened* (Sect. 4.3).

It should be pointed out that a new document frame is *not* created if the document has been opened already. In this case, it is simply activated. Note that there is no *New Window* command in the Window menu, as is customary in Windows applications. Every document is shown in just one document frame.

The opened document contains a link (a pointer) to the previous opened document. Hence, it is possible to return to the previous document, and to activate it by using the *Last* command on the *Document* menu. The same command is also found in the frame window Control menu (the Windows menu in the title bar, to the left of the title).

An opened document can be closed in the usual way for Windows. There is also a *Close* command on the *Document* menu. Because every opened document contains a link (pointer) to the previous document, it is necessary to close all up-per-level documents first. Thus, a *close* command invokes a dialogue asking if the user would like to close all upper-level documents. If the answer is no, closing is aborted. Otherwise, all upper-level documents are closed before the current document. When closing a document, the state of the corresponding component object changes to the *Normal*.

There are two other *close* commands in the *Document* menu. The *Close Component* command is used if, in the current window, there is an opened component. To close it, it is necessary to select the component first, and invoke the *Close Component* command. There is also a *Close All* command in the *Document* menu. This command is used to close all currently opened documents and, thus, also the modelling project.

When closing, an alert box appears to advise the user if the document about to be closed has not been saved. The document then may be saved or not. If it is not saved it stays open. It is currently implemented this way because changes in a document often involve changes in other documents as well, e.g. its component documents. This way to close the document it must be saved.

There is also a *Save* command on the *Document* menu. This saves the current document. When saving, the corresponding document method is called to save the document to a file. In the process, all objects contained in the documents, such as components, ports, bonds, and others, are called to save themselves in a document archive before the document is saved to the file.

We now return to the modes in which the document frames can be opened. As already stated in Sect. 4.5, there are two such modes: *Normal* and *Scale to Fit*. The first is set by default if the project root document is opened in the *Edit* mode (Sect. 4.2). When the document frame is open it can be switched between the *Normal* and *Scale to Fit* modes using the appropriate commands from the View menu. The check mark next to the command indicates the current mode. The other mode is used by default when the project root document is opened in the *Read Only* mode. The document window opened in this mode cannot be switched back to the *Normal* mode.

4.7.2 The Copy, Cut, Insert, and Delete Operations

To simplify model editing, several commands are implement for deleting, copying, cutting, and inserting. Thus, a component port can be deleted and removed from the model by selecting it first, then using the *Delete* command from the *Edit* menu. The *Delete* key may also be used for this purpose. The port can be deleted if it is not externally and/or internally connected (Sect. 3.4). Deleting a port does not only remove the port in the component, but also in the corresponding document. This means that the accompanying document is opened in the background, the document port removed, and the document saved and again closed before the component port is removed. In a similar way, it is possible to delete a bond. This removes it from the document, and the bond data are removed from the ports that it connects.

Dealing with components is a little more involved. The component that is disconnected from the other can be deleted. To delete a component it first must be selected, then the *Delete* command is chosen. A dialogue then appears asking if we wish to move the component to the Waste Bin. If the *Yes* button is clicked the component is removed from the document and moved into the *Waste Bin* buffer (Sect. 4.2). This action does not really remove the component object, but its pointer; that is, the pointer is removed from the list maintained in the document and is inserted in the *Waste Bin* component list. The next time the screen is painted the components disappear from the screen. The Waste Bin buffer can be accessed from the document level, as well (see Sect. 4.2). On the other hand if we select the *No* button in the dialogue the components and its document files are removed as already explained in Sec. 4.2.

Three other commands—*Copy*, *Cut*, and *Insert*—work between the document and an internal *Component Buffer*, which the application maintains (Sect. 3.8). This buffer plays a similar role to that of the *Clipboard* in Windows. Thus, to cut a component not interconnected to others, we first select it, then use the *Cut* command from the *Edit* menu. This removes the component from the document—in the same way as the *Delete* command—and puts it into the *Component Buffer*.

We can also copy a component from the document to the *Component Buffer*. The component may be connected, because we are not moving it, but its copy. To copy a component to the buffer we first select it, and then apply the *Copy* command. Copying means, as has been already stated; creating a new component object that is a replica of the one being copied. The copy has a new *id* and filename. Thus, the document of the original component, as well as those of any contained components, must be copied. The pointer of the newly created component is put into the *Component Buffer*.

The components in the *Component buffer* can be inserted back into a document with the *Component Buffer* command in the *Tools* menu. This command has three subcommands: *Insert*, *Open*, and *Delete*.

The Insert command inserts a component into the current document. Choosing *Insert* opens a dialogue with a list box containing the components in the buffer. Selecting a component from the list sends a message to the active view, which changes its state to the *Insert Component* mode. The status bar displays a message informing the user to pick a place in the document area for the insertion. In the same way as when creating a component (Sect. 2.6.2), as soon as the cursor is moved into the document area it changes to a cross. Clicking with the left mouse button inserts a copy of the component into the document.

The *Open* command can be used to review components placed in the buffer. Selecting a component from a list opens a separate window that displays its document. A copy of the component object is inserted into this document and serves as the interface to the underlying documents. Editing is not permitted in this mode.

The components in the buffer can also be removed when, for example, they are no longer needed. The *Delete* subcommand is used to remove a component from the buffer and place in the *Waste Bin*. In the same way as the *Delete* command describe previously.

All the above commands can also be applied to a set of selected components (Sect. 4.5). The *CSelSet* object enumerates the selected components. We create a separate *CCompSet* object to move a selected set of objects from the document to a buffer, and back (Sect 3.8).

The components in the set can be *deleted* or *cut* from the document if they are not connected to other components not belonging to the selected set. In this case, we simply proceed as explained in Sect. 3.8. Hence, we create a component set object, remove the pointers of the selected objects from the document, and add them to this object. The object can then be added to a buffer (the *Waste Bin* or the *Component*). When adding the component set to the buffer, we must append a suitable reference name. It is not necessary that this name be unique. Any name conflict is resolved easily by adding a version indicator, e.g. *Body*, *Body.1*, *Body.2*. (Whether or not the names are the same, they represent different objects.)

We copy a selection in a similar way. In this case, the components in the set need not be disconnected from the other components. We again create a component set object and add to it copies of all selected components and bonds (Sect. 3.8). Copies of its documents and of all contained documents are made, as well. The new component set object is then added to the buffer, as in the previous case.

Inserting a set of components from the buffer is made in the same way as when inserting a component Sect. 3.8. Copies of all components and bonds in the set are created and added to the document. Their positions are changed to correspond to the position chosen on the screen. Copies of all documents are also made. Next, a temporary selection set object is created, pointers to objects inserted in the document are added to the selection object, and all of the objects in the set are selected. The selected components can now be moved across the screen, if required.

As with components, a set of components placed in the buffer can be reviewed by using the *Open* command. Selecting the component from a list, a document window is opened that is used to review the component set. Copies of the objects from the set are made and inserted into this document in a similar way as when inserting a component set described above. This time, however, these components are just pure replicas of the components in the set and no documents are copied. The components are used only as interfaces to documents of the component set.

The component sets can also be removed from the buffer, just like any other component, by the *Delete* command. Removed components are placed in the *Waste Bin* buffer, or deleted completely.

4.7.3 Library Operations

Library operations at the document level are implemented following the guidelines given in Sect. 3.10. To put a component into the library, it is first selected, then the *Put to Library* command from the *Tools* menu is invoked. This copies the component document and all associated component documents. A library object with all of the relevant data—such as the component name and its root document filename—thus is created and added to the library index.

This operation depends on the type of word model object selected. Common word models are stored in the *Word Model Component Library*. Components represented by electrical circuit symbols are moved into the *Electrical Component*, section and those with a mechanical symbol into the *Mechanical Component Library*.

To insert a component into a document from a library, we chose between the three aforementioned component libraries—*Electrical*, *Mechanical*, and *Word Models*—by the appropriate commands in the *Tools* menu. We then select the *Insert* command. This presents a dialogue box, from which a component can be selected (Figs. 4.14 and 4.15).

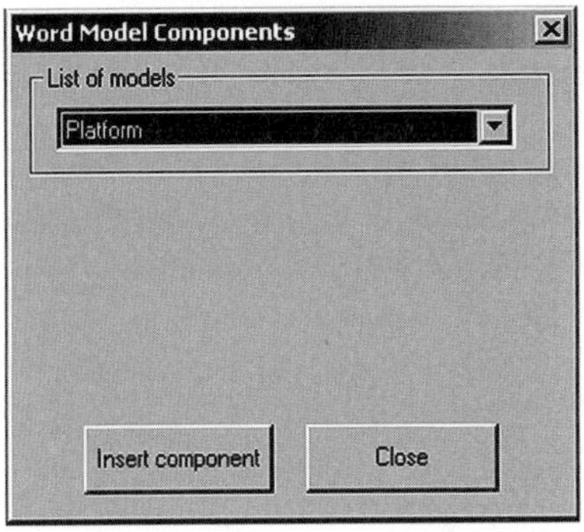

Fig. 4.14. The insert word model component library dialogue

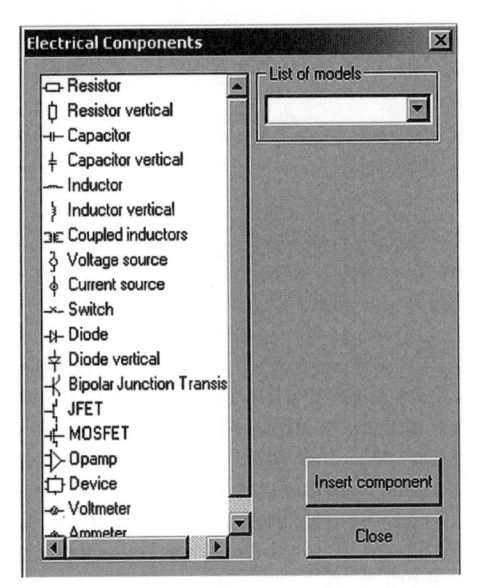

Fig. 4.15. The insert electrical component dialogue

This operation is similar to that employed when inserting a component from the buffer (Sect. 4.7.2). This time, the message is sent to the active document view to change its state to the *Insert Library Component*. When a place is chosen in the document, a component object is created based on the data held in the component library root document, then added to the document. The component root document is copied, as well as the documents of all contained components.

There is also an *Open* command that can be used for reviewing components in the library. This command presents a dialogue that is similar to that used when inserting a component (Figs. 4.14 and 4.15), but with an *Open Component* button instead of the *Insert Component* button.

4.7.4 The Page Layout and Print Commands

Document orientation is important for printing, but it also influences the maximum size (Sect. 4.6.4). The default document is oriented horizontally, but this can be changed to the vertical by the *Page Layout* command on the *Document* menu. This command invokes the dialogue shown in Fig. 4.16.

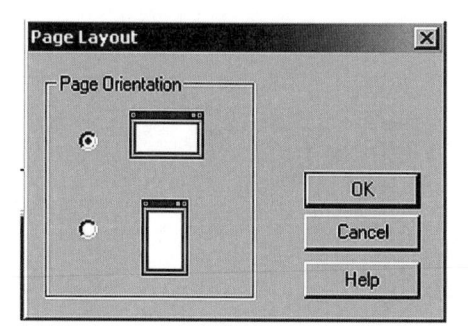

Fig. 4.16. The page layout dialogue

The program also implements printing and print previewing of the documents based on the MFC Library support. The first page of the printout contains the bond graph diagram displayed in the current document. The next page contains all parameters defined at the document level, as well as the constitutive relations and parameters of the elementary and block diagram components. Fig. 4.17 shows a printout of the model data of the LinRot component of Fig. 2.21. Values of parameters a and c are not shown, as they are defined in a lower-level document.

There also is a command for printing graphic data (bond graphs) in the *Enhanced Window Metafile* format (emf). This command invokes methods very similar to those used for drawing to the screen, but uses the MFC Library metafile context for drawing to a file [1]. The emf file can be imported by most word processors and graphic programs, such as MS Word, CorelDraw, Adobe Illustrator.

```
                                    2

COMPONENT : LinRot

Parameters :
x=c
y=a
Component :TF
   Power-input-port constitutive relation :
   M2x=(-x*sin(phi)-y*cos(phi))*F2x
   Power-output-port constitutive relation :
   vc2x=(-x*sin(phi)-y*cos(phi))*OmegaL
Component :TF
   Power-input-port constitutive relation :
   M2y=(x*cos(phi)-y*sin(phi))*F2y
   Power-output-port constitutive relation :
   vc2y=(x*cos(phi)-y*sin(phi))*OmegL
```

Fig. 4.17. Printout of the component LinRot of Fig.2.21

4.8 Editing The Component Constitutive Relations

4.8.1 Component Port Dialogues

Port variables and the constitutive relations of elementary components are stored in their port objects (Sect. 3.4) and, ultimately, in the container document. The same holds for block diagram components, with the difference that the constitutive relations are stored in their output ports; input ports only store input variables. In addition, there are parameters that are defined at the document level, or locally in the component. The constitutive relation can have different forms depending on the type of component, as discussed in Sects. 2.5 and 2.6. The corresponding expressions should conform to the rules of Sect. 3.5.

Default variable names and constitutive relations are defined at the component port creation. These correspond to simple linear relations. The necessary parameters also are defined and stored in the component. Hence, it is necessary to implement methods that enable changing variable names, parameters, and the constitutive relations. We address the problem of the editing of variables and constitutive relations first. The matter of parameters is discussed later.

Suitable dialogues (Fig. 4.18) are used to support changing variables and the constitutive relations of the element ports in a user-friendly way. These are invoked by double-clicking the ports of elementary or block diagram components. In response, the component method is called to open the component port. The method constructs a dialogue object and transfers data to it, which are then displayed on the screen. Variable names and constitutive relations are edited using fields provided by the dialogue. When editing is finished and the data are accepted (by clicking the *OK* button), the data are stored in the component ports and the dialogue is closed.

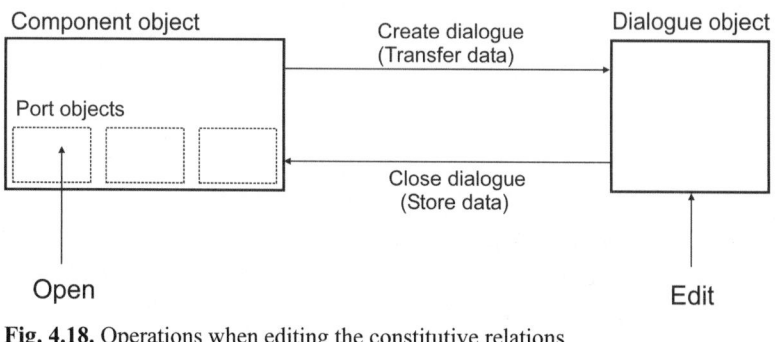

Fig. 4.18. Operations when editing the constitutive relations

To enforce specific types of variables and constitutive relations, a separate dialogue is defined for every type of component port. Fig. 4.19 shows a dialogue corresponding to an inertial component power port. Similar dialogues are used for the other types of ports. The top part has fields used for defining the constitutive relation, which is of the form discussed in Sect. 3.5. Next are fields that define port variables. On the left are the current port variables, and on the right is a list of variables of the component's other ports. There is also a button that can be used to change the values of, or define new, parameters (constants).

Most fields can be changed simply by selection followed by editing using the keyboard. The fields are instances of MFC's *CEdit* class, which implement common text editing operations. Fields that should not to be changed are disabled, e.g., the *Effort* field (Fig. 4.19).

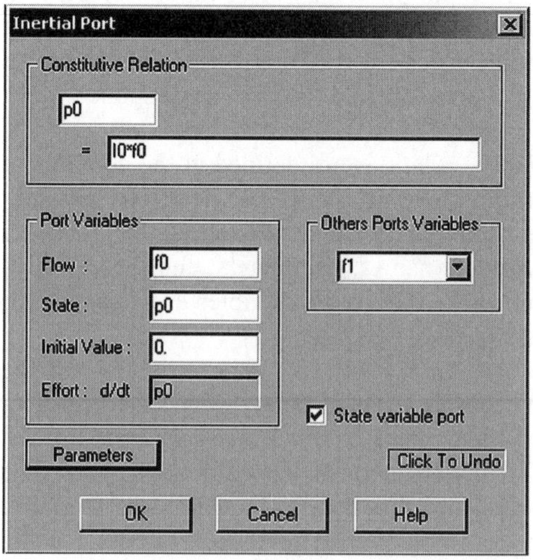

Fig. 4.19. The inertial port dialogue

When editing is finished, the changes are accepted by clicking *OK*, or rejected by clicking *Cancel*. If the OK button is chosen, the data are validated before being sent back into the component.

Validation is done by parsing and syntactical analysis of the expressions. A valid variable starts with a letter, its length is not limited, and is case sensitive. Symbols used in the constitutive relation can be any defined variable, the time symbol t, or numerical and symbolic constants. The syntax of the constitutive expression should conform to the rules discussed in Sect. 3.5. If an error is found, a message box appears with information about the error. In this case, the dialogue is not closed. The error must be corrected before the data in the dialogue are accepted, or else the user must dismiss the dialogue by clicking the *Cancel* button.

Another check is made before the data are transferred back to the component port: When port variable names are changed, these new names are checked against the names used in the constitutive relations of the other ports contained in the component. If a match is found, a dialogue is opened that prompts the user to use a unique variable name. When all such corrections are made, the new data are accepted and the data stored back into the component ports

4.8.2 Defining the Parameters

Parameters used in the port constitutive relation can be defined in the component that contains the port, in its container document, or in a lower-level document. Parameter definitions are stored in the document or components as lists.

A parameter is defined by its name and an expression (Fig. 4.20). The expression can contain literal constants, as well as other parameters already defined.

```
a=0.010
b=a^2+k
c=b+5.0e-3
...
```

Fig. 4.20. An example of a parameters list

To define a parameter in a component, the *Parameters* button of the port dialogue is used (Fig. 4.19). This invokes a dialogue that displays parameters defined in the component (Fig. 4.21). To define a new parameter, its name is typed in the edit field and inserted by pressing the *Insert* button. The parameter definition can be added at the tail of the list box, or before some parameter already in the list. A parameter already defined in the list can be changed easily by selecting it in the list first, then pressing the *Edit* button. Similarly, the parameter can be defined at the document level. In this case, the *Model Parameters* command on the *Edit* menu is used. In response to invoking this command, the same dialogue appears; but this time it displays parameters defined in the document.

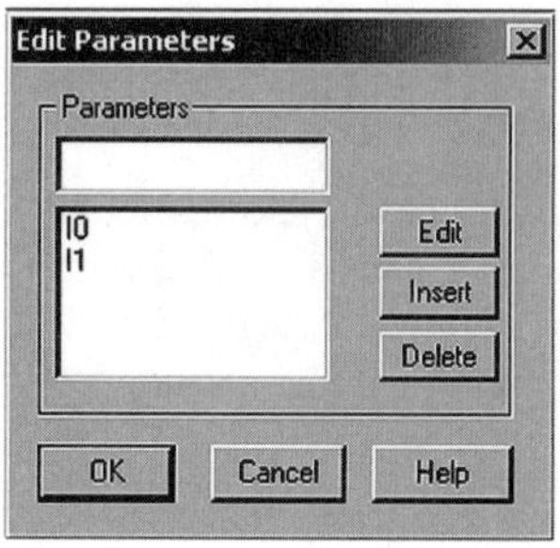

Fig. 4.21. The parameters dialogue

When inserting a new parameter definition or when editing an existing one, a dialogue appears with the parameter name on the left side and a field on the right in which the parameter can be defined as a value or expression (Fig. 4.22). When the *OK* button is pressed, it is parsed to determine if it is syntactically correct.

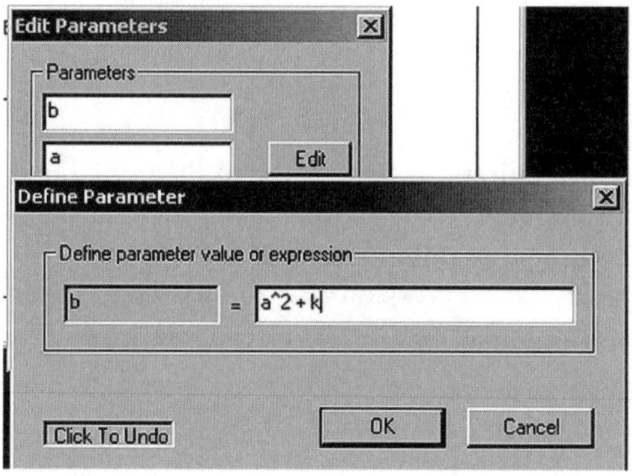

Fig. 4.22. Editing a parameter expression

The parsing and syntactical analysis is similar to that implemented for the constitutive relations. The parameter expression is accepted if all parameters are already defined, i.e. lies in the parameter list before the current parameter, or is defined in some lower-level document.

These parameter lists are ordered in the same way as the components and documents. Thus, the parameter definitions can be looked at as the "list of lists", or as a tree of the parameters. Each node of the lists corresponds to a list contained in a component, or in a lower level document (Fig. 4.20). Some of nodes may be empty.

Lists are searched from the tail of the list towards its head. The search starts at the component or document list and proceeds in the direction of the root document list (Fig. 4.20). Thus, a parameter defined at some place in the list is taken into account, even if the same parameter has been defined previously. This usually is termed parameter hiding, and is similar to hiding local variables in the C/C++ language. It is implemented here in this way to circumvent multiple definition of the same parameter used in components, e.g. gravity, masses etc. The parameters can be defined in some lower-level document and used freely. In the same vein a parameter can be defined in some document by a default value. A specific value of the parameter can defined in the component contained in the document. This is a useful properties as is shown in the second part of the book.

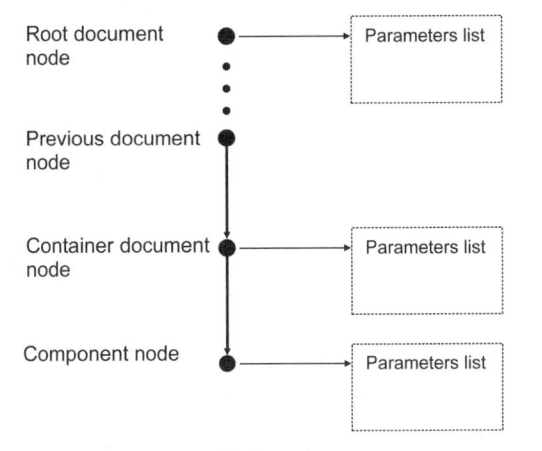

Fig. 4.23. Parameter list branch

Finally, it is appropriate to add a few words about deletion of a parameter from a list. This is accomplished by selecting a parameter in the list, then clicking the *Delete* button (Fig. 4.21). This removes the node at which the parameter is stored, and the parameter definition, as well. But this is potentially dangerous, as already discussed in Sect. 3.9: The parameter can be used somewhere else. Thus, before accepting such a deletion, a warning message dialogue box is displayed. It is the developer's responsibility to decide whether or not to delete a parameter. In any case, final parameter validation is postponed until the building phase, before the start of simulation.

4.9 Collaboration Support

This section describes the application's support for collaborative work. The support is designed, as explained in Sect. 3.10, to enable the exchange of models between different applications. The basic exchange mechanism is the e-mail service with the model files attached.

A typical model contains a number of files because, as explained earlier, it is described by a tree of document files. Thus, it is not convenient to send a model's documents file-by-file, and then re-assemble them. A more convenient approach is to create a package file that contains the documents for the projects, components, and user-defined function documents that are to be exchanged. This file is created as a *compound file*, i.e. a physical file containing a number of document files. Each individual document file in a package can be accessed as if it were a single physical file. The compound package files are created as Microsoft compound files that are implementations of the *Active Structured Storage Model* and are part of the *Object Linking and Embedding* (OLE) technology [1]. These files are supported by MFC's COleDocument class. To create a compound document file, a CBondSimPackage class is derived from the ColeDocument class (Fig. 4.24).

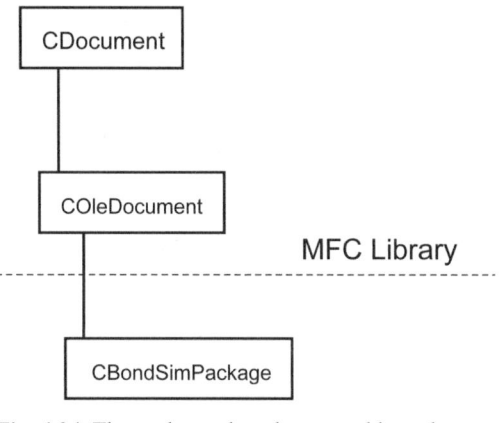

Fig. 4.24. The package class document hierarchy tree

This implementation uses *stream* objects to store data-like files and *storage* objects that, like directories, contain other storage objects and stream objects. Consistent methods exist that support serialization of the storage, stream, and file objects. Thus, compound files behave like a file system with directories and files contained inside a physical file. The CBondSimPackage class is designed to support the creation of a package file with the structure given in Fig. 4.25.

The package file has a header, which gives information about the author of the model and the date when the package was created. Then, for every one of the main categories of models, a separate storage object is created, e.g. Projects, Word model components, Electrical components, Mechanical components, and user-defined Functions. Storage objects contain an index of the contents, which gives information on the projects, the components, and the functions contained in the package. Corresponding document files are stored in the package in the same way as documents are saved in the files, but using streams. CBondSimPackage supports the component model architecture developed so far by working closely with other BondSim classes, notably CBondSimDoc.

An export package is created using the application-level item *Export* on the *Projects* menu (Sect. 4.2). This menu item contains three submenus: *Project*, *Component*, and *Function*. *Component* has its own submenus: *Word Model*, *Electrical*, and *Mechanical*, from which the kind of model to be exported can be chosen. Once a submenu is chosen, a dialogue window appears with a list of models or functions from which an item to be exported can be selected. This is used to determine the name of the item (project, component, function) that will be exported, as well as its root document file from which the exporting will start.

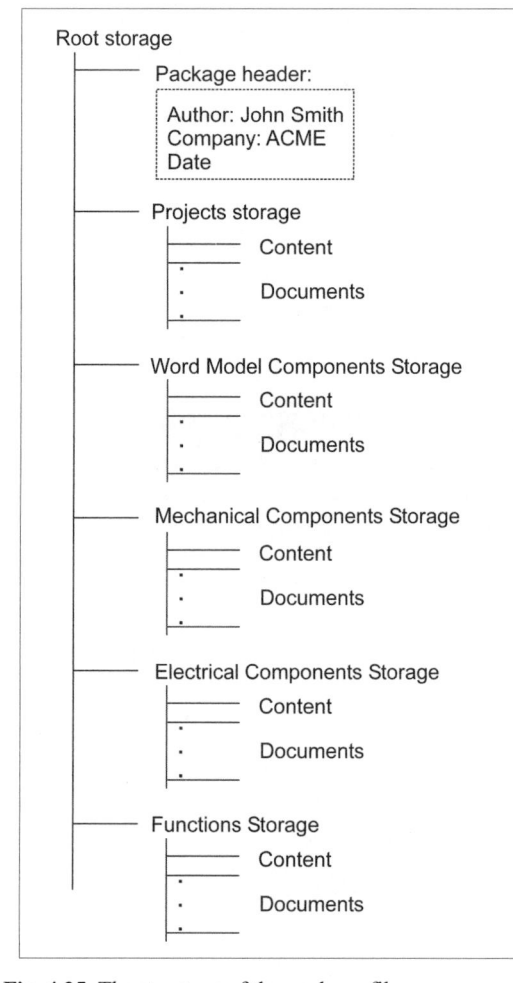

Fig. 4.25. The structure of the package file

If no package file has been created, a new package object is created and the user is asked to supply data on the author name and company. The date is taken directly from the machine. These data are stored in the newly created package and saved during package file saving operations. On the other hand, if the package file already exists, it is simply opened.

Exporting projects, components, or functions is very similar. The major difference is their storage locations (Fig. 4.25). The index of contents consists of maps that contain their names and the names of their root document streams, in the same way as was done with normal documents, but using streams instead of files.

Stream names are based on document filenames (in fact they are just the filenames stripped of the extensions). Thus, to export a project its name and root document filename are used to create a root document stream name. These are stored in the index of contents of the projects. Next, the root document file is opened and is stored using the corresponding stream. Similarly, as in copy operations, the document file of every component contained in the document is opened and stored by its stream. This is continued until all documents are stored. Before the package is closed, it is saved to disk. In a similar way, it is possible to export other objects, such as word model components, electrical components, etc.

Before sending the package, it is useful to preview it. This is accomplished using the *Preview Package* command on the *Project* menu. This command displays a dialogue (Fig. 4.26) in which is contained a list of the storages in the package, i.e. Project, Word model, etc. By selecting a storage, a list of the corresponding items is displayed showing the content of the storage. This dialogue can also be used for removing a project, component, or function from the package.

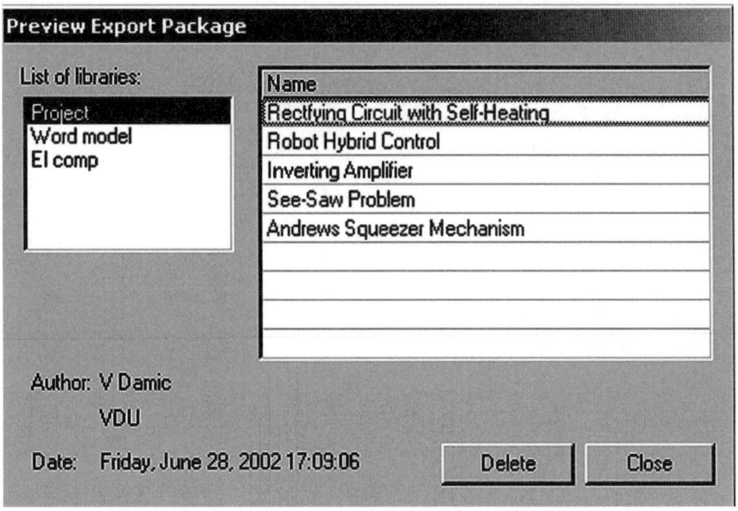

Fig. 4.26. The package preview dialogue

The package is sent by e-mail as an attachment via the user's mail host, if any, using the *Send by Mail* command in the *Project* menu. This command uses support already implemented in MFC's CDocument and COleDocument classes [1]. The package file is not sent immediately, but is first zipped using the ZLib (see Sect.4.1 and footnote 1). After zipping, the base class method that attaches the package file to the message is called. Then, an empty *New message* dialogue (Fig. 4.27) is displayed. Thus, as with any new message, the recipient's address must be typed, along with the subject and text of the message, if any. The message can then be sent.

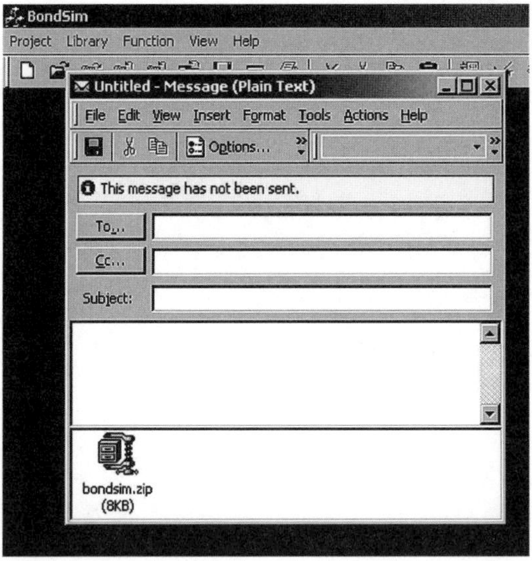

Fig. 4.27. The new message window with the package file attached

The operations on the recipient side are similar. When the recipient receives the message, she can save the attachment to any suitable directory. Next, from within BondSim, the *Get Mail* command on the *Project* menu is selected. This command uses the standard windows browser to get the zipped package file. The file is then unzipped into a directory from which importing is done.

To import models from the package, the *Import* command from the *Project* menu is used. This command displays a dialogue similar to the one associated with the *Preview Package* command described above. This time, the dialogue is used for the import. The type of model (storage) in the file is selected first, then the name of the model. Clicking the *Insert* button begins the importing.

Importing from the package is very similar to inserting from libraries (Sect. 4.7.3). The basic difference is that reading is not from document files, but from document streams. These operations are supported by the package class of Fig. 4.24. This is, in effect, a systematic copy operation. During the copy operation the project, component, and function unique identifiers (*ids*) are created, as well as the corresponding file names that identify where the document will be saved. The models are moved to the project, component, and function libraries, from which they can be reused. This way the possible conflict between imported models and the application database is resolved.

References

1.   Microsoft Corporation (1998), Microsoft Visual C++ 6.0 Reference Library, Microsoft Press, Redmond

Chapter 5 Generation of the Model Equations and Their Solution

5.1 Introduction

In the previous two chapters the systematic component-based approach was developed that enables development of mechatronic system models in a formal way. An important part of this is the description of the element constitutive relation symbolically using a relatively simple language. Thus, not only non-linear relation ships, but also piecewise expressions can be used. This is important in modelling discontinuous mechanical processes and in electronics. This makes simulation of complex systems not only feasible, but also a challenging task. This chapter describes the generation of the system mathematical models and their solution.

We start by describing methods used to generate the mathematical models implied by a component's structure. The equations are machine generated in the form of differential-algebraic equations (DAEs). We have already discussed some of the features of these equations. We continue here with the problem of their numerical solution. We chose the well-known backward differentiation formula (BDF) as the solution method. This is suitable for index-1 problems, but can also be extended to index-2 problems. For higher index problems the solution is much more demanding; we have left this for further study. The problems of starting values and discontinuities are also discussed.

The methods we use depend to a great extent on computational algebra support. This adds flexibility to the modelling. It also is used in the numerical solver e.g. for evaluation of functions, generation of matrixes of partial derivatives, and their subsequent solution. It enables generation of feedback to the modeller supplying information on the generated models. These capabilities can be extended.

This chapter closes the first part of the book, which deals with the fundamentals of a component-based bond graph approach to modelling mechatronic systems. This is implemented in the BondSim program provided with this book (see Appendix).

5.2 General Forms of the Model Equations

In Sects. 2.5 and 2.6 we gave a fairly general description of the constitutive relation of components used as the building blocks for developing models of systems.

Using the techniques of Chapts. 3 and 4, such models are generated and stored as a tree of component models. These models consist of relations defined in elementary components that are the leaves of the model tree and which are constrained by relations describing the interconnections of components that are the branches of the tree. The models are made more general and, unfortunately, more difficult to solve by allowing the mixing of bond graph based models with operations on signals. Here we discus the general procedure for the generation of the mathematical model and some of its characteristics.

5.2.1 System Variables

The first step in generating the mathematical model is defining a unique set of system variables. As explained in Sect. 3.4, at every port of the elementary components two variables are defined: effort and flow. The connection of ports by a bond means that the respective efforts and flows of connected ports are the *same*. In addition, there is an internal state variable (Sect. 2.5 and 2.6), i.e. the momenta at inertial element ports, displacements at the capacitive ports, and the outputs of integrator ports. Hence, we need to develop a technique for defining a unique set of system-wide model variables and their correspondence to port variables. We do this by means of a symbol table.

The symbol table consists of a pair of port and system variables. To create it we first need to define global names for the port variables. We create these relatively simply. Remember that to every component a unique label is assigned, and every port has a label that is unique to a particular component. Thus, the global effort variable of the component of the label complabel belonging to the port of the label portlabel is described by the unique string e_complabel_portlabel. This also holds for flows and internal state variables, i.e. f_complabel_portlabe and s_complabel_portlabe. System variables can be designed as y(1), y(2), etc. It is enough to store the indexes, i.e. 1, 2, etc. The symbol table is simply a map (dictionary) of port and system variable pairs (Fig. 5.1). We used the map structure because of the short search times, which is very important during mathematical model creation.

...	...
e_00100-0	102
e_00100-1	102
f_00012-5	103
s-00051-0	104
...	...

Fig. 5.1. The pairs of variables in the symbol table

The procedure for the creation of the symbol table consists of traversing the components, starting from the lowest level and going up. During this procedure a port variable is put into the table along with its corresponding system variable index. We start with the effort and flow junctions, and then continue with other elementary components.

The effort junctions are characterised by the fact that all of their port flows are the same variable, which is the common junction variable (Sect. 2.5.7). A similar situation holds for the flow junction, with the common junction variable being the effort. The procedure is as follows:

1. Increment the index of the system variables by one. Insert in the symbol table the junction flow (effort) variable symbol as the key, and the current index of the system variable as the value.
2. For every junction power port, find the port of the elementary component to which it is connected. If the component is of the same type, i.e. if the effort junction port is connected to another effort junction, or if the flow junction port is connected to another flow junction, treat the port as internal and the junction as the same junction, and proceed further. If, on the other hand, it is a port of the other component type, generate a unique port flow (effort) symbol and insert it into the symbol table using the current index as the variable index.
3. If the port is a control-out port, the procedure is slightly different. If the symbol of the connected port variable is already in the symbol table, break the search and continue with the next effort (flow) junction port. If it is not, add it to the table. The search is stopped if the connected port is an input port of an Integrator, a Differentiator, or an Output component. Otherwise, if it is a Function or a Summator component input port, continue the search with its output port. We create a symbol corresponding to the output port variable and insert it into the symbol table, together with the new value of the system variable index, if it is not already created. In this case, the search is stopped. Then, find the other port connected to it and continue the search. In the case where it is a Node input port, find the port connected to each of its output ports and continue the search as described at beginning of this step (3), because all node ports share the same variable.
4. Continue with the search and assignment of variables until all power or control ports of the junction and of other junctions of the same type that are connected to it are traversed.
5. Continue with the next junction until all junctions are visited.

The procedure outlined above ensures spreading the junction variables to all ports connected directly to it, or across other junction components of the same type (Fig. 5.2). It also creates variables of input-output components connected to it. It is important to do this during effort or flow junction variable assignment, for it is one of the required connection points between bond graphs and control signal paths. We stress that this procedure ensures that all directly connected junctions of the same type behave as single junction.

After finishing with the effort and flow junctions, we proceed to the other components. The procedure is similar to the one described above, but is slightly simpler. For every unprocessed elementary component port variable—i.e. the effort, flow, or control—we increment the system variable index by one and insert a symbol-index pair into the symbol table. We then find the port to which it is connected and add its symbol variable of the same type and the current value of the system variable index to the symbol table. In the case in which the port is an output control port, we proceed as explained above (item 3). The elementary component control input ports, e.g. resistive component inputs or transformer input ports, are dealt with when assigning input-output (block-diagram) component port variables. Internal variables of inertial and capacitive components are also added to the symbol table.

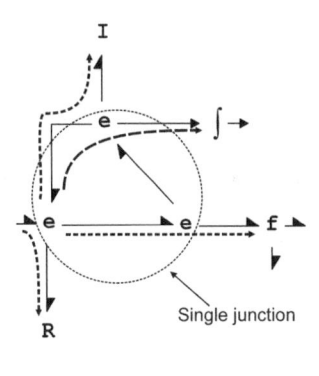

Fig. 5.2. The assignment of the junction variable

During the generation of the system variables and filling of the system table, the correctness of the component model is checked. First, if some ports are not connected the procedure is stopped and the reason is conveyed to the user by a suitable message. Similarly, if the connection on each side of a component does not fit—e.g. because of incorrect dimensions of the bonds (Sect. 3.7)—the procedure is stopped and a message is posted. Finally, the elementary components are checked to ensure that they have valid connections (Table 5.1). Elementary components that are not junction components—i.e. effort or flow junctions, Transformers, or Gyrators— can only be connected to junction components. Junction components, on the other hand, can be connected to any component. We rule out direct interconnections of Transformers and Gyrators, as these combinations are not necessary, e.g. TF-TF is just TF, as is GY-GY, etc. Regarding input-output components, there is no restriction other than an output port should be connected to the input port of another component.

Table 5.1. Permited interconnection of elementary components

Components	Permitted connection to
I, C, R, SE, SF, Sw	e, f, TF, GY
e, f	All
TF, GY	all but TF and GY

5.2.2 Generation of the Equations

Once the system variables are defined and their correspondence to port variables established, the next step is to generate equations that describe the constitutive relations of the elements. There are no built-in relations; every relation is either the default created when the component port is created, or edited later by the user. Such relations are described in a manner similar to that of the usual mathematical expressions, as explained in Sect. 3.5. Relations are not translated to a subroutine or function, but are, instead, translated to another string that enables their efficient evaluation at simulation time. This form is called *byte form*. Thus, the constitutive relations are retained in a symbolic form that can easily be translated back to the more familiar mathematical expression, if required (Sect. 5.4). Without going into too much detail, we now explain how the mathematical model equations are created.

To deal with model equations in symbolic form, a supporting computational algebra program has been developed. During model building the strings, which are held in the elementary component ports, are analysed, translated to byte form, and stored in an array that is accessible for further processing.

The method used is an extension of the approach of [1]. The byte string form is based on the affix operator form

$$\text{Operator operand1 operand2} \tag{5.1}$$

Thus, parameter constants in the expressions are represented by bytes consisting of the operator symbol C and the real value of the parameter as the operand

$$\text{C} real\_value \tag{5.2}$$

Similarly, a system variable y(i) is represented by the operator symbol V and the integer value of its index

$$\text{V} int\_value \tag{5.3}$$

The index values of variables run from 1 to N, where N is the number of variables in the model. The index value of zero (0) is retained for the *time* variable. The time derivative of a variable of an index int_value is coded as

$$\text{D} int\_value \tag{5.4}$$

Various operators are used to describe operations between variables and parameters in constitutive expressions. These include unary, binary, ternary, and some special operators. Unary operators are designated by U, plus a symbol that describes the operation in question. Thus, U+ and U- describe the usual unary plus and minus operations, e.g. −F where F is the force (effort) variable is described as

$$\text{U} - \text{V} index\_value \tag{5.5}$$

where index_value is the value of the system variable index that corresponds to effort F.

Elementary mathematical functions are also treated as unary operations with the second byte used to designate the function, and the operand corresponding to the function argument. We also introduce single- and two-variable operators that currently are used only for coding functions defined by tables (see Sect. 7.5 for an example).

Binary operators B are used to describe addition (+) and subtraction (-), multiplication (*), division (/) and the mod operation (%), as well as for relational operators >, <, >=, <=, and logical operators AND and OR.

The ternary operator ?: is coded as

$$T ? \text{conditional\_part Expression1 Expression2} \tag{5.6}$$

The operator ?: is borrowed from the C language and is used for *if then else* constructions. The constitutive relationship describing some physical process often cannot be described by a single expression valid throughout the range of the variables, but, instead, by two or more expressions valid in different parts of the range. This is the case, for example, with dry friction in mechanics, and in semiconductor models in electronics. To describe such relationships we use ternary ?: operators. The conditional part is described using relational or logical operators, and the expressional parts are constructed in the usual way. The expression is coded as given in Eq. (5.6). Thus, for example, the relation

$$e > 0 ? e : e + 1 \tag{5.7}$$

reads as e if e > 0 and as e+1 otherwise. This will be coded as

$$T ? B > V1C0V1B + V1C1 \tag{5.8}$$

where 1 is used as the index of the variable e. When evaluating such an expression, the first two bytes signify that it is a ternary ?: operator. The string following this is the conditional expression. Reading and evaluating this expression, we find if it is true or false. If true, we read and evaluate the expression that follows, i.e. the *if then* part. If the conditional expression evaluates to false, we also read the next expression, but do not evaluate it, and then read and evaluate the expression that follows, i.e. the *else* part.

The element constitutive relations are described as

$$\text{var} = \text{expression} \tag{5.9}$$

Here var stands for the port variable and the expression is the constitutive relation of the element. Most processes are described by a single relationship of this form (Sects. 2.6 and 2.7). This is the case for processes at Resistor or Transformer ports. Such relations are treated as expressions of the form

$$\text{expression} - \text{var} = 0 \tag{5.10}$$

and coded as

$$B - \text{expression V var\_index} \tag{5.11}$$

with the understanding that the last expression is equal to zero.

In the case of Capacitive and Inertial components, there also is a derivative part of the form

$$\frac{dx}{dt} = y \tag{5.12}$$

where x is the internal state variable displacement or momentum of the process and y is a port effort or flow. The Integrator input-output component is described in a similar form. Such equations are also treated as

$$\frac{dx}{dt} - y = 0 \tag{5.13}$$

and coded by

$$B - D\,var\_index\ V\,var\_index \tag{5.14}$$

Further, it is understood that this expression should evaluate to zero.

Now that we have explained how the constitutive equations are coded, we can describe the procedure for generating the mathematical model of the system based on the information held in the component ports. For components that are not effort junctions, flow junctions, nodes of signals, or summators, we systematically visit every component in the model tree and, for every one of its power ports and control-out ports, generate equations that correspond to the constitutive equations. This includes defining expressions of the form in Eq. (5.12). Every such expression is parsed and translated to byte form, as explained previously. Because the original expressions are given in terms of symbols corresponding to the port variables, every occurrence of such variables is substituted with a system variable using the symbol table of variables (Sect. 5.2.1). Parameters defined by symbols are evaluated and coded as constants. Parameter evaluation means that the complete hierarchy of parameters is searched, starting with parameters in the particular component, until the complete parameter expression is evaluated (Sect.4.8.2). If the complete parameter definition is not found, the translation process is stopped and a message explaining the cause is displayed. If a syntactical error is found during translation, the process is stopped. This normally should not happen because, when constitutive expressions are edited, the syntactical analysis has already been executed and the expressions are accepted and stored only if the checks have been passed.

Constitutive relations corresponding to effort or flow junctions are generated, as explained in Sec. 2.5.7, based on the sense of power of the ports. These are translated to the byte forms of Eq. (5.11). This also holds for summators, where the sign of the input ports define the output relationships. The signal nodes need no constitutive relation at all, as their roles are taken care of during the generation of model variables (Sect. 5.2.1).

During generation of the model equations, every byte form expression corresponding to the left side (the right side is zero) is stored in an array. This array is used to evaluate expressions during simulation. Evaluation is done using a routine for reading and evaluating byte form expressions. Hence, model equations are not compiled and linked to the rest of the executable code, but instead are processed by interpreter routines. The byte form of the expression enables reading and evaluating expressions in a single pass, without looping. We return to this point in Sect. 5.7.

We now summarize the forms of the model equations generated for the simulation of system behaviour. The model consists of three groups of equations. We introduce notation that permits their description in a compact way. All variables appearing in a time-derived form, as in Eq. (5.12), are called *differentiated* variables described by a vector $\mathbf{x}$ of dimension n_d. These variables represent displacements of Capacitive elements, momenta of Inertial elements, outputs of Integrators, and inputs of Differentiators. All other variables are *algebraic* variables and are denoted by a vector $\mathbf{y}$ of dimension n.

The first group of model equations has the form of a differential equation, as in Eq. (5.12), and can be compactly described in vector form as

$$\frac{d\mathbf{x}}{dt} = \mathbf{B_1 y} \tag{5.15}$$

Here $\mathbf{B}_1$ is $n_d \times n$ dimensional incidence matrix having in each row a 1 in the column of the corresponding algebraic variable, and a 0 elsewhere.

The next group comes from equations describing the element constitutive relations of Eq. (5.10). These can be put in the form

$$\mathbf{f(x, y, t)} = 0 \tag{5.16}$$

where $\mathbf{f}$ is vector function of the corresponding dimension.

The final group of equations describes the balancing of efforts and flows at effort and flow junctions, respectively. To this group we also add summator equations. This group constitutes a system of linear equations with coefficients +1 or − 1. It can be described in vector form as

$$\mathbf{B_2 y} = 0 \tag{5.17}$$

Matrix $\mathbf{B}_2$ is a rectangular matrix of corresponding dimension and is of full row rank.

Eqs. (5.15) – (5.17) constitute a system of *differential-algebraic* equations. Because of Eq. (5.15), such a form is usually called *extended*. Thus, direct application of the element constitutive relation, as described in Sect. 2.5 and 2.6, leads to a mathematical model of the system in the form a system of differential-algebraic equations in extended form. In the next subsection we discuss the numerical solution of the model.

5.2.3 The Characteristics of the Model

Systems composed of both differential and algebraic equations generally differ significantly from ordinary differential equations and are much harder to solve [2]. DAEs may be characterised by indices that measure the difficulties encountered when solving them. Several types of indexes are defined. We do not give here their precise definition; this is found in [3-6].

The *differentiation index* is defined as the number of times the DAEs should be differentiated with respect to time in order to reduce them to a system of ordinary differential equations [3, 4]. We illustrate this on an example of semi-explicit equations

$$\left.\begin{array}{l} \dot{\mathbf{x}} = \mathbf{f}(\mathbf{x}, \mathbf{y}, t) \\ \mathbf{g}(\mathbf{x}, \mathbf{y}, t) = 0 \end{array}\right\} \tag{5.18}$$

Differentiating the second equation with respect to time yields

$$\left.\begin{array}{l} \dot{\mathbf{x}} = \mathbf{f}(\mathbf{x}, \mathbf{y}, t) \\[2mm] \dfrac{\partial \mathbf{g}}{\partial \mathbf{x}} \dot{\mathbf{x}} + \dfrac{\partial \mathbf{g}}{\partial \mathbf{y}} \dot{\mathbf{y}} = -\dfrac{\partial \mathbf{g}}{\partial t} \end{array}\right\} \tag{5.19}$$

Thus, if the partial derivative matrix $\partial \mathbf{g}/\partial \mathbf{y}$ is not singular, this is a system of ordinary differential equations and, hence, the DAEs of Eq. (5.18) have a differential index of 1.

If, on the other hand, the partial derivative matrix is singular, we proceed as described in [4]. Supposing that $\partial \mathbf{g}/\partial \mathbf{y}$ is of constant rank, we can transform Eq. (5.18) at least locally to the form

$$\left.\begin{array}{l} \dot{\mathbf{x}} = \mathbf{f}_1(\mathbf{x}, \mathbf{y}, t) \\ \mathbf{g}_1(\mathbf{x}, t) = 0 \end{array}\right\} \tag{5.20}$$

This can be achieved by expressing the y–variables that appear in some of the algebraic equations of Eq. (5.18) as functions of the x– and other y–variables, eliminating them from the system. As a result, the new algebraic equation is independent of the y-variables.

As an illustration, we return to the see-saw problem of Sec. 2.7 and show that Eq. (2.69) – (2.72) and (2.95) – (2. 99), which describe the system, can easily be converted to the form of Eq. (5.20). Thus, forces F_{m1x} and F_{m1y} can be evaluated from the last two equations in Eq. (2.69) of **Body 1** motion and substituted into the first two. We can also eliminate the gravity force G_{1y}. In a similar way, we can eliminate the corresponding variables in Eq. (2.70) of **Body 2** motion. In the same way we can eliminate components F_{Cx}, F_{Cy} and G_y from Eq. (2.98) of the platform mass center motion, and the moment M_C in Eq. (2.99) of the platform rotation. Next, we can eliminate moments in Eq. (2.99) by use of the relations in Eq. (2.95) – (2.97), (2.71) and (2.72). The angular velocity in Eq. (2.72) is eliminated using Eqs. (2.71) and (2.99). In this way, we get the first part of the system equations, which involves derivatives of variables with respect to time, as shown in Eq. (5.21). We can now eliminate the velocity components from the system. Thus, the velocity components in Eq. (2.69) can be eliminated using Eqs. (2.95), (2.97), (2.71), (2.72) and (2.99). We handle the velocity components in Eq. (2.70) and (2.98) in a similar way. Finally, we obtain the system of algebraic equations in Eq. (5.22), which contains only differentiated variables: momenta and the rotation angle.

$$
\left.
\begin{aligned}
\dot{p}_{m1x} &= -F_{1x} \\
\dot{p}_{m1y} &= -F_{1y} - m_1 g \\
\dot{p}_{m2x} &= -F_{2x} \\
\dot{p}_{m2y} &= -F_{2x} - m_2 g \\
\dot{p}_x &= F_{1x} + F_{2x} - F_{Fx} \\
\dot{p}_y &= F_{1y} + F_{2y} - F_{Fy} - mg \\
\dot{K}_C &= (a\sin\phi + c\cos\phi)F_{1x} + (-a\cos\phi + c\sin\phi)F_{1y} + \\
&\quad (-a\sin\phi + c\cos\phi)F_{2x} + (a\cos\phi + c\sin\phi)F_{2y} + \\
&\quad (b\cos\phi)F_{Px} + (b\sin\phi)F_{Py} \\
\dot{\phi} &= K_C / I_C
\end{aligned}
\right\}
\tag{5.21}
$$

$$
\left.
\begin{aligned}
p_{m1x} &= (m_1 / I_C)(a\sin\phi + (b+c)\cos\phi)K_C \\
p_{m1y} &= (m_1 / I_C)(-a\cos\phi + (b+c)\sin\phi)K_C \\
p_{m2x} &= (m_2 / I_C)(-a\sin\phi + (b+c)\cos\phi)K_C \\
p_{m2y} &= (m_2 / I_C)(a\cos\phi + (b+c)\sin\phi)K_C \\
p_x &= (m / I_C)(b\cos\phi)K_C \\
p_x &= (m / I_C)(b\cos\phi)K_C
\end{aligned}
\right\}
\tag{5.22}
$$

In this way, the equations of the see-saw motion can be reduced to eight equations in differential form and six algebraic equations. These contain eight differentiated and six algebraic variables. The differentiated variables are constrained by the algebraic equations, hence only two of them are independent. This is to be expected, for the see-saw is a single-degree-of-freedom system and, as such, its dynamics can be described by only two variables, the angle of rotation and its angular momentum (or angular velocity).

We now return to Eq. (5.20) and differentiate the last equation with respect to t

$$
\frac{\partial g_1}{\partial x} \dot{x} = -\frac{\partial g_1}{\partial t}
$$

Substituting from the first Eq. (5.20), we obtain the expression

$$
\frac{\partial g_1}{\partial x} f_1(x, y, t) = -\frac{\partial g_1}{\partial t}
\tag{5.23}
$$

This last equation constitutes a *hidden constraint* that the solution of the system must satisfy. Thus, if the matrix

$$
\frac{\partial g_1}{\partial x} \frac{\partial f_1}{\partial y}
\tag{5.24}
$$

is invertible, then the first Eq. (5.20) and Eq. (5.23) constitute an index 1 problem. Differentiating with respect to time, Eq. (5.23) gives a differential equation with respect to variable y. Hence, the system of Eq. (5.20) is of differential index 2. Algebraic variables that need two differentiations in order to express them as differential equations often are called index-2 variables.

Returning to Eq. (5.21) and (5.22), we see that they constitute a system of DAEs of index-2. The hidden constraints can be found by differentiating the momentum relations of Eq. (5.22) and by substituting from Eq. (5.21). We do not give them here because of their length. It should be noted that reaction forces between bodies and the platform, as well as between the frame and the pivot (Fig. 2.15), are index-2 variables. These variables correspond to the Lagrangian multipliers of constrained body mechanics. In the original formulation of the see-saw problem all the variables are not of index-2 type. Thus, for example, the velocity components v_{1x} and v_{2x} of Eq. (2.69) are of index-1 because we only need one differentiation to get the corresponding differential equations, e.g.

$$\dot{v}_{1x} = \dot{p}_{m1x} / m_1 = F_{m1x} / m_1$$

Eqs. (5.15)-(5.17) are semi-explicit differential-algebraic equations. To simplify the notation we define a variable

$$\mathbf{z} = \begin{pmatrix} \mathbf{x} \\ \mathbf{y} \end{pmatrix} \tag{5.25}$$

The equations can now be represented as

$$\mathbf{g}(\dot{\mathbf{z}}, \mathbf{z}, t) = \mathbf{0} \tag{5.26}$$

We define the *leading coefficient* matrix by

$$\mathbf{A} = \frac{\partial \mathbf{g}}{\partial \dot{\mathbf{z}}} \tag{5.27}$$

In our case this matrix is extremely simple, i.e.

$$\mathbf{A} = \begin{pmatrix} \mathbf{I} & \mathbf{0} \\ \mathbf{0} & \mathbf{0} \end{pmatrix} \tag{5.28}$$

where $\mathbf{I}$ is the identity matrix of dimension equal to the number of differentiated variables.

Another type of index introduced for the detailed study of DAEs is the *tractability index* [5, 6]. It is not based on differentiation, but uses the underlying vector space. The vector space of a column vectors is denoted as $\mathbf{R}^m$, where m is the (constant) dimension. Important roles in the *tractability* approach to DAEs are played by two subspaces. The *image* of a matrix $\mathbf{A}$, denoted as im($\mathbf{A}$), is a space consisting of vectors $\mathbf{z}$ such that $\mathbf{z} = \mathbf{A}\mathbf{u}$ for some $\mathbf{u} \in \mathbf{R}^m$. The *null-space* of $\mathbf{A}$ is denoted as ker($\mathbf{A}$) and is a space of vectors $\mathbf{u}$ such that $\mathbf{A}\mathbf{u} = 0$. Using projectors onto the ker($\mathbf{A}$) it has been shown that DAEs of tractability index 1 can be transformed to ordinary differential equations in state-space form. Such systems are called transferable (to state-space form). The others—those of higher index—are not transferable. Unfortunately, many DAEs of engineering interest are not transferable of index ≥ 2. Characterization of an important class of index-2 systems has been developed. Particular attention has been paid to DAEs arising in the modelling of electrical systems [7,8].

Another important DAEs index, the *perturbation index* [3], is based on the behaviour of the solution under the influence of perturbations. This index can be used to explain why higher index DAEs are very sensitive to perturbations caused by, for example, numerical inaccuracies, discontinuities, and even changes in step size.

After this short introduction to some of the important concepts of DAEs, we now return to our problem, the numerical solution of the system represented in Eq. (5.15) – (5.17). A basic characteristic of such a system is that the equations are in semi-explicit form. They could be converted to a fully implicit form by eliminating variables on the right hand side of Eq. (5.15). The index of the resulting system will be one degree lower.

An important characteristic of the original equations is that the leading coefficient matrix of Eq. (5.26) has an extremely simple form. Thus, its null-space is constant. On the other hand, the reduced system can, in general, contain non-linear functions of the derivatives of variables with respect to time. As a result, the leading coefficient matrix depends generally on the system variables. The corresponding null-space is not constant, but changes during the solution. It has been shown that a system of tractability index 1, in which the leading coefficient matrix null-space changes with the solution, behaves analytically and numerically as index 2 tractable DAEs with a perturbation index of 2 [12]. This confirms the assertion in [2] that semi-explicit DAEs behave like fully implicit ones of one index lower.

To show the similarity of Eqs. (5.15) – (5.17) to some other formalisms, we consider models from two fields important to mechatronics. These are the Lagrangian formulation of equations of constrained rigid bodies and the charge/flux formulation of modified nodal equations (MNA) from electrical circuit analysis.

The classical approach based on the Lagrangian formulation leads to equations of motion of constrained bodies in terms of the generalized coordinate vector $\mathbf{q}$ of the form [9]

$$\left.\begin{aligned}
\dot{\mathbf{q}} &= \mathbf{v} \\
\mathbf{M}(\mathbf{q})\dot{\mathbf{v}} &= \mathbf{Q}(\mathbf{q},\mathbf{v}) - \mathbf{G}^{\mathsf{T}}\boldsymbol{\lambda} \\
\mathbf{g}(\mathbf{q}) &= 0
\end{aligned}\right\} \tag{5.29}$$

where is $\mathbf{M}$ is the mass-inertia matrix, $\mathbf{Q}$ is a vector of generalized forces, and λ is a vector of Lagrange multipliers. The last equations represent the position constraint that the coordinates must satisfy. Note that $\mathbf{G} = \partial\mathbf{g}/\partial\mathbf{q}$. (The superscript T denotes matrix transposition.) The first equation is of form of Eq. (5.15) because in the original formulation Eq. (5.29) it is of second order with respect to the generalized coordinates. This normally leads to equations in extended form with respect to the position coordinates. Such equations are known to be of index 3. Different approaches have been used to lower the index of such equations [2, 3, 9]. In Chap. 9 we solve a problem of this type using the bond graph modelling approach that naturally leads to equations with velocity constraints, i.e. of the form

$$\left.\begin{array}{l} \dot{\mathbf{q}} = \mathbf{v} \\ \mathbf{M(q)}\dot{\mathbf{v}} = \mathbf{Q(q,v)} - \mathbf{G}^{\mathsf{T}}\boldsymbol{\lambda} \\ \mathbf{G(q)v} = 0 \end{array}\right\} \tag{5.30}$$

Such equations are of differential index 2. The general form of the equations thus is similar with the velocity constraint replacing the positional constraints.

Analysis of electrical circuits is usually based on the classical modified nodal analysis (MNA). There is only one type of junction in circuits, i.e. the node. The governing equations are developed by applying the Kirchhoff current law to every node in the circuit. The variables consist of nodal potentials and currents in the voltage-controlled elements (inductors and sources). The constitutive relations of voltage-controlled elements are appended to the system equations. The charge/flux oriented MNA introduces charges and fluxes also as system variables. Equations describing inductors are used in the same way as those defined by inertial elements of bond graphs, i.e. using the extended form of equations

$$\left.\begin{array}{l} \dot{\phi} - u_2 + u_1 = 0 \\ \phi - L \cdot i = 0 \end{array}\right\} \tag{5.31}$$

Here u_2-u_1 is the voltage across the inductor, ϕ is the flux, i is the current through the inductor, and L is the inductance parameter (for linear inductors). The equations are added along with the constitutive relations for capacitors to the nodal equations. The constitutive relations of current-controlled elements are set directly into the nodal equations. As an illustration of equations formed using the charge/flux based MNA approach, consider the simple of circuit in Fig. 5.3.

The equations read

$$\left.\begin{array}{l} -i_1 + (u_1 - u_2)/R = 0 \\ -(u_1 - u_2)/R + i_2 = 0 \\ -i_2 + \dot{q} = 0 \\ u_1 = V \\ \dot{\phi} - (u_2 - u_3) = 0 \\ \phi = L \cdot i_2 \\ q = C \cdot u_3 \end{array}\right\} \tag{5.32}$$

The first three are written by applying the Kirchhoff current law to the three nodes. The fourth equation gives the constitutive relation of the voltage source, the next two describe the inductor, and the last is the constitutive relation of the capacitor.

Charge/flux based MNA is used to describe charge and flux accumulations in circuits in a better way. It has also been shown that circuit equations based on this formalism are better suited to simulation [7, 8]. The explanation for this is found in the leading coefficient matrix null-space that, for the charge/flux based formulation, does not change during the solution.

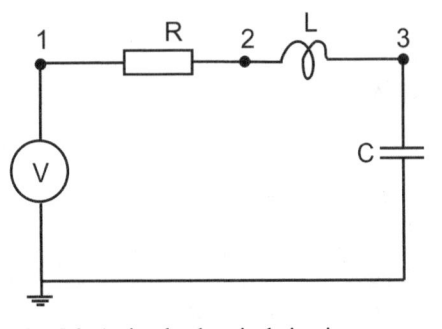

Fig. 5.3. A simple electrical circuit

There are few papers in the bond graph literature dealing with the analysis of the DAEs formulation of bond graph models. Important papers in this respect are [10, 11]. These show that the index of the system model can be larger than 2. A similar analysis is made for electrical circuits in [12]. We believe that most models of engineering systems have indices not greater than two. Higher index systems at the current state of solver techniques are difficult to solve by a general-purpose modelling and simulation system. Model reformulation and some specialized approaches are then necessary.

In [13] an approach based on tearing variables for dependent storage elements has been proposed. It suggests changing the model by introducing Langrange multipliers. Structural type analysis, such as the causality analysis already discussed in Sec. 2.10, is generally of limited value, as the structure of real problem models generally change during solution. The problems can be treated more generally using the projectors of [5, 6]. But, unfortunately, this is not an easy task.

The conclusion that can be drawn from these discussions is that the form of the model generated by the program is suited to simulation because of the constant leading coefficient matrix. We believe it serves as an acceptable frame for the successful modelling and simulation of the mechatronic systems in Part 2.

5.3 Numerical Solution Using BDF Methods

This section describes the methods and strategy for the solution of the generated differential-algebraic equations. Among the possible candidates for solving DAEs, two methods have attracted most attention: those based on backward difference formula methods, known as BDF methods; and various kinds of the Runge-Kutta methods [3, 4]. Among the implementations of BDF methods, perhaps the best-known software is DASSL, of which a detailed description can be found in [3]. It is freely available from the NETLIB web repository. Similarly, among the implementations of the Runge-Kutta methods, possibly the most useful is Radau5 of [4], which also is freely available from those authors.

We use an implementation of the BDF method for the solution of the generated DAEs. One of the reasons for this choice is that we have had experience with BDF methods from the time of the famous DIFSUB program [14]. BDF codes are widely used in electronic circuit simulators and continue to attract attention as capable, general-purpose methods for solving DAEs. Part 2 shows that it also is a method with which it is possible to solve mechatronics problems based on the bond graph modelling approach.

5.3.1 The Implementation of the BDF Method

The solver used for solving DAEs of the model system is based on the *variable coefficient* version of BDF [3, 15]. In comparison, DASSL uses a fixed coefficient implementation of BDF [3]. The variable coefficient form is perhaps the most stable implementation of the BDF methods, though it is less efficient because it requires frequent re-evaluation of the partial-derivative matrix. The reason for using it here is that the solution of DAEs is not an easy problem, and the higher level of stability of this method is welcome. Unfortunately, the authors are only aware of one study only that compares these two BDF implementations [15]. The frequent re-evaluation of the partial-derivative matrix is less expensive in our approach, as we use it in an analytical form and not by a numerical approximation. This improves the stability of the BDF code. In the following section we describe in some detail the method we use. We use a similar notation as in [3].

The system we solve numerically is described by Eq. (5.26). The numerical solver generates approximations z_i to the true solution $z(t_i)$ (Fig.5.4). The BDF method is a variable step and variable order predictor-corrector method. Because of the stability requirement, the order k of the method is limited to 5.

The method predicts values of the solution at the next instant of time t_{n+1} by evaluating a polynomial that interpolates the last k+1 values, i.e.

$$z_{n+1}^0 = P(t_{n+1}) \tag{5.33}$$

where P are interpolating polynomials defined by

$$P(t_{n-i}) = z_{n-i}, \ i = 0,1,...,k \tag{5.34}$$

The predicted value of the time derivative at time t_{n+1} is similarly found by evaluating the time derivative of the interpolating polynomial, that is

$$\dot{z}_{n+1}^0 = \dot{P}(t_{n+1}) \tag{5.35}$$

In the DIFSUB code of [14], the interpolating polynomials were originally represented by Taylor series. The past history is held in the form of Nordsieck vectors of the current values of the variables and their scaled derivatives up to the order k. It was found that it is much more efficient to use interpolating polynomials based on the Newton interpolating polynomial with modified divided differences [16, 17]. DASSL follows this approach and it is also used in our code. A detailed description of the implementation details can be found in [3, 17].

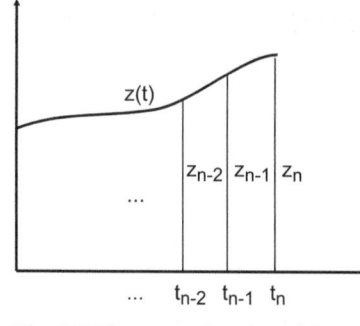

Fig. 5.4. The approximation of the solution at discrete times

The approximation to the solution value is determined using corrector polynomials $\mathbf{Q}$

$$\mathbf{z}_{n+1} = \mathbf{Q}(t_{n+1}) \tag{5.36}$$

such that

$$g(\dot{\mathbf{Q}}(t_{n+1}), \mathbf{Q}(t_{n+1}), t_{n+1}) = 0 \tag{5.37}$$

There are several ways correction polynomials can be defined [14]. The method we use defines it as a polynomial that interpolates through the same last k points as the predictor, i.e.

$$\mathbf{Q}(t_{n+1-i}) = \mathbf{z}_{n+1-i}, \ i = 1, \dots, k \tag{5.38}$$

Continuing in the same way as in [3, 15], we arrive at expressions for approximating the derivative, known as the *differentiation formula*

$$\dot{\mathbf{z}}_{n+1} = \dot{\mathbf{z}}_{n+1}^0 - \frac{\alpha^0 (n+1)}{h_{n+1}} (\mathbf{z}_{n+1} - \mathbf{z}_{n+1}^0) \tag{5.39}$$

Here

$$h_{n+1} = t_{n+1} - t_n \tag{5.40}$$

is the attempted step size and

$$\alpha^0 (n+1) = -\sum_{i=1}^{k} \alpha_i (n+1) \tag{5.41}$$

where

$$\alpha_i (n+1) = \frac{h_{n+1}}{t_{n+1} - t_{n+1-i}}, \ (i \geq 1) \tag{5.42}$$

The coefficient α^0 in Eq. (5.41) depends on the ratio of the current and accumulated previous step sizes and changes each time the step size or order changes. Such a formula is known as a *variable coefficient* formula. In DASSL, a simplified form of this formula is used in which the coefficient is independent of the step-size (denoted in [3] as α_s). The formula was developed using corrector polynomials that interpolate at equi-distant time points.

After substituting in Eq. (5.37), we get

$$\mathbf{g}(\dot{\mathbf{z}}_{n+1}^0 - \frac{\alpha^0(n+1)}{h_{n+1}}(\mathbf{z}_{n+1} - \mathbf{z}_{n+1}^0), \mathbf{z}_{n+1}, t_{n+1}) = 0 \tag{5.43}$$

This is an implicit vector equation that can be solved iteratively using the predicted values as the starting point. The modified Newton method is used. If we denote by $\mathbf{d}_m$ the corrections at the m-th iteration step, the corrections can be found by solving the linear equation

$$\mathbf{J}\mathbf{d}_{n+1}^m = -\mathbf{g}(\dot{\mathbf{z}}_{n+1}^m, \mathbf{z}_{n+1}^m, t_{n+1}) \tag{5.44}$$

where $\mathbf{J}$ is the partial derivative (Jacobian) matrix

$$\mathbf{J} = \frac{\partial \mathbf{g}}{\partial \mathbf{z}} + \alpha \frac{\partial \mathbf{g}}{\partial \dot{\mathbf{z}}} \tag{5.45}$$

with

$$\alpha = -\frac{\alpha^0(n+1)}{h_{n+1}} \tag{5.46}$$

The values of the variables and the derivatives are updated for the next iteration by using the formulas

$$\left.\begin{array}{l} \mathbf{z}_{n+1}^{m+1} = \mathbf{z}_{n+1}^m + \mathbf{d}_{n+1}^m \\ \dot{\mathbf{z}}_{n+1}^{m+1} = \dot{\mathbf{z}}_{n+1}^m + \alpha\mathbf{d}_{n+1}^m \end{array}\right\} \tag{5.47}$$

The solution Eq. (5.44) is accomplished by the method of LU decomposition and back substitution. Because of the sparse structure of the matrix, we use a sparse package for the LU decomposition of the matrix and the solution of the corresponding systems of triangular equations. As a practical implementation of the method, we use the Sparse Linear Equation Solver (version 1.3a) of the University of California, Berkeley. [1] The solver was originally developed for circuit simulators. We used its basic functionality only and, in particular, the possibility to decompose the matrix at a fixed pivotal order once the complete decomposition is done. This way, if the matrix has not changed too much, then decomposition can be performed quite efficiently. The strategy used for solving Eqs. (5.44) – (5.47), including the stopping criterion, is the same as described in [3] for DASSL.

We use a similar strategy for accepting an integration step, the step size selection, and change of order, as in [3]. This has its roots in [17]. The basic difference is a slightly different expression for the estimated principal part of the truncation error. Other differences are described later. The estimation used for the truncation error is of the form

---

[1] The code and documentation are freely available through the NETLIB repository. We express our thanks to the University of California, Berkeley, Department of Electrical Engineering, and to the authors of this really sophisticated software.

$$\boldsymbol{\theta}_{n+1} = \alpha_{k+1}(n+1)\left\|\mathbf{z}_{n+1} - \mathbf{z}_{n+1}^0\right\| \tag{5.48}$$

where $\|\ldots\|$ is a weighted mean square root norm. The expressions are slightly different from DASSL because of the variable coefficient form of the differentiation formula in Eq. (5.39).

5.3.2 The Generation of the Partial Derivative Matrix

Solving Eq. (5.44) requires evaluating the partial derivative matrix of Eq. (5.45). This matrix is often approximated numerically. We generate analytical expressions for this matrix by symbolic differentiation of the model equations.

The matrix is of sparse structure, having a small number of nonzero elements per row and column. We generate and store only the expressions of nonzero elements. One of the often-used methods for storing sparse matrices is based on the coordinate scheme. It is used here because it is very easy to implement (Fig. 5.5).

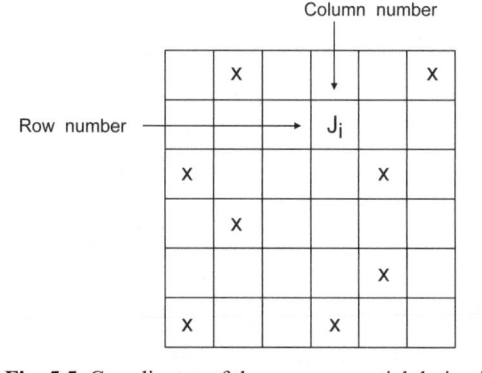

Fig. 5.5. Coordinates of the nonzero partial derivative matrix entry

Thus, for every nonzero element of the partial derivative matrix we generate an analytical expression and record its row and column number. The procedure is straightforward and is as follows

1. Take the left side of the expression of the first system model equation
2. Find the list of variables and the derivatives of variables that appear in the expression
3. For every variable evaluate, in turn, the partial derivative of the expression by symbolic differentiation. Evaluated expressions are stored successively in the corresponding string array. Store, in separate arrays, the index of the equation (row index) and of the variable (column index).
4. Repeat step 2 with the next expression until all equations have been processed.

The values of the partial derivative matrix are evaluated during the simulation using routines for reading and evaluating byte form expressions, as was done with the mathematical model expressions.

5.3.3 The Error Control Strategy

We now describe the strategies used for error control. We do not apply any projections on the model equations, but try to solve them directly. The BDF code, as used here, can handle DAEs of index 1 and 2, if some precautions are taken. Putting aside the problem of initialisation that we consider in Sect. 5.5, one of the most sensitive points is error control. This concerns the strategy for step acceptance and how the next step and/or the order of the method is chosen.

Error control in BDF codes is based on testing the truncation error estimates against predefined error tolerances. The truncation error is estimated by Eq. (5.48), which is similar to that in DASSL, with some difference in the leading coefficient. The error is thus proportional to the corrections made in the Newton iterative loop. If the error test is not satisfied a new, smaller step size is chosen until the test is passed. This works well for index-1 DAEs. In index-2 equations there is a repeatable failure of the error tests and the procedure is finally stopped. The problem is that errors in some algebraic variables do not approach zero as the step size approaches zero. As a result, these cannot be made less than the tolerance (in some norm) with decreasing step size. Thus, solving higher-index DAEs requires changing the way the error is tested.

The simplest way to do this is suggested in [3]: All the algebraic variables are removed from the error tests, the next-step-size test, and the order-selection tests, and these tests are applied to the differentiated variables only. This has the drawback that the algebraic variables are partially out of control. They, however, satisfy the corrector equation solving loop and are accepted if the differentiated variables pass the integration error test. The reasoning behind such a strategy is that errors in the differentiated variables do not influence directly the future state of the system. In this way, integration can be executed successfully, though with the price of lower accuracy in the algebraic variables. In many cases this is acceptable, as long as high accuracy in such variables is not required.

We also scale the algebraic equations by the factor 1/h, where h is the current step-size. This scaling, as discussed in [3], improves the ill conditioning of the iteration matrix at very small step sizes. This, in turn, improves the accuracy of the differentiated variables and, to a lesser extent, of the algebraic variables. In any case, scaling is recommended. We have not applied scaling directly in the model equations because these become quite difficult for the user to understand (Sect. 5.4). Instead, we simply modify the values of the left side of the equations and of the partial matrix elements when they are evaluated during iteration.

We have also implemented another strategy based on *local error* control. This has its roots in a study of the local error control of DAEs of index-1 and 2 in [18]. Using the notation of Sec. 5.3.1, the proposed estimation of the local error is given by

$$\mathbf{S}_{n+1} = \mathbf{J}^{-1}\mathbf{A}(\kappa\mathbf{I} - \frac{(\alpha^0)^2}{\alpha^0(n+1)\cdot h^2}\mathbf{J}^{-1}\mathbf{A})\mathbf{\theta}_{n+1} \qquad (5.49)$$

Matrix $\mathbf{J}$ is the last evaluated and decomposed partial derivative matrix defined in Eq. (5.45), $\mathbf{A}$ is the leading coefficient matrix of Eq. (5.27), and θ_{n+1} is the truncation error estimate of Eq. (5.48). The α^0 is the value of the BDF parameter in Eq. (5.41) used in the last evaluated partial derivative matrix, and h is the value of the step size used at that time. Parameter κ is a user-supplied factor used to weight the inherent differentiation in index-2 DAEs. This factor can be set from 1 to a value of the order of $1/h$.

In the implementation of the formula the inverse of the Jacobian is not required because the matrix is already evaluated and decomposed into LU factors. Corresponding terms in Eq. (5.49) can be evaluated by back substitutions from the corresponding linear system. The cost of the formula is, in essence, these two back substitutions. We also use this formula for step size and order selection, as suggested in [18]. The basic strategy, as described in [3], is retained but, instead of the estimates of the truncation errors, the corresponding local errors are evaluated according Eq. (5.49) and used for the next step size and order selection. This increases the computational costs, but the BDF solver works better.

In [18] it was shown that the error control based on the local error is satisfactory for index-1 and index-2 DAEs. It is expected that this control strategy is an improvement over the error control of using the first strategy. We have found this to be true at least for some of the problems treated in this book (Chapt.9). We also found that the local error control strategy works much better if the algebraic equations are scaled by the step-size.

To summarize the integration control used for solving the model equations during simulation, we describe the parameters and options at the user's disposal. First of all, after the mathematical model is built and suitable data are created, the user has to select the

1. Simulation interval
2. Output interval
3. Maximum step size

After the simulation is finished, it is possible to continue or to restart the simulation. Output is generated at every output interval value. The maximum step size that the integrator selects is limited. By default, both the output interval and maximum step size are set to $1/100$ of the simulation interval. For many problems, however, quite a short maximum step size is often necessary in order to finish the simulation successfully.

The BDF method "likes" relatively tight error tolerances. Thus, the default values of absolute and relative error tolerance are set to 10^{-6}. These can be changed, up or down. Lower values are, in general, not advised. On the other hand, some problems are not easy to solve with too tight a tolerance, i.e. 10^{-12} or 10^{-15}.

Currently only one integration method is implemented, the variable coefficient BDF method. It is possible, however, to scale the algebraic equations (this is the default behaviour), or to disable scaling. Both error control strategies described above are implemented:

1. Control of errors in differentiated variables only (default)
2. Local error control

In the last option it is possible to input the differentiation weight parameter value, too. It is set to 1 by default.

5.4 Decompiling of the Model Equations

The system model equations, including the partial derivative matrices, are generated automatically by the BondSim software. They are generated in byte form, as explained in Sect.s 5.2 and 5.3, and hence are not easily readable. We think, however, that it is important for the user to have feedback on the model generated, so we have implemented routines that *decompile* the equations to a readable form. We here give the information currently available and how it can be interpreted.

In Sec. 5.2 a procedure for model equation generation is given. It is possible for equations assembled in byte form to be converted back to a form that can be easily read. However, the symbols used in the decompiled form are not the original symbols used when defining the constitutive relations. Instead, system variables in the form Y(1), Y(2), etc. are used, and t is used for time. Because the parameters have already been evaluated, they are converted to numbers written in the usual floating-point form. The typical form of the decompiled equations is shown in Fig. 5.6.

Equations of the model :
    EQ(1) = ((Y(2)-Y(18))+Y(35)) = 0
    EQ(2) = (Y(35)-((t<1E-009)?(2.5*(1-EXP (-t/1E-011))):(2.5*(1+SIN
    (((2*3.1415927)*20000000)*(t-1E-009)))))) = 0
    EQ(3) = (((((-Y(1)-Y(63))-Y(33))-Y(24))-Y(22))+Y(31)) = 0
    EQ(4) = ((-Y(14)+Y(11))-Y(7)) = 0
    EQ(5) = ((Y(6)-Y(38))-Y(36)) = 0
    EQ(6) = (Y(36)-Y'(37)) = 0
    EQ(7) = (Y(37)-(((Y(7)/0.75)<0.5)?(((5E-015*0.75)/(1-0.5))*(1-((1-(Y(7)/0.75))^(1-
    0.5)))):((5E-015*0.75)*((0.585786437626905+(1.4142135623731*((Y(7)/0.75)-
    0.5)))+(0.707106781186548*(((Y(7)/0.75)-0.5)^2))))))) = 0
    EQ(8) = (Y(38)-(((Y(7)/(1*0.0258))>50)?((1E-014*((EXP
    (50)*((1+(Y(7)/(1*0.0258)))-50))-1))+(Y(7)*1E-012)):((1E-014*(EXP
    (Y(7)/(1*0.0258))-1))+(Y(7)*1E-012)))) = 0

Fig. 5.6. The decompiled form of the model equations

In addition to the model equations, other information is also collected and displayed, such as

1. Conditional expressions (switch functions)
2. A list of the differentiated variables
3. A list of the algebraic equations.

The problem is how to interpret these equations and, in particular, the system variables in terms of the original model variables, such as. efforts and flows. To that end, the symbolic table generated during the system variables assignment (Sec. 5.2.1) is used. The symbol table is not destroyed after the model equations are built, but is retained in the memory.

Using the symbol table every system model variable Y(l) can be correlated to a port variable. It is clearer to the program what variable it is than it is to the user, because the port variables are internally coded. Hence, to show the correspondence between these two sets of variables a visual technique is used. In simple terms, when the user puts the cursor over a port of an elementary component the port variables are shown using the system variable symbols, i.e. effort Y(4), flow Y(18), etc. How this works is shown in the Part 2.

In a similar way, data for the partial derivative matrix can be decompiled (Fig. 5.7). For every nonzero element of the matrix the expression, accompanied by row and column index, is given. Data for the leading coefficient matrix of Eq. (5.27) also is given.

Matrix of partial derivatives :
  J(1) = 1
  IRow(1) = 1   JCol(1) = 2
  J(2) = -1
  IRow(2) = 1   JCol(2) = 18
  J(3) = 1
  IRow(3) = 1   JCol(3) = 35
  J(4) = (1-((t<1E-009)?0:0))
  ...
  J(20) = -(((Y(7)/0.75)<0.5)?(7.5E-015*-(((1-(Y(7)*1.33333333333333))^0.5)*(0.5*(-1.33333333333333/(1-(Y(7)*1.33333333333333))))):(3.75E-015*(1.88561808316413+(0.707106781186548*(2.66666666666667*((Y(7)*1.33333333333333)-0.5))))))
  IRow(20) = 7   JCol(20) = 7

Fig. 5.7. The decompiled form of the partial derivative matrix

5.5 The Problem of Starting Values

In this and the next section we describe the approaches used for solving two important issues of simulation: the generation of starting values and integrating across discontinuities.

The problem of values from which the simulation starts is one of the difficult problems of DAEs that has attracted much attention. In the equations in state-space form, such as

$$\left.\begin{array}{l} \dot{\mathbf{x}} = \mathbf{f}(\mathbf{x}, t) \\ \mathbf{y} = \mathbf{g}(\mathbf{x}, t) \end{array}\right\} \tag{5.50}$$

this is not much of a problem. To start solving the equations it is enough to specify the initial values of the state variables, e.g. by specifying the values at initial time t = 0

$$\mathbf{x}(0) = \mathbf{x}_0 \tag{5.51}$$

In this way, the time derivatives of the variables needed by the integration routine can be evaluated directly from the first equation

$$\dot{\mathbf{x}}(0) = \mathbf{f}(\mathbf{x}_0, 0) \tag{5.52}$$

and the integration can start. The output values are generated using the second equation. The problem thus is decoupled into solving the differential equation and generating the outputs. This is well known and is why much attention in bond graphs, and in other approaches, has been given to setting the model equations in this form. In modelling real systems, however, models in state space form generally are too restrictive. This makes it necessary to deal with DAEs.

Because there is no decoupling of equations and variables in DAEs, we have to cope with the complete system simultaneously. In DAEs of index 1, such as of Eq. (5.18), some of the similarity with state-space form is retained. Because the Jacobian matrix $\partial \mathbf{g}/\partial \mathbf{y}$ is not singular, it is possible to find values of the algebraic variables $\mathbf{y}$, given values of the differentiated variables $\mathbf{x}$. Hence, if values of the latter are known at the initial time, values of the former can be found by solving the algebraic equations

$$\mathbf{g}(\mathbf{x}_0, \mathbf{y}, 0) = 0 \tag{5.53}$$

This can be accomplished using iterative methods, typically Newton type methods, for which we need only an initial guess at their values to start the iteration. Once values are found to the prescribed accuracy, the starting values of variables are known. Hence, in index-1 DAEs, the differentiated variables play the role of state variables, in the sense that they are independent and completely define the values of all other variables.

Index-1 DAEs are typically solved using BDF methods. To start the solution we need also the time derivatives of the variables. For the semi-explicit DAEs of Eq. (5.18), derivatives of the differentiated variables with respect to time can be found easily by evaluating the right-hand side of the first equation at initial time. For algebraic variables, on other hand, we first need to find the time derivative of the algebraic equations. We can do this by symbolic differentiation. The resulting equations are linear in the time derivative of the algebraic variables, as can be seen from the second equation in Eq. (5.19) and, hence, are solved readily for the time derivative of the variables.

Values of the variables and their time derivatives constitute a set of *consistent* starting values for the system. If we find such values with sufficient accuracy, integration continues smoothly. Otherwise, we can expect wide fluctuations, which hopefully converge to the solution.

For higher index DAEs the problem is more complicated. Looking at Eq. (5.20), which is often used as a prototype of semi-explicit index-2 DAEs, we see that the differentiated variables are constrained by the algebraic equations. Hence, all of their values cannot be set independently. The independent part plays the role of system state variables. These can be set initially to appropriate values. The others then are found by solving the second equation. Even for these we need starting values that are used to initialise the iteration.

An independent set of differentiated variables traditionally is found in bond graphs by the causality assignment procedure (Sec. 2.10). More generally, they are found using the subspace structure of the underlying equations [5, 6]. This way, only part of the first equation in Eq. (5.20) is an ordinary equation that involves integration. The other constitutes algebraic equations that involve differentiation of the variables. But this is only part of the story and, perhaps, the easier part. What about the algebraic equations? To find them we need to differentiate the second equation in Eq. (5.20). After substituting from the first equation, we get the hidden constraint of Eq. (5.23). If the corresponding Jacobian given in Eq. (5.24) is not singular, we solve the hidden constraint for the algebraic equations. For this, however, we again need starting values.

Determination of the initial values of the time derivatives of variables is accomplished in a similar way as that used for index-1 DAEs. Initial values of the time derivatives of the differentiated variables are found from the first equation in Eq. (5.20). But for the algebraic variables we again need to differentiate the hidden constraint with respect to time and find the corresponding values of the variables from the resulting linear equations.

In DAEs of index higher than 2, the problems are even more difficult, as more differentiation is necessary and the subsequent equations are more difficult to solve. Different approaches have been reported for solving the problem of starting values [19 - 23]. Many are concerned with index-1 DAEs, but some also treat index-2 equations. In general, they resort in some way or other to differentiation. A good survey of these is found in [3, 23]. One of the problems is how to determine the index of the system. Approaches based only on structural information, as in [20], are often not feasible, as the system structure can change because, for example, some of the constitutive relations are defined conditionally, depending on values of variables. As a simple example, we return to the see-saw problem (Sec. 2.7 and Sect. 5.2.3). The model generated for this system is of index-2. If we assume, however, that there is dry friction—as there is!—between the see-saw seat and the children, when the children stick to the seat this becomes an index-2 problem. When they slip even for a short time, however, the model changes to index–1, because the differentiated variables become independent.

The problem of consistent starting values is not only important at the start of the simulation, but also after every discontinuity. In addition, any approach to the problem of starting values assumes that the user supplies the necessary data. Even under the best conditions this is often too much to ask. The modelling approach in this book has been developed to help the user design or analyse mechatronic equipment. For this we need a simpler approach to the starting of the simulation.

The approach we use is close to that of [19, 21]. It was prompted by how we start real equipment. Typically, we switch the power on and, after some time, the system settles down to the appropriate operating state and we start using it. We thus assume that the system starts off un-energized, i.e. with efforts and flow equal to zero. We also assume that the time derivatives of all variables are equal to zero. By default, the starting values of all differentiated variables defined in the corresponding elements (capacitive, inertial, and integrator elements) also are set to zero, but can be changed to some other value if required. These are taken as an estimation of consistent starting values.

To find a starting value for the simulation we use a version of the BDF solver that employs a first-order method (implicit Euler) and a fairly small and constant step size. To simulate the transients that can be quite intensive until the system settles down, we simulate the system for several steps without error control. We then advance the simulation time for one step more and check the error. If the error test is passed, we integrate back to the initial time and use the state reached as the starting state for the simulation. If, on the other hand, the test is not passed, we repeat the starting procedure with a smaller step. The starting procedure is stopped and failure reported if either the step was reduced to the minimum without success, or if too many initialisations have been attempted. During initialisation the corrector equations are solved at every step to the prescribed accuracy. The partial derivative matrix is updated and decomposed at every simulation or iteration step. If, for any reason, it is not possible to solve these equations, initialisation is stopped with the appropriate message. In Part 2, in which different mechatronic problems are solved by bond graph modelling, all initialisations are done using this approach quite successfully. The reader can test this for herself.

5.6 The Treatment of Discontinuities

Discontinuities give reality to models. There are numerous examples of engineering systems where such features are necessary. This is the case, for example, when dealing with dry friction and impact in mechanical systems. In electronics, processes are often described by different expressions, depending on the range of variable values. This, of course, complicates the problem of equation solving.

We analysed several schemas for dealing with discontinuities. During model generation sub-expressions contained in the constitutive equations are extracted and stored in a separate array. These sub-expressions are decompiled and shown below, along with the model equations discussed in Sec. 5.4. They can be used to analyse model behaviour. Fig. 5.8 shows conditional expressions (switch functions) used in the model of the CMOS inverter of Sec. 7.4.2 (Fig. 7.86).

We use these functions to generate state transition tables. The location time instant of switching was found by a binary search using both interpolating polynomials that the BDF solver maintains or by repeating the time step. After the discontinuity, the integration was reinitialised using a method similar to that used for simulation initialisation (Sec.5.5). We found, however, that this approach did not work as expected: Many times there was a conflict during the Newton step and the corresponding step-advancing.

We thus do not change the way that the solver behaves at a discontinuity. In the current implementation of BondSim, we do not use discontinuity location and transition tables to go over the discontinuity; we simply let the solver handle the discontinuities. As seen in Figs. 5.6 and 5.7, the conditional expressions are retained in the model expressions and those of the partial derivative matrix elements. When the discontinuity is encountered, the solver tries to go over it. If it cannot—owing to the error control requirements—it shortens the step-size and repeats the step. This perhaps is neither the most efficient nor the best technique that might be used, but it works. It works well in all the examples treated in this book, as well as with both error control strategies discussed in Sec. 5.3.3.

```
The model switch functions :
    Switch(1) = ((t<1E-009)?0:1)
    Switch(2) = (((Y(7)/0.75)<0.5)?0:1)
    Switch(3) = (((Y(7)/(1*0.0258))>50)?0:1)
    Switch(4) = (((Y(9)/0.75)<0.5)?0:1)
    Switch(5) = (((Y(9)/(1*0.0258))>50)?0:1)
    Switch(6) = ((Y(12)<=(1+(0*(SQRT (0.6-Y(13))-SQRT
    (0.6)))))?0:((Y(42)>=0)?((Y(42)<((Y(12)-1)-(0*(SQRT (0.6-Y(13))-SQRT
    (0.6)))))?1:2):((-Y(42)<((Y(12)-1)-(0*(SQRT (0.6-Y(13))-SQRT (0.6)))))?3:4)))
    Switch(7) = (((Y(23)/0.75)<0.5)?0:1)
    Switch(8) = (((Y(23)/(1*0.0258))>50)?0:1)
    Switch(9) = (((Y(25)/0.75)<0.5)?0:1)
    Switch(10) = (((Y(25)/(1*0.0258))>50)?0:1)
    Switch(11) = ((Y(19)<=(1+(0*(SQRT (0.6-Y(3))-SQRT
    (0.6)))))?0:((Y(55)>=0)?((Y(55)<((Y(19)-1)-(0*(SQRT (0.6-Y(3))-SQRT
    (0.6)))))?1:2):((-Y(55)<((Y(19)-1)-(0*(SQRT (0.6-Y(3))-SQRT (0.6)))))?3:4)))
```

Fig. 5.8. Conditional expressions of the CMOS inverter model

To help the solver to deal with discontinuities it is advisable to ensure continuity of sub-expressions at their boundaries, if at all possible. In many cases discontinuities appear because of the simplification of the model. Thus, complicated non-linearities present fewer problems than an on-off discontinuity. This is particularly true at the start of a simulation.

5.7 Pros and Cons of the Combined Compiled/Interpretative Approach

We end this section with a discussion of the advantages and disadvantages of the combined compiled and interpreting code approach used in the current implementation of BondSim.

The conventional approach in simulation software is based on the compiled code approach. This is motivated mainly by its better efficiency, as measured in terms of execution time. But it has its drawbacks, too. For example, every change to the model requires that new functions or subroutines be generated, compiled, and linked with the rest of the code. Hence, the program should be part of a suitable programming environment, such as Microsoft Developer Studio; or a suitable compiler and linker should be shipped with the program. This is but one—and perhaps a minor—problem. More importantly is the relatively long process involved in developing the model to the point at which simulations can be run. This was one of the main reasons why another approach was taken.

As described in several places in this chapter, instead of functions or subroutines, the constitutive relations of the model are packed into byte codes that can be read and evaluated in one pass by a suitable interpreter. Thus, when the model is developed as a tree of components, or taken from a library, the tree is analysed and the model built. This operation needs no external program, such as in the compiled approach. After the model is built, the simulation can start. Later, if some changes are made, i.e. if some local parameters are changed, it is not necessary to rebuild the complete model, but only update part of it. Currently, this is implemented for parameters defined locally in the sources. For parameters defined in other components and at lower levels, e.g. in the documents, the complete model is rebuilt. It is implemented in this way because such parameters can be used by any upper level components. It is also possible to implement a more sophisticated approach, one that is analogous to the incremental linking technique. This has been left for further research.

Another advantage of this approach is that it opens up the possibility of implementing computational algebra support. Symbolic differentiation currently is implemented only for the generation of Jacobian matrices in analytical form, which the BDF solver needs. Much more, however, can be done, particularly in the field of DAEs solving and initialisation. There are other interesting topics, as well.

In our opinion, execution speed remains the only drawback to the combined compiled/interpretative approach. This is not, however, a problem of such great concern. A greater part of the time during simulation, we suppose, is used on matrix decomposition, and this is performed using compiled subroutines. Function evaluation and sparse matrix evaluation are responsible for a smaller part. Thus, we expect that the difference between compiled and interpretative routines is not too great. This assertion is based on comparing problems solved in Part 2 with those reported in other sources. Such an approach is at least partly encouraged by recent advances in scripting languages and, in particular, JAVA. Now even numerical libraries are implemented in interpretative languages, such as JAVA.

References

1. Alain Reverchon and Marc Ducamp (1993) Mathematical Software Tools in C++. John Wiley, Chichester
2. L Petzold (1982), Differential/Algebraic equations are not ODEs, SIAM J. Sci. Statist. Comup., 3:367-384
3. KE Brenan, SL Cambell and LR Petzold (1996) Numerical Solution of Initial-Value Problems in Differential-Algebraic equations, Classics in Applied Mathematics. SIAM, Philadelphia
4. E Hairer and G Wanner (1996) Solving ordinary Differential Equations II, Stiff and Differential-Algebraic Problems, 2nd Revisited edn. Springer-Verlag, Berlin Heidelberg New York

5. E Griepentrog and R März (1986) Differential-Algebraic Equations and Their Numerical Treatment. BSB Teubner, Leipzig

6. R März (1992), Numerical Methods for Differential-Algebraic Equations, Acta Numerica, 141-198

7. C Tischendorf (1995) Solution of Index-2 Differential Algebraic Equations and Its Applications in Circuit Simulation, PhD thesis Humboldt Univ. Berlin. Logos Verlag, Berlin

8. R März and C Tischendorf (1997), Recent Results in Solving Index-2 Differential-Algebraic Equations in Circuit Simulations, SIM J. Sci. Comput., 18:139-159

9. EJ Haug (1989) Computer-Aided Kinematics and Dynamics of Mechanical Systems, Vol. I: Basic Methods. Allyn and Bacon, Needham Heights, Massachusetts

10. J Van Dijk and PC Breedveld (1991) Simulation of System Models Containing Zero-order Causal Paths – I. Classification of Zero-order Causal Paths, J. of The Franklin Institute 328:959-979

11. J Van Dijk and PC Breedveld (1991) Simulation of System Models Containing Zero-order Causal Paths – II. Numerical Implications of Class 1 Zero-order Causal Paths, J. of The Franklin Institute 328:981-1004

12. DE Schwartz and C Tischendorf (2000) Structural Analysis of Electrical Circuits and Consequences for MNA, Int. J. Circ. Theor. App. 28:131-162

13. W Borutzky F Cellier (1996), Tearing in Bond Graphs with Dependent Storage Elements. Proc. Symp. on Modelling, Analysis and Simulation, CESA'96, IMACS Multi-Conference on Computational Engineering in Systems Applications, Lille, France 2:1113-1119

14. CW Gear (1971) Numerical initial-value problems in ordinary differential equations, Prenice Hall. Englewood Cliffs

15. KR Jackson and R Sacks-Davis (1980), An Alternative Implementation of Variable Step-size Multistep Formulas for Stiff ODEs. ACM Trans. Math. Software, 6:295-318

16. FT Krogh (1974) Changing Step Size in Integrations of Differential Equations Using Modified Divided Differences. Proc. Conf. Num. Solution of ODEs, Lecture Notes in Mathematics 362, Springer-Verlag, New York

17. LF Shampine and MK Gordon (1975) Computer Simulation of Ordinary Differential Equations. FH Friedman and Co., San Francisco

18. J Sieber (1997), Local Error Control for General Index-1 and Index-2 Differential-Algebraic Equations. Humboldt University Berlin, Preprint

19. RF Sinovec, AM Erisman, EL Yip and MA Epton (1981) Analysis of Descriptor Systems Using Numerical Algorithms, IEEE Trans. Aut. Control. AC-26:139-147

20. CC Pantelides (1988) The Consistent Initialization of Differential-Algebraic Systems. SIAM J. Sci. Comput.9:213-232

21. A Kröner, W Marquardt and ED Giles (1992) Computing Consistent Initial Conditions for Differential – Algebraic Equations, Computers & Chemical Engineering, 16: S131-S138

22. PN Brown, AC Hindmarsh and LR Petzold (1998) Consistent Initial Conditions Calculations for Differential-Algebraic Systems. SIAM J. Sci. Comput. 19:1495-1512

23. DE Schwarz (2000) Consistent Initialization for Index-2 Differential Algebraic Equations and Its Applications to Circuit Simulation. PhD thesis Humboldt Univ. Berlin.

PART 2

APPLICATIONS

Chapter 6 Mechanical Systems

6.1 Introduction

In this chapter we start the study of problems from Mechatronics using the bond graph modelling approach developed in Part 1. The general procedure consists of

1. Analysis of the problem under study
2. Development of the corresponding model in terms of bond graphs, and
3. Analysis of the behaviour by the simulation

Bond graph modelling and simulation will be effected by the *BondSim* research package included as a supplement to this book. Readers are advised to use it when reading material given in this and subsequent chapters (see also the Appendix).

In this chapter we study simple, mostly one-dimensional mechanical problems. One reason for this is to familiarize the reader with using the *BondSim* software on relatively simple problems, though these problems are of interest in their own right.

We start with the well-known Body Mass Damper problem already discussed in Sect. 2.6.1. After that we continue with the study of the influence of dry friction, which introduces discontinuities in the model equations. We then study the Bouncing Ball problem. This also involves discontinuities, but is better known for its chaotic behaviour. The section concludes with a discussion of higher index problems. It is not the intention to show that it is possible to solve all such problems using the methodology developed; we simply will show that a class of such problems of interest in Mechatronics can be solved in an acceptable way from an engineering point of view. This is particularly true for some problems in multibody dynamics. In this chapter we will study only the simple pendulum problem. Further discussion of solving problems in multibody dynamics is left to Chapt. 9.

6.2 The Body Spring Damper Problem

6.2.1 The Problem

We start by analysing a well-known problem from engineering mechanics, the simple Body Spring Damper system (Fig. 6.1), which has already been discussed in Sect. 2.7.1. The system consists of a body of mass m that can translate along the ground, is connected to a wall by a spring of stiffness k and a damper having the linear friction velocity constant b. An external force F acts on the body parallel to the floor. The effect of dry friction is not included in the model. The modelling of dry friction is a topic of Sect. 6.3.

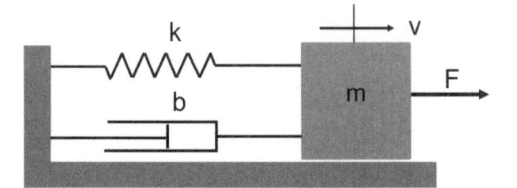

Fig. 6.1. The body spring damper problem

We will analyse the transient behaviour of this simple system, and also its behaviour under sinusoidal forcing. We first develop a bond graph model using the BondSim program. After that we analyse the dynamical behaviour of the system by simulation. The explanations of the procedures used are somewhat detailed and serve as an introduction to using the BondSim program.

6.2.2 The Bond Graph Model

Model Development

Before we start with model development, we must launch BondSim. This can be done in the usual way for a MS Window environment, e.g. by double clicking the program button on the computer desktop; or using the Windows *Start* button at the left corner of the computer screen, then choosing the *Program* command from the menu that appears, and then *BondSim*. A welcome window appears on the screen which, after a few second, disappears. The main program screen then appears (see Fig. 4.2). There is a menu bar just bellow the program title showing the main menu commands—*Project, Library, Function, View* and *Help*—from which program commands are invoked (Sect. 4.2). There is also a row of toolbar buttons used to invoke some of the more often-used commands. Some of these buttons are familiar to Windows users, but other are specifically designed for BondSim. By moving the mouse cursor over a button, a tool tip appears that contains a short description of the corresponding command. Some of these buttons are disabled.

To begin modelling the *Body Spring Damper* system, we need to define a new project. This can be done using the *New* command in the *Projects* menu, or by using the *New Project* toolbar button. In the dialogue that appears (Fig. 4.3), we type a suitable project name, e.g. "*Body Spring Damper Problem*", then click the OK button. The command is accepted if there is no existing project with the same name. The dialogue closes, and a new empty document appears having the title of the new project as shown in Fig. 6.2. The menu also changes to show the commands that can be used at the document level (Sects. 4.6 and 4.7).

The document window is used to define the overall structure of the system model. The basic tool for the bond graph model development is the *Editing Palette* (Fig. 4.9). We can open the palette by clicking the *Show Palette* toolbar button, or by the *Bond Graph Palette* command on the *Tools* menu. Once the palette is opened, we move (dragging it by its title bar) it to some suitable place where it does not cover the document window (Fig. 6.2). Similar to the toolbar buttons, a short description of various buttons on the palette can be found by moving the mouse cursor over them. Clicking the *Keep visible* ensures that the palette doesn't close after the first component is created.

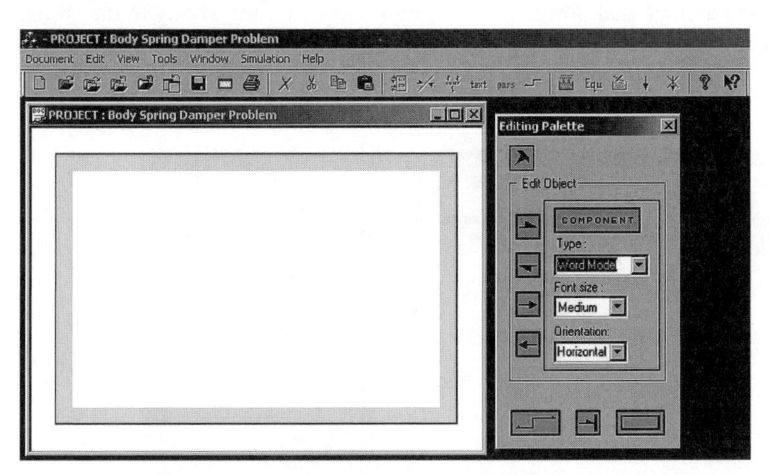

Fig. 6.2. The project document window with the Editing Palette opened

Model development starts by decomposing the system into components. Looking at the system schematics of Fig. 6.1, we identify the following components:

1. The wall used to connect the spring and damper, and the floor over which body can move. We represent the wall and the floor by a single two-port component called Wall, the ports being the places where the spring and the damper are connected.
2. The spring of stiffness k, which can be represented by a two-port component Spring, in which the ports represent the spring ends used for connection.
3. The damper, of linear viscosity coefficient b, represented by a two-port component Damper. The ports again represent the ends used for connection.
4. The body itself, which is connected to the spring and the damper, and on which there is an applied force. The body can be modelled by a three-port component Body.
5. Finally, there is a force acting on the body in the direction parallel to the floor. This effect can be represented by a source effort component.

As a first step in the model development we create these components in the document working area, i.e. inside the rectangle (Sect. 4.6.2). We start with the Body word component model. Looking at the *Editing palette* we see that the *Word*

Model component type is already selected by default, as are the medium font size and the horizontal orientation of the component text. We accept this and click on *Component* button. Next, we move the mouse cursor inside the document working area. When in this area it changes its shape to a cross. We can position the cross anywhere inside the document working area, but not too close to its boundary, and then click the mouse. A vertical textual cursor (caret) appears and we can start editing the component name simply by typing in "Body". To finish text editing of the component name, we click anywhere outside of the text.

To complete the Body word model it is necessary to create the corresponding ports. We will assume that power flows from the force through body to the spring and the damper and then to the wall (Fig. 6.1). Hence, we create a power-in port on the right side of the Body, where the force acts; and two power-out ports on the left side, where the spring and the damper are connected (Fig. 6.3).

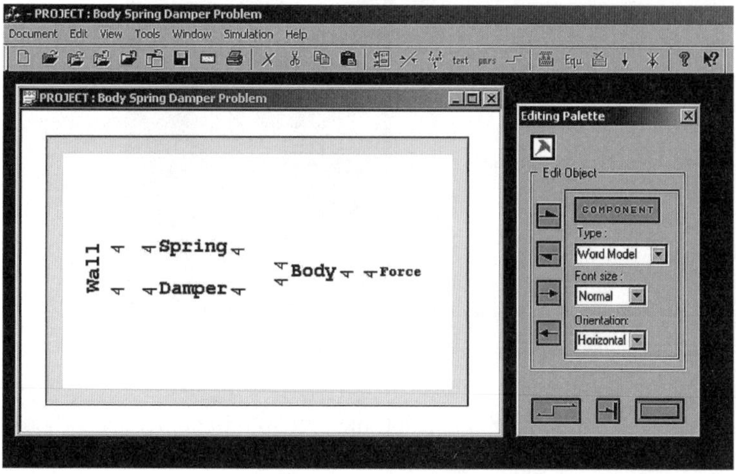

Fig. 6.3. Creating the components for the Body Spring Damper problem

In a similar way, we create the Spring and the Damper components. To each we add two power ports, a power-in port on the right and a power-out port on the left side. The wall component we create a little differently, with the name running vertically. Before clicking the component button in the palette, we select *Vertical* in the *Orientation* combo box on the palette. We also add two power-in ports to the Wall for the connection of the spring and the damper.

The final component that we create is a source effort, which describes the force applied to the body by the environment. To create such a component we select the *Source Effort* type from the list of component types on the palette, and change the font size to small.

By default, the small sizes are pre-set for elementary components. Similarly, the medium sizes are used for word models, i.e. the components consisting of other components. In a similar vein, large fonts can be used for some of the main word model components. This is not, of course obligatory; the user can choose the most appropriate size.

The elementary components are created with the standard bond graph names (Fig. 2.6). We will change the name of the source effort to "Force" by selecting it and then clicking the *Text* toolbar button. When the caret appears, we delete the old name, type the new name, then click outside to end the text editing. By changing the name we don't change the component type, just its name.

To complete the Force component, we add a power-out port because it is a power-generating component. This completes creating of the components. Before we proceed with further model development, we rearrange the component positions to correspond to the scheme of Fig. 6.1 by moving them to the appropriate positions. In this way, we arrive at the arrangement as shown in Fig. 6.3.

Next, we draw bond lines between component ports by connecting them in the same way as the components are connected in Fig. 6.1. To do this, we click on the *Connect ports* button on the palette. Now we can start to connect corresponding ports by the bonds (Sect. 4.6.3). Thus, to connect the Spring to the Body, we click the right spring port, then draw the mouse cursor to the corresponding Body port. As the mouse is moved, a bond line is drawn, starting at the first port. When the target port is reached and clicked, the ports disappear and only the bond line connecting the components remains. When all bonds are thus drawn we click again on the Connect port button to deselect. Thus, we have created the bond graph representing the basic level model of the *Body Spring Damper* system (Fig. 6.4).

Note

A component can be created anywhere inside the document working area, but not too close to its boundary. It is better to create it near the centre and then move it. Similarly, a port cannot be created too close to another ports. It is better to create it away from other ports, then drag it. When drawing a bond, it is also possible to create intermediate points by clicking on to it and then continuing. A bond line can be dragged to change its shape. Bond lines often have a zigzag appearance. There is no grid that can be used as guide. To create a nice looking bond graph, it is usually necessary to move components or their ports a bit before drawing the bonds. Drawing of a bond can be cancelet by double-clicking!

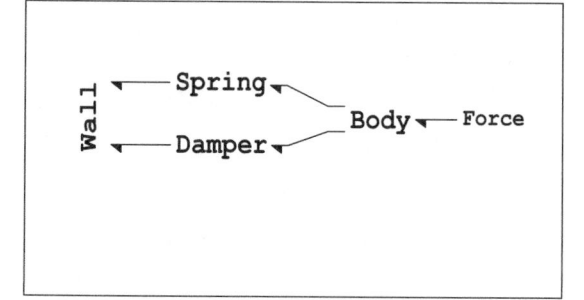

Fig. 6.4. The basic level of the Body Spring Damper system model

To complete the model it is necessary to develop models of all of the word model components, and to modify the default models of the elementary components created. To define the **Body** model, we double-click its component. Because its model has not been defined previously, a dialogue appears asking if we would like to create a new component model. We select *Yes*, and a new empty document window appears where the component model will be created (Fig. 6.5).

We may drag the document window, if we wish, to uncover its word component model in the previous document. The document has three document ports. These correspond to the **Body** component ports. Comparing the **Body** document and **Body** word model, it can be easily seen which document port corresponds to which component port.

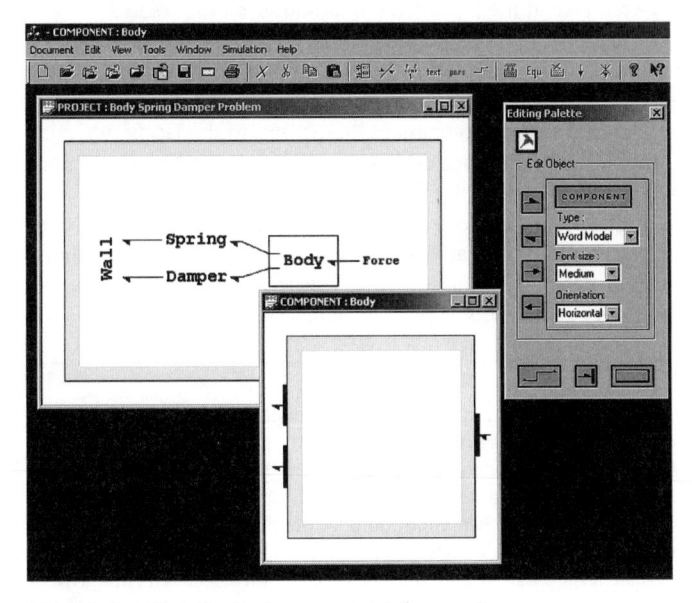

Fig. 6.5. Creating the **Body** component document

The dynamics of body translation is governed by the equation of balance of all forces acting on the body, including the inertia. We define the body inertia first. Thus, we select the *Inertial* in the list of component types on the *Editing palette* and create an inertial component I in the **Body** document. We add a power-in port, assuming that part of power transferred to **Body** is used to overcome its inertia. To balance all forces on the body, a source effort junction e is created by selecting the corresponding component type from the list on the palette. To this junction are added a power-in and three power-out ports. These ports are joined to the document ports and to the inertial component port, as shown in Fig. 6.6 on the left.

To complete the model of the body, its inertial properties are defined (Sect. 4.8.1). Double-clicking the inertial port opens *Inertial Port* dialogue (Fig. 6.6 on the right). By default, there are predefined names of the port variables and the linear constitutive relation between the port momentum and flow. This relationship is valid for body inertia in this problem, hence we simply change the names of the variables and the parameter. We use the p as the momentum variable, the body velocity v as the flow variable, and mass m as the inertial parameter. The momentum relationship thus reads p = m*v. We cannot simply click OK, because the equation parser won't understand what m is. Hence, we define it using the *Parameter* button (Sect. 4.8.2), insert a parameter m and assign to it a value of 1.0 (kg).

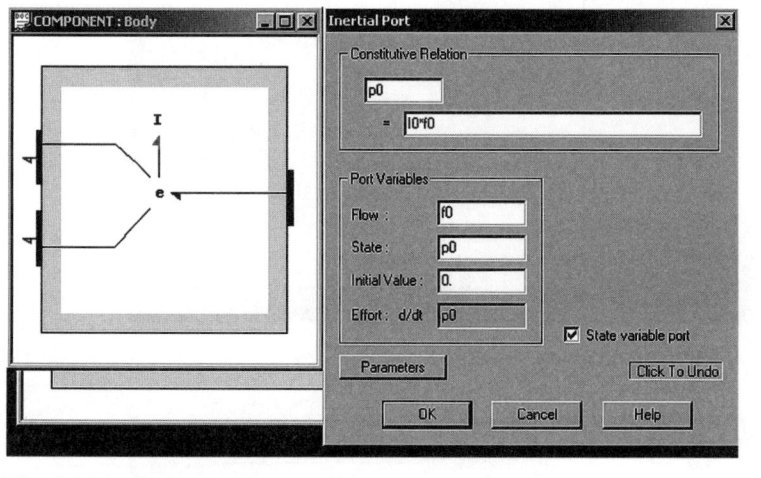

Fig. 6.6. The Body component model

In a similar way we define the Spring model (Fig. 6.7 on the left). The deformation of the spring depends on the relative velocities of its ends. Hence the Spring component can be described by a flow junction f whose two ports are connected to the document ports (spring ends); the third is connected to a capacitive port C used to model the process in the spring. These components are created similarly to the other components by first selecting the component type from the list on the palette.

The properties of the spring can be defined by double-clicking the capacitive component port and using the dialogue shown on the right side of Fig. 6.7. We change the state variable q0 to spring deformation xs and the effort to spring force Fs. The default constitutive relation corresponds to that of the electrical capacitor This will be changed to read Fs = k*xs. The spring stiffness k is inserted in the capacitive component parameters list and assigned a value of 1e4, i.e. 10 kN/m.

The Damper model is defined similarly (Fig. 6.8 on the left). The damper force depends on the relative velocities of its ends. Thus, the Damper model can be described by a flow junction f with two ports connected to the document ports (damper ends) and a third connected to the port of a resistive component R.

The properties of the damper can be defined using the dialogue shown on the right of Fig. 6.8. We change the effort variable to the damper force Fr and change the flow variable to the damper extension velocity vr. We accept the linear constitutive relation, but the resistive parameter R0 we change to read b. Because of this change, we have to define b as a new parameter. We accept the temporary default value of 0, i.e. no friction. This can be easily changed later.

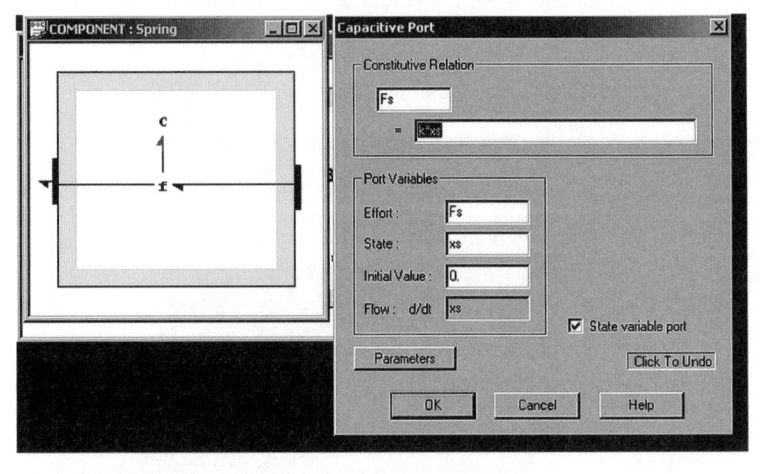

Fig. 6.7. The Spring component model

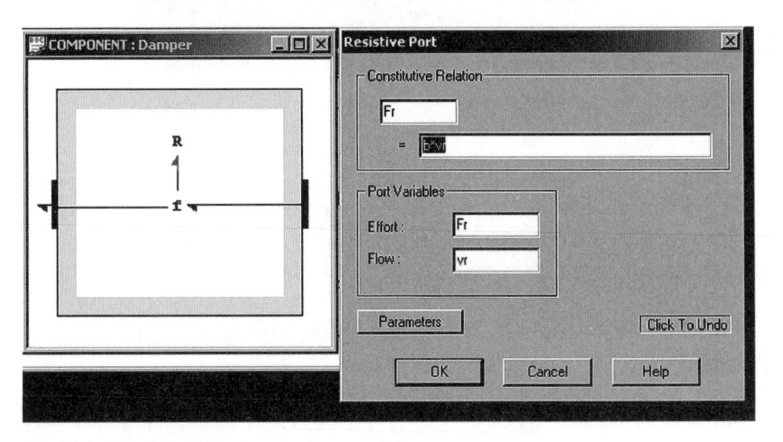

Fig. 6.8. The Damper component model

We create the Wall model in a similar way. The wall fixes one side of the spring and one side of the damper. This implies zero velocity at the spring and the damper ends. Thus, in the Wall document (Fig. 6.9), we create an effort junction e connected to a zero-flow source SF. The other two junction's ports are connected to the Wall document ports and, thus to the spring and the damper ports (Fig. 6.4). The effort at the flow source port is equal to the sum of the forces at the other two junction ports, i.e. it represents the force transmitted to the Wall.

The value generated by the Force component (a source effort) is at creation set to zero, i.e. there is no force. We change only the name of the effort to F (Fig.

6.10). The constitutive relation for the force will be changed later when analysing the system behaviour.

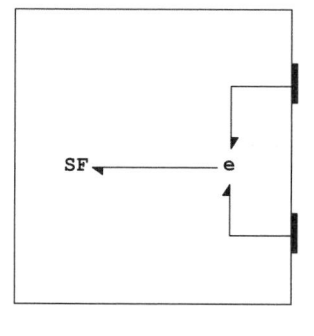

Fig. 6.9. The model of the Wall

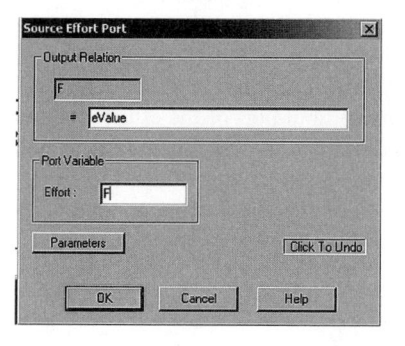

Fig. 6.10. The applied force *Source Effort Port* dialogue

The model of the **Body Spring Damper** system is nearly finished. It is still necessary to define the variables to monitor the system behaviour. These could be the force acting on the body and the body position. We also might be interested in the total force transmitted to the wall. We thus create three **Output** components with a single input control port each, to generate of the x-t plots. [1] We rename these components **Position, Force** and **Wall force**, respectively (Fig. 6.11).

We next define the signals fed to these display components by creating two control-out ports at the periphery of the **Body** component. These are connected to the **Position** and the **Force** output components (Fig. 6.11). We also create a control-out port at the **Wall** and connect it to the **Wall Force** component. To define the labels used for the x axes of the plots, double-click the input ports of the output components and in the dialogues that appear, change the default symbols to **x, F**

---

[1] It is also possible to create a single Output component with several input ports.

and Fw respectively. To define these signals, we have to open the Body and Wall components again and make the necessary changes there.

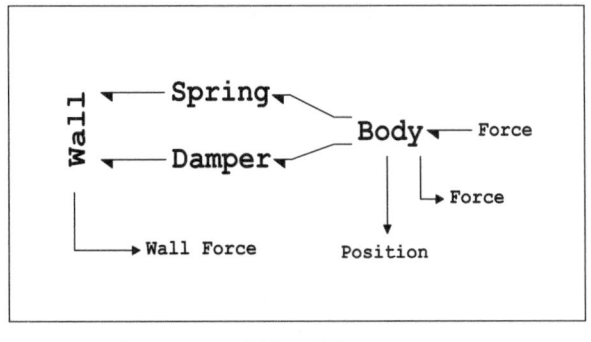

Fig. 6.11. The output variables of the system

The new model of the Body is shown in Fig. 6.12 (compare to Fig. 6.6). Integrating the velocity provides information on the body position. Hence, to get the velocity we create a control-out port at the effort junction. Next, we create an Integrator component, depicted by the usual integration symbol, and add input and output control ports to it. We connect the control output port of the effort junction to the integrator input, and the integrator output port to the document port, which on the other side, is connected to the Position output component (Fig. 6.11).

The initial body position is defined by double-clicking the integrator output port (Fig. 6.13). By default, the initial value is 0.0.

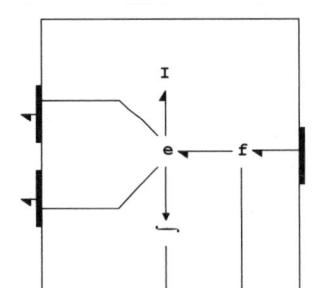

Fig. 6.12. Creating access to the Body output variables

Finally, to create access to the force applied to the body, we insert a flow junction between the Body document port, where the force is applied, and the effort junction port. Next, we create a control-out port at the junction and connect it to the other Body output document port (Fig. 6.12), which is connected on the other side to the Force display component.

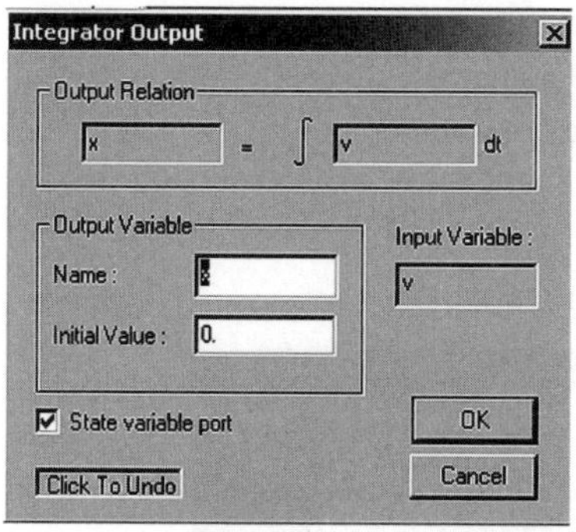

Integrator Output ☒

Output Relation

$$x = \int y \, dt$$

Output Variable
Name : x
Initial Value : 0.

Input Variable :
y

☑ State variable port

[Click To Undo]   [OK]   [Cancel]

Fig. 6.13. The *Integrator Output* port dialogue

In a similar way we create access to the signal of the total force transferred to the wall (Fig. 6.14).

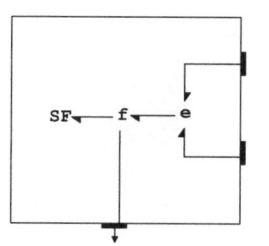

SF ◄── f ◄── e

Fig. 6.14. Generating access to the force transmitted to the Wall.

In this way, the bond graph model of the system is developed as a two level model, i.e. at system level and at the level of component models.

Before we finish with the modelling stage, we check the connections. To check the connection of the Position output component, we select its port and click the *Joined Port* toolbar button. The connection bond between the Position port (at the system level) and the Integrator output port (at the Body component level) is shown in blue. By clicking the end ports we can walk between these two components.

The Use of Mechanical Symbols

The Body Spring Damper System can also be described by the familiar mechanical symbols (Sect. 4.6.5). The procedure is quite similar to that of the previous subsection. This time we create a new project entitled "Body Spring Damper – Mechanical Model". Next we open the *Mechanical Component* palette using the command *Use Icons/Mechanical* on the *Tools* menu.

This palette is used to create the components represented by the familiar symbols for springs, dampers, bodies etc. in a way similar to using the *Editing Palette* (Fig. 6.15 on the right). We simply click on the component icon in the palette and then click at a position in the document. The components created already contain the ports and also the appended text labels. These are word models but represented by graphically!

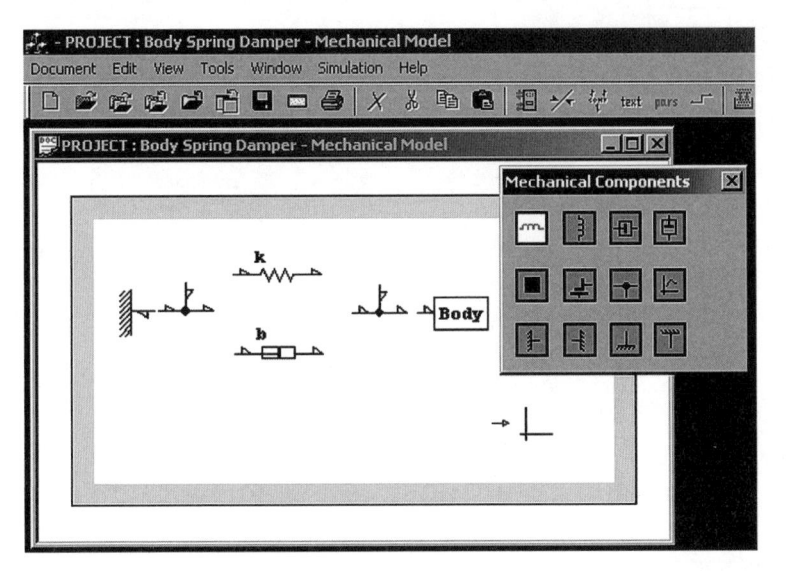

Fig. 6.15. Use of the mechanical symbols for modelling of Body Spring Damper system

There are small differences from the usual word models. Firstly, the ports are fixed in place. Also, the text is limited in the length, the font size and the orientation.

We now develop a bond graph model using this palette. Starting from the left (see Fig. 6.1) we create the wall and ground component using the *Ground vertical left* button in the palette. Next, we use the *Node* button to create a node that is used for joining the left spring and damper ends,. Then we create spring and damper objects using the corresponding *Spring horizontal* and *Damper horizontal* buttons. An additional node for joining the right spring and damper ends is also created. Finally we create the body component using the *Body* button. We also create a display (output) component, represented by the x-y axes symbol, using the *X-Y Plotter* button (Fig. 6.15). This component is used for displaying all the output signals. The force generator will be created later.

In this case, the default sense of power flow is different from the anticipated power flow through the system. We change it by changing the power ports from power-in to power-out or *vice versa*. We do this in the usual way: using the toolbar *Change Port* button. We can also change all ports by a single command. Thus, to change the ports of the spring component, we select it, and then chose the command *Change Ports All* on the *Edit* menu.

It would be nice to move the right port at the left node to bottom, as well as the left port of the right node. But the node ports are fixed and cannot be moved. Thus, we remove them and create corresponding node ports using the *Editing palette*. When the palette opens, the *Mechanical Components* palette disappears, for it no longer is needed.

We continue with model development using the palette in the usual way. We also create a power-in port on the right side of the body and the Force component (source effort) in the same way as we did it in the previous subsection (Fig. 6.16).

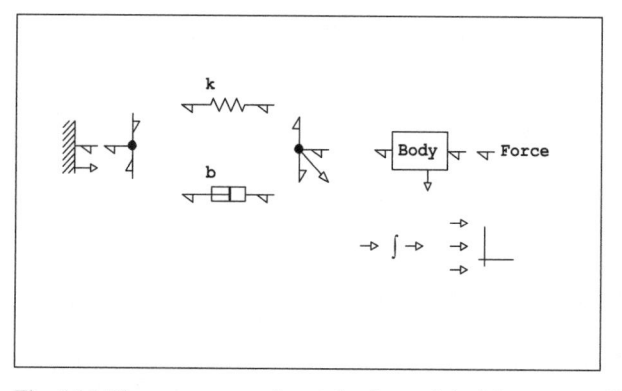

Fig. 6.16. The components that make the model of the system of Fig. 6.1

We need to generate the outputs, too. The x-y display component has been created already (Fig. 6.15), so we must create the ports that will be used to feed in the signals. The body velocity is available at the right node. We thus create the control-out port at the bottom of the node. In addition, we create the integrator component with input and output control ports (Fig. 6.16). The applied and the wall forces are not directly available. A technique similar to that used in the last subsection is used.

Connecting all of the components we get the basic model (the first level) of the Body Spring Damper system (Fig. 6.17). The models of the components that constitute the system are developed in a way similar to the last section. There are small differences between Figs. 6.17 and 6.11.

Because of the nodes in Fig. 6.17, the forces applied to the body and to the wall are the sum of the damper and the spring forces. Thus, the wall component of Fig. 6.17 has only one power input port and there is no effort junction, as in Fig. 6.14. Similarly, the body component has only one power-out port, in comparison with Fig. 6.12 and the integrator is put outside of the body component. The spring and the damper are represented by the same models in Figs. 6.7 and 6.8.

There is another difference. We use a single output (display) component with several input ports. By default, all variables connected to output component are plotted on the y-axis and time on the x-axis. We can change this easily by double clicking the port. A dialogue (Fig. 6.19) appears, in which we can assign variables to the x or y-axis. We can also choose *None* if we wish to remove it from the plot.

There are also some restrictions. Only one of the variables can be plotted on the x-axis and all others are y-axis variables.

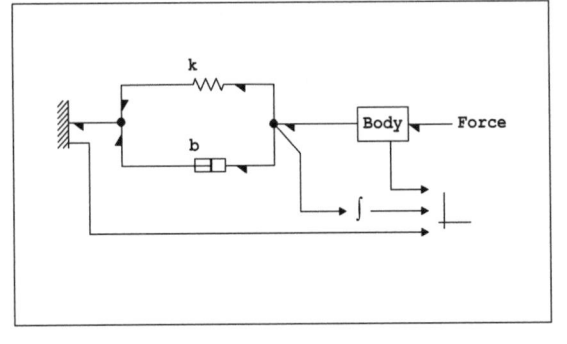

Fig. 6.17. Alternative model of the Body Spring Damper system

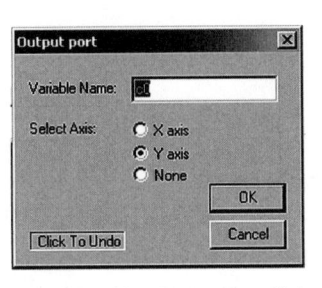

Fig. 6.18. The *Output Port* dialogue

We suggest that the reader to compare this model with that of the previous sub-section. The models are stored in the BondSim project library and can be taken from the library by using the command *Get From* in the *Project* menu.

6.2.3 Analysis of the System Behaviour by Simulation

In previous sections the bond graph model of the Body Spring Damper system was developed. The model closely resembles the real system. By running simulations under different conditions, we can observe the time behaviour of different variables in a way that is analogous to observing behaviour of the real system *via* the outputs of instruments. To analyse the dynamical behaviour of the system, we thus change the inputs and parameters, and study the effects.

Building the Model

We must *build* the mathematical model that will be solved during the simulation runs. To build the model, the modelling project must be opened; the *Build* button on the toolbar then is clicked, or the *Build Model* command on the *Simulation* menu selected. We analyse the Body Spring Damper system using the Body Spring Damper Problem project developed in Sect. 6.2.2 We set the parameters of the model to the following values:

1. The body mass m= 5 kg
2. The spring stiffness k = 112.5 kN/m
3. Linear damping value b=150 N s/m
4. The input force F=500 N

Denoting the natural frequency of the system as

$$\omega_n = \sqrt{\frac{k}{m}} \tag{6.1}$$

we get ω_n= 150 rad/s = 23.87 Hz. Similarly, the damping ratio

$$\zeta = \frac{b}{2\sqrt{mk}} \tag{6.2}$$

has a value of 0.1.

After the equations are built, we display them in the message window at the right lower corner of the main window (Fig. 6.19) using the toolbar button *Equations*, or using the command *Show Model/Model Equations* on the *Simulation* menu.

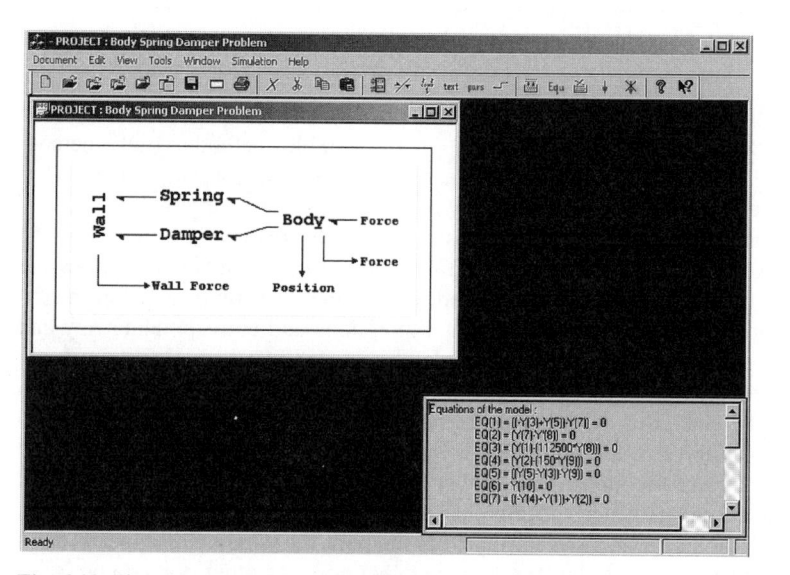

Fig. 6.19. The creation of the model equations

The list of machine-generated equations of the system's mathematical model is shown in Fig. 6.20. In the first part, the model equations are written in implicit form. Next is the list of variables that appear in the model as time derivatives. The others are purely algebraic. The equations are not actually used in this form internally in the program, but rather in byte form that permits much more efficient evaluation. To display them in a familiar mathematical form, the model equations are first decompiled and simplified. The equations can also be printed to a file in ASCII format by clicking on the window using the right mouse button, then selecting the command *Print* from the menu that drops down.

```
Equations of the model :
      EQ(1)  =  ((-Y(3)+Y(5))-Y(7))  = 0
      EQ(2)  =  (Y(7)-Y'(8))  = 0
      EQ(3)  =  (Y(1)-(112500*Y(8)))  = 0
      EQ(4)  =  (Y(2)-(150*Y(9)))  = 0
      EQ(5)  =  ((Y(5)-Y(3))-Y(9))  = 0
      EQ(6)  =  Y(10)  = 0
      EQ(7)  =  ((-Y(4)+Y(1))+Y(2))  = 0
      EQ(8)  =  (Y(3)-Y(10))  = 0
      EQ(9)  =  (((-Y(1)-Y(2))+Y(6))-Y(11))  = 0
      EQ(10) =  (Y(11)-Y'(12))  = 0
      EQ(11) =  (Y(12)-(5*Y(5)))  = 0
      EQ(12) =  (Y'(13)-Y(5))  = 0
      EQ(13) =  (Y(14)-Y(5))  = 0
      EQ(14) =  (Y(6)-500)  = 0
List of differentiated variables :
      Y(8),  Y(12),  Y(13)
```

Fig. 6.20. Listing of machine generated model equation

The mathematical model is in the form of a system of differential-algebraic equations that is solved during simulation by suitable methods (BDF method, Chapt. 5). During model building the matrix of partial derivatives (Jacobian) of the system equations is also symbolically generated and stored internally.

This matrix of partial derivatives is also available for reviewing by selecting the *Show Model/Matrix* command on the *Simulations* menu, which decompiles the matrix and display it in the message window. The matrix is represented in the coordinate form, i.e. as the matrix element expressions and corresponding row and column index values (Fig. 6.21). It can also be printed to a textual file similar to the model equations, i.e. by clicking on the window using the right mouse button.

The model equations are expressed using the machine-generated variables. To see which variables in the equations correspond to which variables in the bond graph model, place the mouse cursor over an elementary component port. A small window then appears giving information on the variables stored at the port (Fig. 6.22). When the cursor is moved away from the port, the window disappears. Such a window appears only at elementary component ports, as other ports do not store variables.

```
Matrix of partial derivatives :
      J(1) = -1
      IRow(1) = 1      JCol(1) = 3
      J(2) = 1
      IRow(2) = 1      JCol(2) = 5
      J(3) = -1
      IRow(3) = 1      JCol(3) = 7
      J(4) = 1
      IRow(4) = 2      JCol(4) = 7
      J(5) = -cj
      IRow(5) = 2      JCol(5) = 8
      J(6) = 1
      IRow(6) = 3      JCol(6) = 1
      J(7) = -112500
      IRow(7) = 3      JCol(7) = 8
      . . .
```

Fig. 6.21. Part of the system matrix of partial derivatives

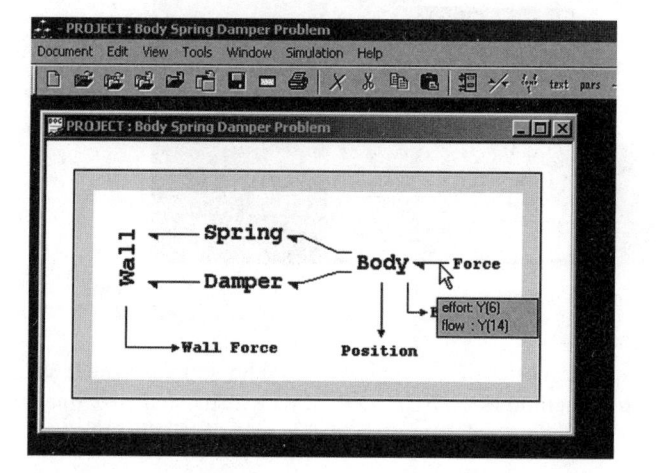

Fig. 6.22. A window showing the variables stored in the port

Running Simulations

The model is created with zero initial conditions for all variables. We changed only the value of the input force. Simulation of such a model corresponds to simulating the system step response to the applied force. To start, we click the *Run* toolbar button, or choose the *Run* command on the *Simulation* menu. The command display a dialogue used to define the necessary simulation parameters, such as the duration of the simulation run, the increment at which output values will be generated, error tolerances etc. (Fig. 6.23).

The natural frequency of the system is about 24 Hz, hence the period is about 0.04s. We can expect the transient to die out after about five periods, or 0.2 s. We use a somewhat greater simulation period, e.g. 0.5 s, to be sure that the transient has settled down. By typing in the simulation period, the output interval and the maximal step-size changes automatically to *one hundredth* of this value, i.e. to 0.005 s. We accept this value. The maximum step-size should be greater than the output interval. It is usually set to the same value as the output interval. We can choose a smaller value for the output interval if we wish to get better resolution and smoother diagrams. This means, however, that there are more calculations and the simulation run needs more computer time and memory resources.

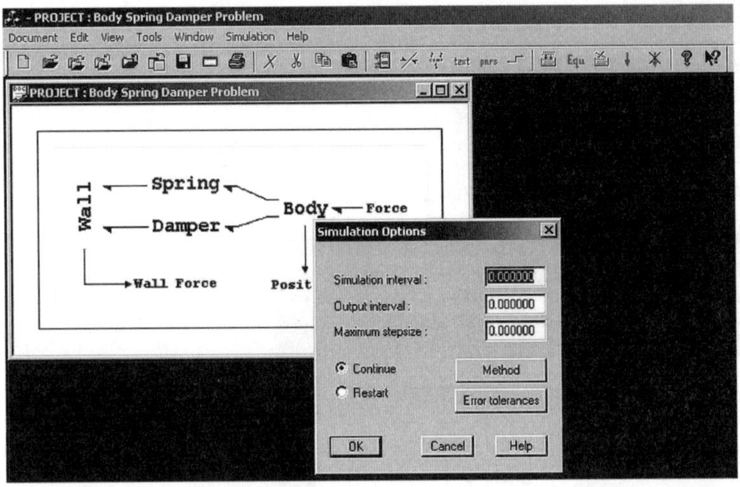

Fig. 6.23. The simulation parameter dialogue

Default error tolerances are set to 10^{-6} for both the absolute and the relative errors. The default integration method is the *BDF* method with a sparse matrix linear equation solver (Sect. 5.3). Accepting the simulation data by pressing the OK button, the simulation starts.

When the simulation starts, a message at the right bottom part of the main screen informs the user of the integration method used. Simultaneously, at the right end of the status bar, a progress bar appears. This shows the progress of the simulation. When the simulation finishes, the progress bar disappears and a message informs the user of this fact. To see the results of the simulation we double-click on the output components. This opens windows that present results as x-y or x-t plots. Thus, opening the Position component of Fig. 6.19 we get the time history of the body position as shown in Fig. 6.24.

We can examine the values of the variables by clicking inside the plot. Horizontal and vertical lines appear, which intersect on the curve. The corresponding values of the variables appear below the plot, as seen in the Fig. 6.24.

It is also possible to get a list of output values by clicking on the plot with the *right* mouse button. A drop-down menu appears from which we can choose the *Show data* command. This command creates a list box containing values of the variables. By scrolling through the list box we can examine values of variable pairs generated by the simulation. It should be noted that the number of value pairs is limited. The values listed are symmetrically with respect to the current value selected by the mouse. If there is a large number of output values, the values throughout the complete range can be examined by a combination of mouse clicking and the use of the list box.

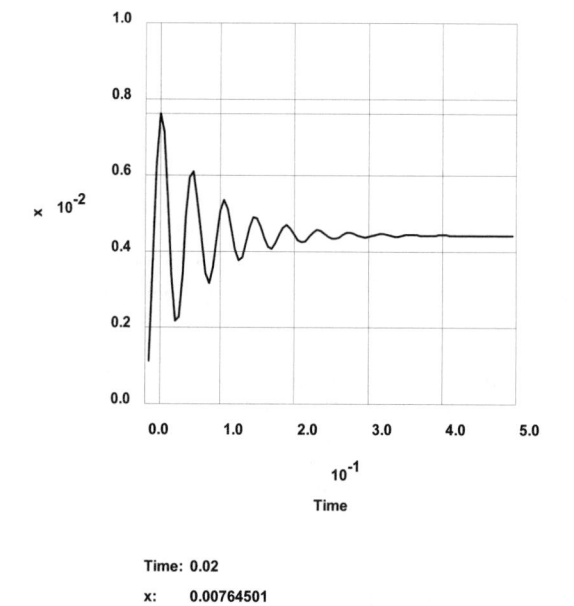

Time: 0.02

x: 0.00764501

Fig. 6.24. The response of the body position to a step in the applied force

There is also another thing worth mentioning: The output windows can be opened before we start the simulation run. In that case, an empty plot appears. Using the *Set axes* command on the menu *Simulation*, we can change the setting of the plot axes used—i.e. quadrants used, the range and the names of the variables plotted. Of these, the range is the least important, as the program automatically calculates and resets the range during the simulation. The setting of axes also can be changed when simulation ends. When the simulation starts the output values are displayed as the simulation advances. This way, we can observe the behaviour of the system during simulation. This makes the simulation somewhat longer, but this usually doesn't matter, as much can be gained by viewing how the processes develop.

In present example, we can compare the values obtained by simulation with the exact solution of the problem under study. Recall that the equation of motion of the Body Spring Damper system can be written as

$$m\frac{d^2x}{dt^2} + b\frac{dx}{dt} + kx = F \tag{6.3}$$

Solving the equation the response to a step F of force can be written as (see e.g. [1])

$$x = x_0 \left[1 - \frac{e^{\zeta\omega_n t}}{\sqrt{1-\zeta^2}} \cos(\omega_n \sqrt{1-\zeta^2} t - \varphi) \right] \tag{6.4}$$

where

$$x_0 = F/k \tag{6.5}$$

and

$$\tan\varphi = \frac{\zeta}{\sqrt{1-\zeta^2}} \tag{6.6}$$

Table 6.1 compare values obtained by simulation and by direct calculation according to Eqs. (6.4)–(6.6). The values are rounded to ten figures. The absolute errors are within the absolute error tolerance (10^{-6}). Better agreement can obtain by increasing the error tolerance.

Table 6.1. Response of the body position to a step in the applied force

Time in s	Simulation	Eq.(6.2)
0.01	0.003761887676	0.003761880557
0.05	0.003448204502	0.003448187707
0.1	0.005076850687	0.005076930584
0.2	0.004465699802	0.004465689113
0.3	0.004406268287	0.004406235937
0.4	0.004455428887	0.004455470625
0.495	0.004444573861	0.004444576945

We also look at the simulation statistics using the *Show statistics* command on the *Simulation* menu. The results are summarized in Table 6.2. The simulation was relatively efficient using the maximal order of the method (5) at the end of the simulation.

Table 6.2. The simulation statistics

Name	Value
The order of the method at the last step	5
The number of the steps taken	431
The number of the function evaluations	901
The number of the partial derivative matrix evaluations	468
The relative error tolerance	10^{-6}
The absolute error tolerance	10^{-6}
The elapsed time s	0.290 (0.391)

Note: The simulation was conducted on a laptop with a Pentium III 650 MHz processor. The elapsed time is for the simulation runing in the backgound. In parenthesis is given the time with the plot window opened.

Response to Harmonic Excitation

The study of mechanical vibration pays great attention to the influence of forces that vary harmonically (see e.g. [1]). We can study such influence by defining a force at the power port of the Force component (Figs. 6.10 and 6.11) as

$$F = F0 * \sin(omega * t) \tag{6.7}$$

The amplitude F0 = 500 N and frequency omega of the applied force are parameters, which can be set to any appropriate value.

We study the effects of the harmonic force on the system by changing its frequency. The simulation interval is chosen as 1.0 s and the output interval is 0.001 s for better resolution. Some typical results are given in Figs. (6.25)–(6.28).

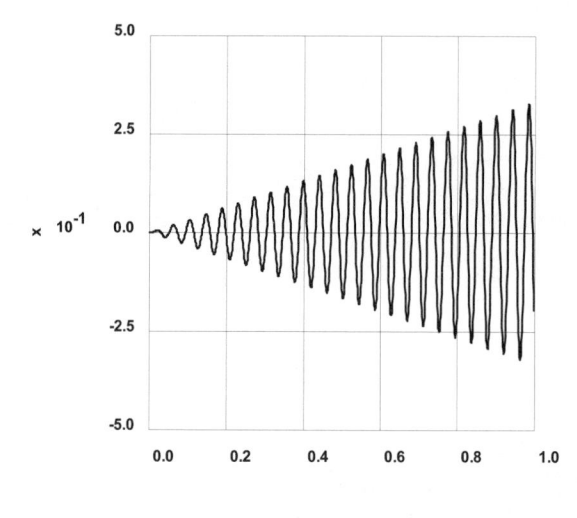

Fig. 6.25. Resonant response of the Body Spring Damper system (ω_n = 150 rad/s, ζ = 0)

In vibrations as well in other fields an important role in design is played by amplitude-frequency and phase-frequency diagrams. Determination of such diagrams is not an easy task particularly for non-linear systems (see [2] for detailed discussion). In the case of linear systems such as the Body Spring Damper system these diagrams can be found from the Fourier transform of the impulse response of the system [1,2]. We confine further attention to the evaluation of the amplitude-frequency diagrams only.

The impulse response of the system can be approximated as the time response to a very short force pulse of unit area (Fig. 6.29). Such a pulse can be described as

$$F = \begin{cases} F_p, & t < T_p \\ 0, & t \geq T_p \end{cases} \tag{6.8}$$

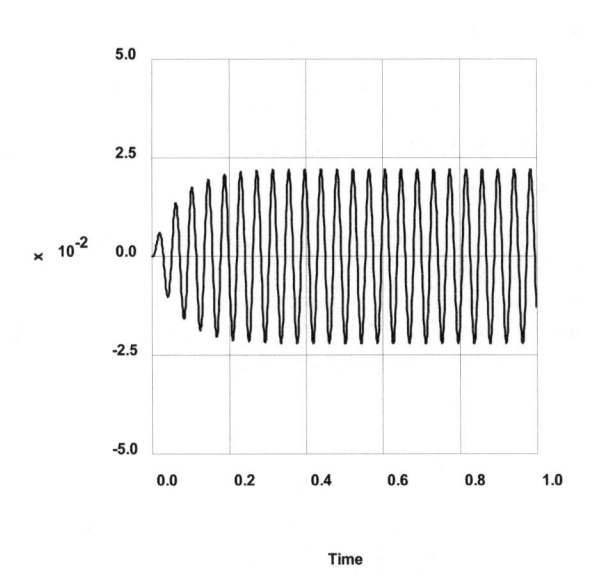

Fig. 6.26. Resonant response of the Body Spring Damper System (ω_n = 150 rad/s, ζ = 0.1)

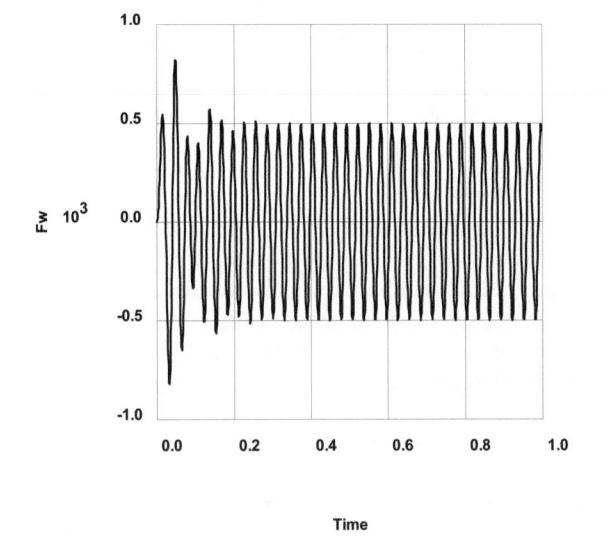

Fig. 6.27. Force at the wall at a frequency of 212.1 rad/s ($\sqrt{2}\omega_n$)

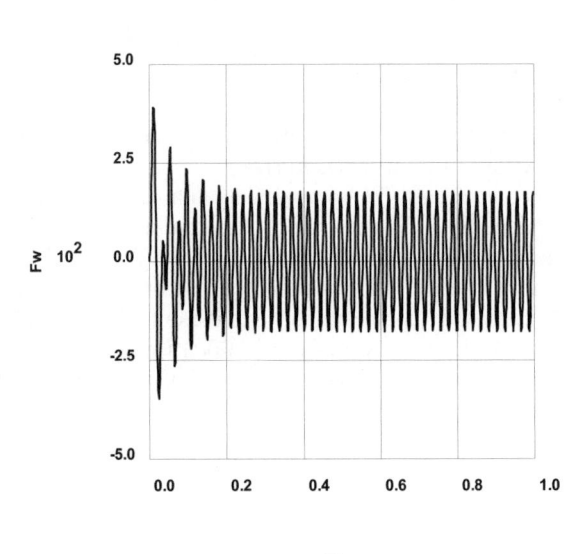

Fig. 6.28. Force at the wall at a frequency of 300 rad/s

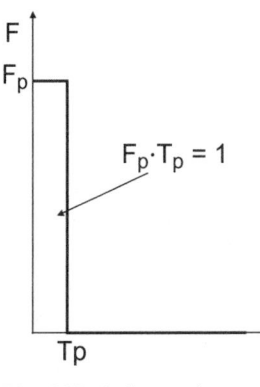

Fig. 6.29. A short unit area pulse as an approximation of a unit impulse

Using the question operator of Sect. 3.5, the relationship for the pulse of force can be expressed as

$$F = t < Tp \; ? \; Fp : 0 \qquad (6.9)$$

The pulse duration parameter Tp should be very short and the strength Fp should be such that their product is equal to one

Let the response to such a pulse be denoted as h(t). Then the frequency response of the system is given by the Fourier transform

$$X(j\omega) = \int_{-\infty}^{+\infty} h(t)e^{j\omega t}dt \qquad (6.10)$$

We evaluate the Fourier transform using the Fast Fourier Transform (FFT) method and find the amplitudes as functions of frequencies in Hz.

The FFT treats the function as periodic. To approximate an aperiodic function, we can extend the function values by adding zeros [2]. As more zeros are added, better resolution in the frequency response is obtained. To create the impulse response of the system, we select a simulation interval of 1 and an output interval of 0.001 s. This is because we expect the impulse response to die out well before 0.5 s (see Fig. 6. 24). The duration of the pulse we take as 0.0001s, i.e. one tenth of the output interval. The impulse response of the body position is given in Fig. 6.30.

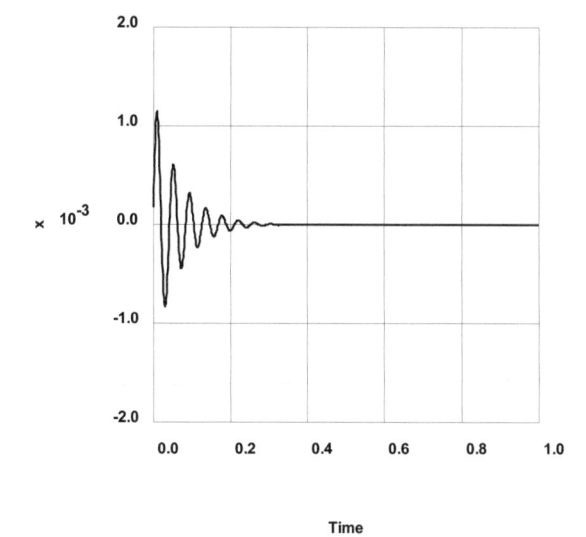

Fig. 6.30. The impulse response of the body position

To create the corresponding frequency response, click the right mouse button on the plot and select from the drop-down menu the *Frequency response* command. The response created covers frequencies up to 500 Hz. We expand a part of it by clicking on the plot using the right mouse button again, selecting the *Expand* command, then setting the range from 0 to 100 Hz. The resulting diagram is shown in Fig. 6.31.

We again compare simulation values with the exact solution. Recall from Eq. (6.3) that the amplitude of the frequency response to the unit impulse force is given by (see e.g. [1])

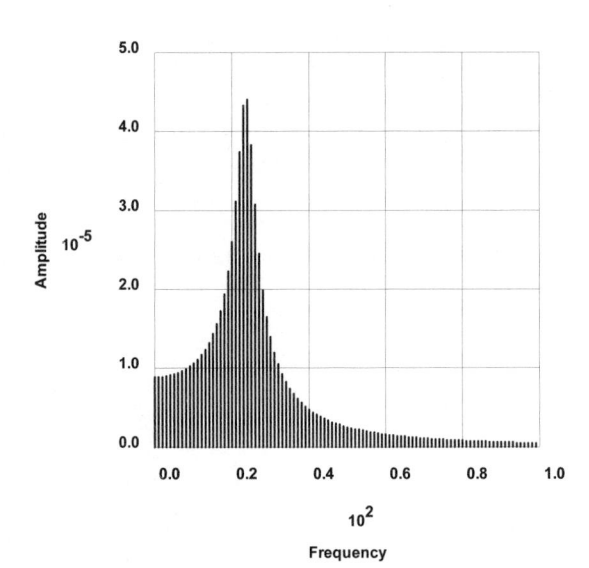

Fig. 6.31. The frequency response of the body position

$$\left|X(j\omega)\right| = \frac{1/m}{\sqrt{(\omega_n^2 - \omega^2)^2 + 4\zeta^2\omega^2\omega_n^2}} \qquad (6.11)$$

After normalization of the frequencies, we get

$$\left|X(j\omega)\right| = \frac{1/k}{\sqrt{\left[1-(\frac{\omega}{\omega_n})^2\right]^2 + 4\zeta^2(\frac{\omega}{\omega_n})^2}} \qquad (6.12)$$

Using the last expression, we can calculate the amplitude at the same frequencies as that obtained by simulation. The amplitudes obtained by simulation and by Eq. (6.12), rounded to six figures, show good agreement (Table 6.3).

Table 6.3. Amplitude of the frequency response of the body position (m)

Frequency in Hz	Simulation	Eq.(6.12)
0.0	$8.87694 \cdot 10^{-6}$	$8.88889 \cdot 10^{-6}$
10.01	$1.07177 \cdot 10^{-5}$	$1.07296 \cdot 10^{-5}$
24.02	$4.40747 \cdot 10^{-5}$	$4.40782 \cdot 10^{-5}$
50.05	$2.61028 \cdot 10^{-6}$	$2.59828 \cdot 10^{-6}$
99.10	$5.59191 \cdot 10^{-7}$	$5.46925 \cdot 10^{-7}$

The impulse response and the frequency response are quite sensitive to the duration of the input pulse. Thus, using the pulse width equal to the output interval, the amplitude obtained agrees with the exact values to one single significant figure. This is to be expected because the Fourier transform of a pulse is not equal to one but is *sinc* function [2].

The Fourier transform of the pulse can be obtained in the same way as the impulse response the system. This time we use a simulation interval of 0.0005 s and the output interval of $1.0 \cdot 10^{-6}$ s and restart the simulation. Then, by opening the Force plot and taking the frequency response command with right mouse button, we get the transform we wish. We also expand only part of the frequency range from 0 to $1 \cdot 10^{5}$ Hz. The results are shown in Fig. 6.33. The well-know *sinc* function is clearly visible.

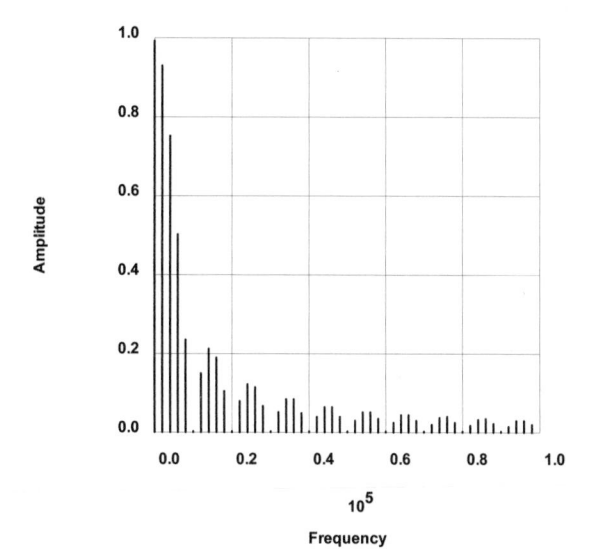

Fig. 6.32. The Fourier transform of the input force pulse

It should be noted that the responses obtained are *not* normalized to the dc value (zero frequency) and are close to the form obtained by direct measurements on the physical system.

6.3 Effect of Dry Friction

We continue with the study of vibration systems by analysing the influence of *dry friction*. This is usually treated by separately studying the motion in the positive and negative directions, and the conditions under which the motion ceases [1]. We will develop a model of dry friction that can be used for the prediction of such a motion. The model can be used for the analysis of more general systems in which this type of friction is important.

6.3.1 The Model of Dry Friction

So-called *dry friction* occurs when one solid slides over another. The laws governing such friction date back to Leonardo da Vinci (1452-1519), but are better known by the work of Coulomb in 1785 (Fig. 6.33). A modern exposition of the theory is given in [3]. In the bond graph literature there were also attempts to model dry friction. Thus in [4] the friction around zero velocity is modelled as a force dependent on motion and, outside of this region, as a function of velocity only. In [5] discontinuous laws of friction were modelled by sinks of fixed causalities. The modelling of friction was analysed also in [6], motivated by the physical theory of [3]. The model proposed consists of capacitors and resistors interconnected by transformers, whose ratios change smoothly from one to zero, depending on the state of the body motion.

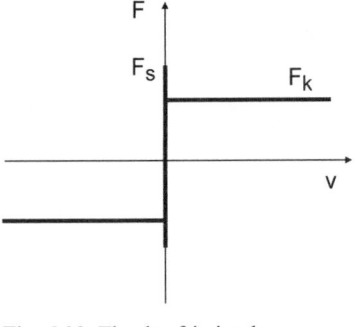

Fig. 6.33. The dry friction law

We also use the theory of [3], but friction will be modelled by a single element that imposes restrictions on the body motion. A suitable element for this is the switch element (Sect. 2.5.8), as it is capable of imposing the zero flow (sliding velocity) condition before motion commences, and also when the motion ceases. It also accounts for constant—or possibly variable—effort (friction force) during the motion.

The contact area between bodies generally is not smooth, but is actually rough Fig. 6.34. Thus, when a body is pressed onto another, the real contact starts at the tips of the highest asperities. These deform until the area of contact is large enough to support the load without yielding. Owing to high pressures and temperatures at the contacts, solid junctions are created at the contact places known as *micro welds*.

To slide one body over another body (ground), it is necessary to apply a force in the direction of the sliding that is large enough to break the micro welds. Until the motion commences, the body is in static equilibrium under the action of the applied forces, including the other body's reaction force (Fig. 6.35).

The component of the reaction force F_n that is normal to the sliding direction is the usual *normal reaction*, and the component in the direction of sliding is the *friction force*. It is denoted simply as F.

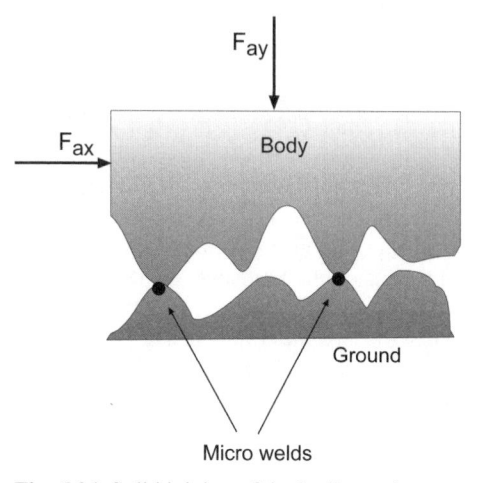

Fig. 6.34. Solid joining of the bodies at the contact points

Thus, the conditions of equilibrium can be stated as

$$\left.\begin{array}{l} F_{ax} + F = 0 \\ F_{ay} + F_n = 0 \end{array}\right\} \tag{6.13}$$

and

$$\left.\begin{array}{l} v_x = 0 \\ v_y = 0 \end{array}\right\} \tag{6.14}$$

where v_x and v_y are the velocity components of the body relative to the other body (ground).

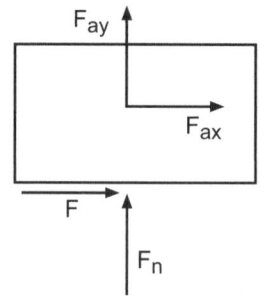

Fig. 6.35. The forces acting on the body

The limiting force value at which the welds break and hence sliding commences is termed the *static friction* F_s. Thus under conditions of Eq. (6.13) the condition that the friction force satisfies is

$$|F| \le F_s \tag{6.15}$$

This limiting value, corresponding to the strength of the micro welds, depends also on the time that the bodies are in the contact. When the bodies are in contact there is migration of particles over the junction by diffusion, and this needs some time. If there is enough time, intimate junctions are established and the two bodies behave as single body. This way, the static force increases to a maximum value that the micro junctions can sustain. When we speak of static friction, we mean this full strength value. The Coulomb law relates this limiting force to the normal reaction by

$$F_s = \mu_s F_n \tag{6.16}$$

The μ_s is the well-known static friction coefficient.

When motion starts, there is deformation of the asperities that are micro welded until they break. Perhaps before complete break down other asperities come into contact and new micro welds form. They, in turn, break and yet other micro welds form and the process continues as the body slides. During the motion there is not enough time for migration across junctions, hence the force of asperities breaking during body motion is somewhat less than the static friction value. According to experimental evidence, this force is more or less independent of the sliding velocity and has the sense opposite to the velocity. During sliding, friction satisfies the relationship

$$\left.\begin{array}{l} F + \mu_k F_n = 0, \quad v > 0 \\ F - \mu_k F_n = 0, \quad v < 0 \end{array}\right\} \tag{6.17}$$

Here the μ_k is the *kinetic coefficient of friction*, which is somewhat less then the static coefficient, typically about $20 - 25\%$.

To complete the model of friction at this stage, we need to define the state at the very moment when motion commences, which ideally occurs at zero velocity (Fig. 6.33). This is not covered by Eq. (6.17) and hence we add this condition to Eq. (6.15). The friction characteristics valid from no sliding until the sliding commences read now as

$$\left.\begin{array}{l} v = 0, |F| \le F_s \\ F - \mu_k F_n = 0, \quad F > \mu_s F_n \\ F + \mu_k F_n = 0, \quad F < -\mu_s F_n \end{array}\right\} (v = 0) \tag{6.18}$$

This equation is compatible with Eq. (6.17) that is it is valid for the sliding. The Eqs. (6.17) and (6.18) are constrains on the body motion imposed by the mechanism of dry friction. We have not tried to find an explicit relationship for the friction force, as this is difficult to find in the general case. We don't know if the force at the contact is known or not, and similarly for the velocity. Hence, the effect of friction is represented by a simple linear implicit equation, different for the body sliding and not sliding.

We dwell a little more on Eq. (6.18). To find out whether it is applicable for a given state of the body motion, we need to test the sliding velocity against the zero value. In computer arithmetic it is, in general impossible to find out the exact moment when some variable attains a definite value. Thus, some tolerance on the velocity around zero is necessary. In parallel with, from the physical side there is no such thing as "rest". The theory of [3] proves that the static friction behaviour holds also when the sliding is very slow. Thus, we reformulate the above relations to be valid if |v|<tol. The tol represents a tolerance, which we take to be much smaller than the error tolerance used for simulating motion and close to the machine *epsilon*.[2] In the BondSim there is a predefined parameter BG_EPSILON, defined as four times the machine epsilon. This can be used if such a low tolerance is needed.

The relationships given by Eqs. (6.17) and (6.18) can be represented in the form that is used when describing the element constitutive relations (Sect. 3.5). The relations can be compactly expressed using the question ('?:') operator. The relationship reads

$$\mathrm{abs}(v) < \mathrm{tol}\,?(\mathrm{abs}(F) <= \mathrm{mus}*Fn\,?\,v : (F >= 0\,?F-\mathrm{muk}*Fn :$$
$$F+\mathrm{muk}*Fn)) : (v >= \mathrm{tol}\,?F+\mathrm{muk}*Fn : F-\mathrm{muk}*Fn) = 0 \qquad (6.19)$$

In the equation instead of the Greek symbol μ, mu is used, and also the indices are written on the same line. The program supports only Latin characters, without any formatting.

The statement can be understood as:

if (the absolute value of v is less than a tolerance) then
    if (the absolute value of F is less than or equal to the static friction) then
        v = 0
    else if (F is greater then or equal to zero) then
        F- muk*Fn = 0
    else
        F+muk*Fn = 0
else if (v is greater than or equal to the tolerance) then
    F+muk*Fn = 0
else
    F-muk*Fn = 0

The left side of Eq. (6.19) looks like a C language statement. It states in a compact form the conditions imposed by dry friction on the rest of the system. The mechanism of friction is not a trivial one. There are five possible states through which the system goes during its motion. Unfortunately, this is not the end of the story. We have to define also the conditions for motion stopping. Otherwise, such a model is not of much use in simulations.

In books on vibrations such as [1,7], the motion stops when the amplitude of the spring force acting on the body is less than the static friction. This is not a precise enough statement. According to [3,6], when the body stops, friction does not achieve the static value again, but at most the kinetic value. To find a more precise

---

[2] By definition the machine epsilon is the smallest positive number ε such that $1+\varepsilon \neq 1$ in machine arithmetic and for the given floating-point number representation. It is for double precision numbers about $2.2 \cdot 10^{-16}$.

statement of the stopping condition, we need to express it in terms of the variables at the interfaces of the two bodies, i.e. the force and the velocity at contact.

We look more carefully at what happens when the body velocity changes its sign. During the motion there is dynamic equilibrium of the forces acting on the body, including the body inertia force. Suppose that the active force is acting in a manner that reverses the body motion. When the velocity changes its sign, the sense of the friction force changes too. Hence, the body acceleration changes abruptly. If the active force on the body is outside the kinetic friction limit, the acceleration will not change its sign and hence the body will continue to move in the opposite direction (Fig. 6.36a).

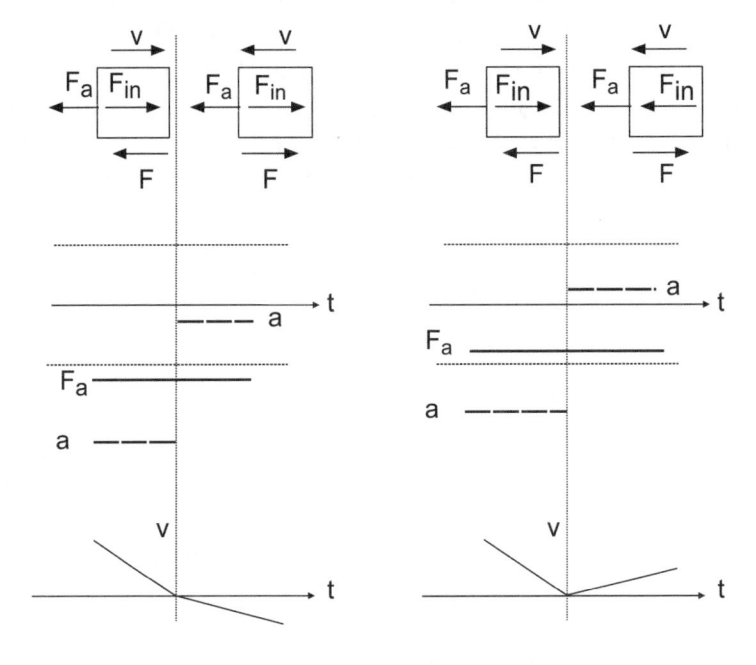

(a)   (b)

Fig. 6.36. Effects when the velocity changes sign and if the body active force is:
(a) outside the friction limit, (b) inside the friction limit

If, on the other hand, the active force is within the kinetic friction limit, the body acceleration will change its sign and the body will try to move back (Fig. 6.36b). This, in effect, means that the friction force tries to push the body back, which is impossible because dry friction is not capable of delivering positive power to the body. The dynamic equation of motion and the characteristics of dry friction as given by Eq. (6.19) thus are incompatible. Looking from the viewpoint of the asperities, there is no abrupt change in the friction force. However, owing to the asperities stiffness and internal damping, the body first slows down and stops. It continues to move in the opposite direction only if the absolute value of the total active force on the body is greater then the asperities breaking force, i.e. the kinetic friction.

Thus, to solve this problem we must apply additional requirements. In particular, if the power delivered by dry friction is *positive*, the velocity of sliding should be equal to zero. This is stated as follows

$$F \cdot v > 0 \Rightarrow v = 0 \qquad (6.20)$$

Adding this requirement, the final form of the dry friction constitutive relation reads (see Eq. (6.19))

$$abs(v) < tol? \ (abs(F) <= mus * Fn? \ v : (F >= 0?F - muk * Fn :$$
$$F + muk * Fn)) : \qquad (6.21)$$
$$(F * v <= 0? \ (v >= 0?F + muk * Fn : F - muk * Fn) : v) = 0$$

To model dry friction we use a Switch element with a constitutive relation given by Eq. (6.21). The model of a ground body that acts on another body by the mechanism of dry friction is shown in Fig. 6.37. There are two branches to the model.

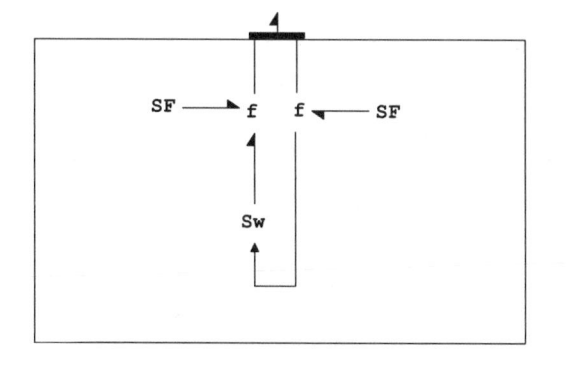

Fig. 6.37. The bond graph model of the ground with dry friction

The left branch corresponds to the interactions in the sliding direction, and the other describes the interactions in the normal direction (Fig. 3.35). The flow junction on the left simply states that the body velocity (the upper port) is equal to the sum of the ground body velocity—defined by the flow source SF on the left—and the body sliding velocity. The flow junction variable is the friction force. Positive power flow at the port corresponds to the force and velocity components taken in the direction of the coordinate axis. This way, power transfer from the ground by the mechanism of dry friction is positive if the friction is positive. Eq. (6.17) clearly shows that it is negative, hence the real power transfer is negative. The other flow junction asserts that the velocities of the bodies in the normal direction are equal. The junction variable is the normal reaction, which is fed back to the control port of the switch element supplying it with the necessary information.

Based on this model we can define a component representing the dry friction interactions (Fig. 6.38). The component implements the model represented by the central part of Fig. 6.37 (without the flow sources), as shown in Fig. 6.39.

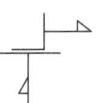

Fig. 6.38. The component representing dry friction

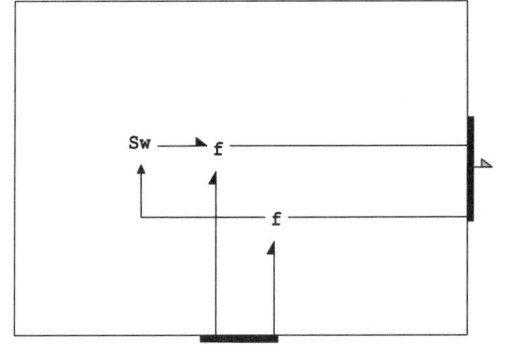

Fig. 6.39. Model of dry friction

We will now test the model on various problems in which the friction is important. We start with vibrations of the single degree of freedom system of Sec. 6.2, replacing the linear friction by dry friction. We then analyse some problems in which *stick-slip* appears.

6.3.2 Free Vibration of a Body with Dry Friction

We return to the problem of Sect. 6.2.1 and analyse the free motion of the body under the influence of dry friction at the contact with the ground (Fig. 6.1). The corresponding bond graph model is given in Fig. 6.40.

The model is similar to that of Fig. 6.11. The main difference lies in replacing the linear mechanical damper with the model of the ground in Fig. 6.37. This includes the effect of dry friction, but the external force is removed. We are no longer interested in the wall force, hence the wall will be represented simply by a zero flow source. The model of the body is somewhat more complicated than that in Fig. 6.12 because we need a model of the body motion in two dimensions, i.e. along the ground and normal to the ground (Fig. 6.41).

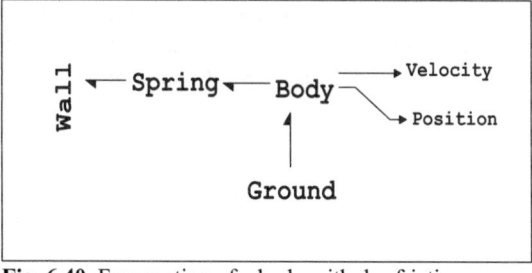

Fig. 6.40. Free motion of a body with dry friction

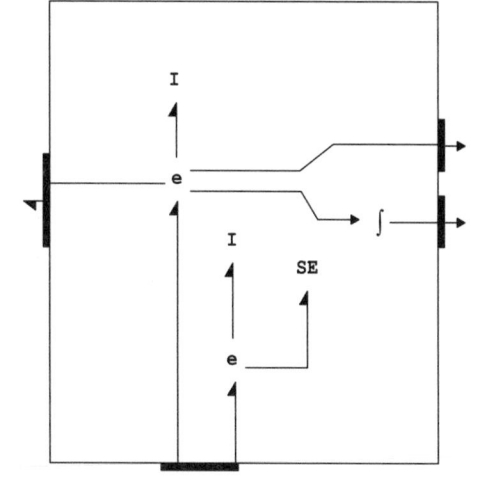

Fig. 6.41. Model of the body motion in two dimensions

The left effort junction of Fig. 6.41 describes the translation of the body along the ground surface. Similarly the other junction represents the summation of the forces in the normal direction. The weight of the body is represented by the source effort SE. The I elements represent the inertia of the body. To monitor the motion of the body, the velocity is taken from the junction and fed out to the corresponding display component Velocity (Fig. 6.40). The position is obtained by integrating the velocity and is fed out and connected to the Position output (display) component.

To compare the behaviour of the two models, the basic parameters used are the same. Thus, the parameters of the model are taken as

1. The body mass m = 5 kg
2. The spring stiffness k = 112.5 kN/m
3. The coefficient of static friction μ_s = 0.5
4. The coefficient of kinetic friction μ_k = 0.4

5. Gravitational acceleration g = 9.81 m/s^2

The initial displacement of the body is $5.0 \cdot 10^{-3}$ m and the initial velocity is 0 m/s^2. The complete model can be found in the project library under the name Body Motion with Dry Friction.

The model is more complex than in Sect. 6.2; it consists of seventeen equations and is of index 2 type (Chap. 5). During the simulation the structure of the model frequently changes, which could pose a problem for the solver. The method described in Chapt. 5, however, copes with this rather well. The simulation was run for 0.5 s with an output interval of 0.001 s.

Results (Fig. 6.42) show that amplitudes of the body position decease linearly until the body settles down after about 0.3 s. The period of vibrations is equal to $2\pi/\omega_n$, where ω_n is the natural frequency of vibration. The theoretical value of the decrease in amplitude per cycle is equal to 4Δ, where $\Delta = \mu mg/k$, is based on the same value for both friction coefficients [1].

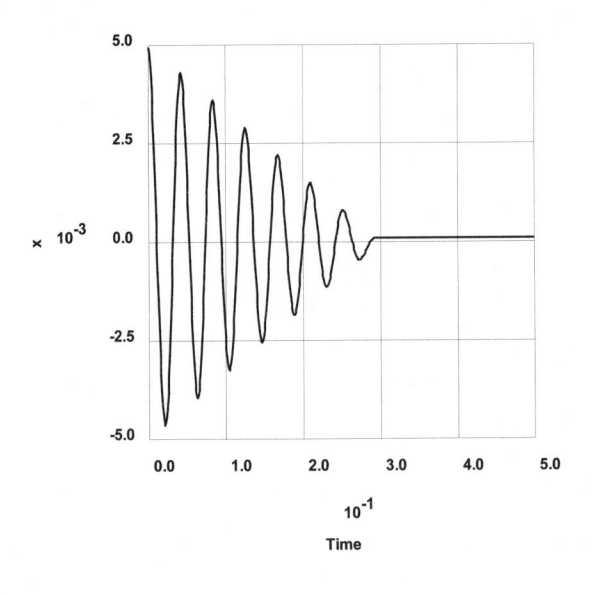

Fig. 6.42. The transient of the body position under the action of dry friction

Using the parameters given above the value of the amplitude drop we obtained by this formula is 0.0006976 m/cycle. By simulation the amplitude drop per cycle of 0.000699 m is obtained. The body settles down at a position of 0.000114159 m, which corresponds to a spring force of 12.84 N, or 65 % of the kinetic friction

The simulation statistics are given in Table 6.4 (for the output interval 0.005 s). In comparison with Table 6.2 there are apparently many more integration steps, and evaluations the function and the matrix. The simulation takes twice as long. What is more important is that, despite of the mathematical model, the simulation is conducted efficiently.

Table 6.4. Simulation statistcs for body with dry friction

Name	Value
The order of the method on the last step	1
The number of the steps taken	915
The number of the function evaluations	2205
The number of the partial derivative matrix evaluations	1305
The relative error tolerance	10^{-6}
The absolute error tolerance	10^{-6}
The elapsed time s	0.631 (0.741)

6.3.3 Stick-Slip Motion

In the previous problem static, friction is important only at the beginning of the motion. Later, when oscillations begin and finally settle down, the kinetic friction is the main influence. This and similar behaviour in engineering equipment often leads to neglecting the difference between static and kinetic friction and thus simplifying the model of dry friction (see e.g. [7]). This is not the case, however, in processes in which intermittent (*stick-slip*) motion appears as a consequence of the difference between these two friction coefficients. This is analysed experimentally and theoretically in detail in the classical reference on friction [3].

The stick-slip motion occurs between the sliding surfaces of two bodies in contact when one body is driven with constant, fairly low velocity, and the other has a certain degree of elasticity. This is the case with the bodies in Fig. 6.40 if the ground body is driven with constant and very low velocity and the other body is initially at rest. At the beginning when the spring force is less than the static friction, the ground drags the body, therby increasing the tension in the spring. This is the "stick" phase. It lasts until the tension in the spring reaches a value at which the spring force overcomes the static friction. At that moment, friction between the body and the ground drops to the kinetic friction value, which generally is lower. Hence, the body under the action of the spring slips quickly back over the ground. The "slip" phase lasts until the two velocities again are equal. A new "stick" phase then begins.

Intermittent motion can be a problem in applications in which smooth motion is important. As discussed in [3], if there is some damping in the system, intermittent motion may not occur at all, even if there is a finite difference between static and kinetic friction. Nevertheless, we will show the behaviour of the system of Fig. 6.40 under very slow ground motion conditions.

In the problem of Fig. 6.40, we assume the ground is driven at a constant velocity $V_0 = 0.001$ m/s and the body is initially at rest. Thus in the left flow source of the Ground component (see Fig. 6.37) we change the value of the flow source velocity to 0.001. Also, in the capacitive element of the Spring (see Fig. 6.8), we set the initial displacement to zero. In the inertial elements of the Body model (Fig. 6.41), we retain zero initial conditions, but change the initial position in the integrator to zero. We build the simulation model and run the simulation for 0.5s with an output interval of 0.001 s, accepting the default values for the other simulation parameters. The results are given in Figs. 6.43 and 6.44.

Fig. 6.44 shows the position of the body and Fig. 6.45 its velocity with time. The body sticks to the ground body until the spring tension is greater than the static friction. This occurs 0.218 s after the start of the motion. The body then detaches from the ground and, at first, continues to move in the same direction. Shortly after, at about 0.219 s, its velocity drops to zero and it then moves in the opposite direction. Reattachment occurs at 0.241 s, when its velocity catches up with the velocity of the ground.

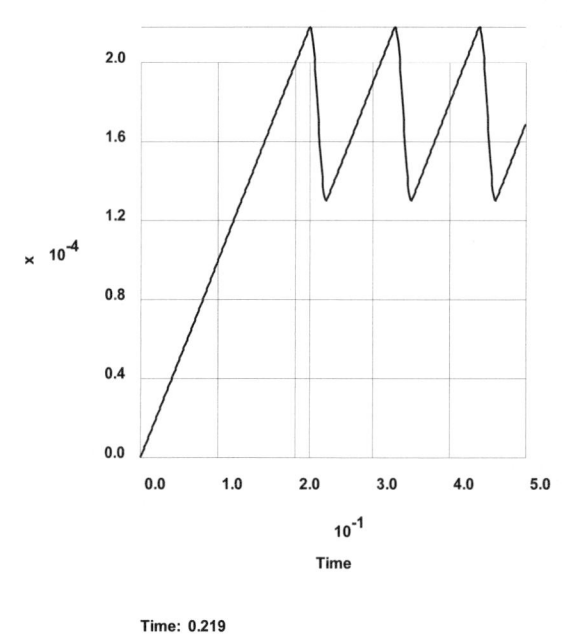

Time: 0.219

x: 0.000218394

Fig. 6.43. The simulation of the stick slip motion of the body

We can easily find the displacement of the body during slip. Let y_1 be the position of the body at the start, and y_2 at end, of slip. During slip the friction force is equal to kinetic friction. Applying the law of kinetic energy change during the slip, we easily find the relationship

$$\frac{1}{2}ky_1^2 - \frac{1}{2}ky_2^2 = F_k(y_1 - y_2)$$
(6.22)

or

$$y_1 + y_2 = 2F_k / k$$
(6.23)

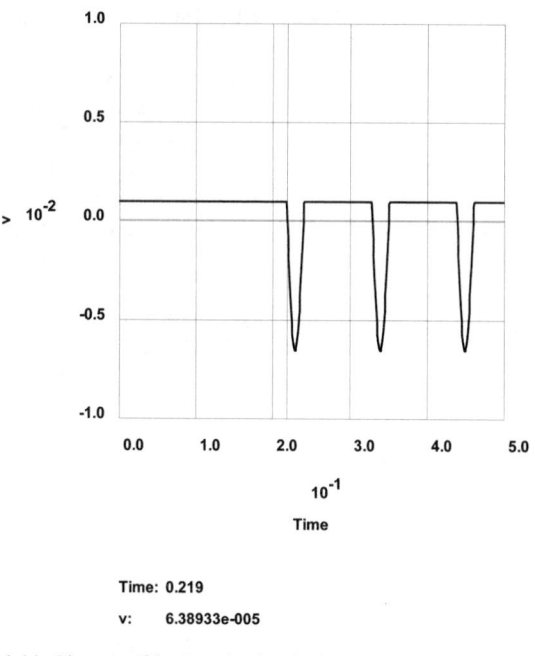

Time: 0.219

v: 6.38933e-005

Fig. 6.44. Change of body velocity during stick-slip motion

On the other hand, at the start of slip, we have $y_1=Fs/k$. Taking into account that $F_s=\mu_s mg$ and $F_k=\mu_k mg$ we get

$$y_1 = \mu_s mg/k$$
$$y_2 = (2\mu_k - \mu_s)mg/k$$

(6.24)

Hence, the slip is given by

$$s = y_1 - y_2 = 2(\mu_s - \mu_k)mg/k \qquad (6.25)$$

This shows that slip is possible if there is a difference between the static and kinetic coefficients of friction. The value shown in Fig. 6.44, as well as the difference between maximal and minimal positions, are somewhat greater then these values. This is because detachment and reattachment of the body to the ground occurs shortly before and after the zero velocity positions (Fig. 6.44).

6.3.4 The Stick-Slip Oscillator

As the next example, we simulate the behaviour of the stick-slip oscillator described in [8]. The system that we analyse consists of a body which has one elastic degree of freedom and which can slide over the ground under the action of an external force (Fig. 6.45). On the body there is placed another body, which can also slide over the first under the action of a force.

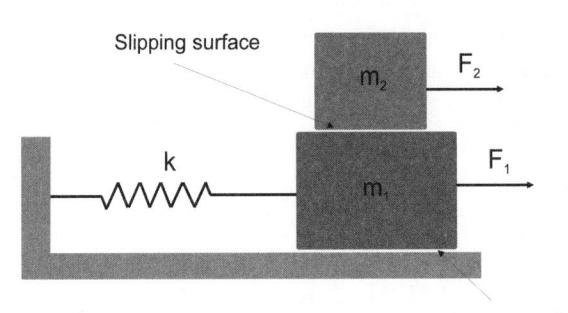

Fig. 6.45. The stick-slip oscillator

The forces change with time harmonically, i.e.

$$F_i = F_{i0} \cos(\omega t + \varphi_i) , \quad (i = 1,2) \tag{6.26}$$

We assume that dry friction exists at both slipping surfaces.

A detailed analysis on the possible states of sticking or slipping, along with phase portraits obtained by numerical integration, is described in [8]. We apply here the more direct approach of bond graph modelling and simulation, as in the previous problems.

The bond graph model of the system in Fig. 6.45 can be developed easily using the model of dry friction developed in Sect. 6.3.1. We develop the model using the mechanical schematics. Dry friction is modelled using the corresponding component from the library (Figs. 6.38 and 6.39). The complete model is shown in Fig. 6.46. The main components are similar to the model of Fig. 6.40. Components F1 and F2 are source efforts described by Eq. (6.26). The component Body 1 differs from Body2 because there are four ports, two for connecting to the spring and the external force source and two for the interactions with the ground and the upper Body2. There are also several signals used for monitoring. Thus, from the first body signals monitoring the inertial force, velocity and position in the sliding direction are taken and connected to the x-y displays shown on the right. Similarly, signals of velocity and position of the upper body motions are taken from the Body2 component and connected to the third display. Details of the model can be found in the library under the project name Stick-Slip Oscillator.

The parameters of the model are as in [8]:

1. The body masses are $m_1 = m_2 = 1$ kg
2. The spring stiffness $k = 150$ N/m
3. The coefficients of friction $\mu_s = \mu_k = 1$ (the same for both surfaces)
4. The driver amplitudes $F_{01} = 60$ N and $F_{02} = 60$ N
5. Driver frequencies $\omega_1 = \omega_2 = 2\pi$ rad/s
6. The drive phase angles $\varphi_1 = 0$ and $\varphi_2 = \pi$ rad
7. The acceleration due to gravity $g = 10$ m/s^2

The simulation was run for **5 s**, which corresponds to five cycles of vibration. To more accurately simulate discontinuities in inertial forces (accelerations here),

a fairly small output interval of 0.001 s is chosen as well as a tighter error tolerance of 10^{-8}. Even with these values, the simulation time is not too long (about 5.9
s on a Pentium III laptop). The results are shown in Figs. 6.47–6.49.

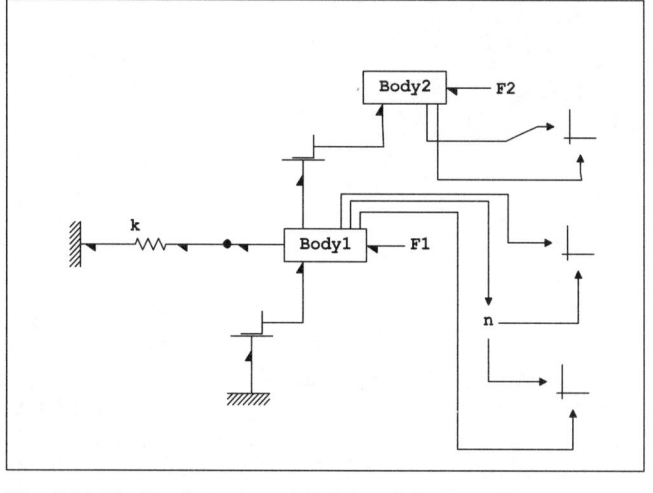

Fig. 6.46. The bond graph model of the stick-slip oscillator

The system establishes stable limit cycles after two or three cycles (Figs. 6.47
and 6.48). The oscillations of Body 1 are fairly symmetrical about the origin and
of Body2 are displaced a little to the right (about 0.04 m).

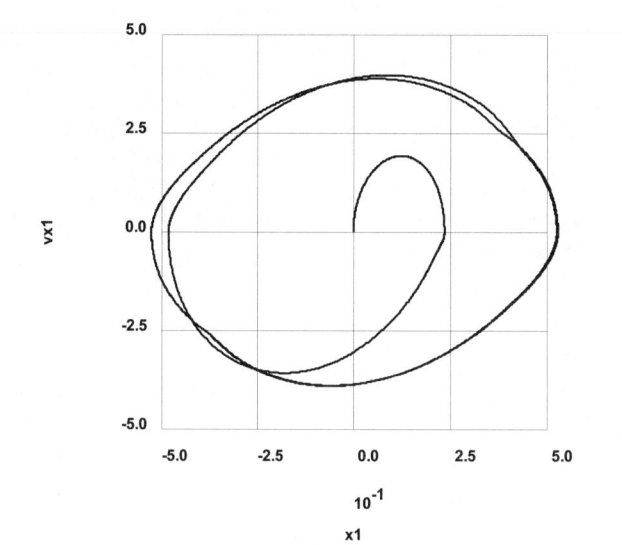

Fig. 6.47. The phase portrait of Body1

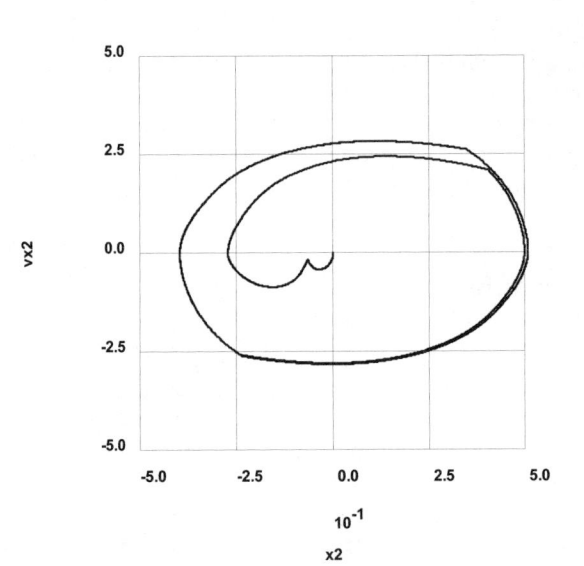

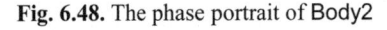

Fig. 6.48. The phase portrait of Body2

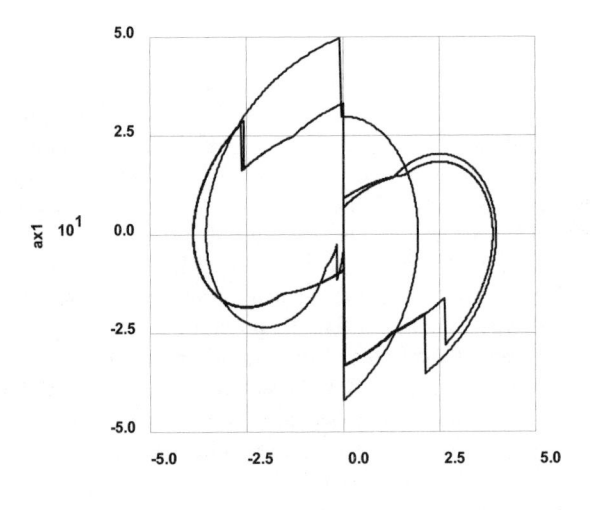

Fig. 6.49. The acceleration–velocity diagram of Body1

Fig. 6.49 gives a plot of inertial force—or acceleration, as the mass is 1kg—against velocity for Body 1. The diagram is not smooth as in the previous case, because there are jumps in the inertial forces when either body change motion from sticking to slipping, or *vice versa*. The bodies are coupled by friction, hence, these changes are reflected in both bodies. By looking carefully, we can see the limit cycle (the inner loop).

Discontinuities in accelerations occur at zero velocity and at a velocity of ±2.626 m/s. The limit cycles closely agree with the phase diagrams of [8]. In addition, Figs. 6.47–6.49 also show how the cycles gradually develop.

6.4 Bouncing Ball Problems

The next type of problem that we analyse is one of the simplest problems dealing with impact. We analyse the impact of a ball dropping on a massive table (ground). We analyse first the case where the table is at rest. But more interesting is the problem when the table vibrates with constant frequency in which case chaotic vibrations can occur [9]. We first develop a simple model of impact, which is then used in bouncing ball problems. This model can be used as bases for solving much more complicated systems involving multibody motion with impact. It also can be combined with the model of dry friction developed in the previous section to model complex interactions at the contacts.

6.4.1 Simple Model of Impact

The dynamics of multibodies with contacts are studied in detail in [8]. This uses an approach based on the classical Newton theory and the Poisson laws of impact. It is applied to solution of a range of problems. We will not follow such an approach here, but use the relatively simple model of Fig. 6.51 in which the contact between two bodies in the direction normal to the contacting surface is represented by a spring-dashpot component. Such a model is also presented in [8], but we analyse it in more detail. The model is convenient because it enables us to treat processes at the impact of two bodies as a component that imposes certain conditions on the rest of the system. This type of model offers the possibility of a more rigorous treatment of the properties of the materials at the contact. It is usually argued that such models lead to very stiff systems, which are not easy to solve. But other approaches, such as in [8], also are not simple because the interactions at impact are rather complicated. The BDF integration method used in the BondSim is capable to solving such stiff models relatively efficiently.

We formulate the equation of motion of the ball and the spring-dashpot component as in [8]. First we define the kinematic relationships

$$v = \frac{dy}{dt} \tag{6.27}$$

and

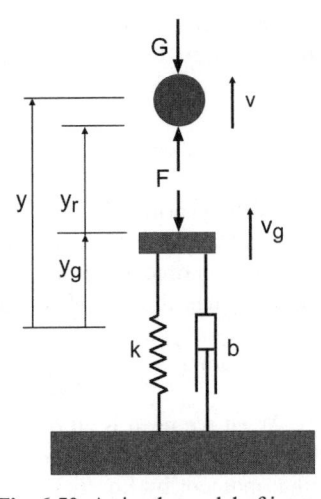

Fig. 6.50. A simple model of impact

$$v_g = \frac{dy_g}{dt} \tag{6.28}$$

The relative velocity of the ball with respect to another body (ground) is

$$v_r = v - v_g \tag{6.29}$$

Hence, the relative distance of the ball to the point of contact is given by

$$y_r = y_{r0} + \int_0^t (v - v_g)d\tau \tag{6.30}$$

If $y_r > 0$ there is free fall of the ball, hence we have the equations

$$\left. \begin{array}{l} m\dfrac{dv}{dt} = -mg + F \\[2mm] b\dfrac{dy_g}{dt} + ky_g = -F \\[2mm] F = 0 \end{array} \right\} \ (y_r > 0) \tag{6.31}$$

The ball hits the other body when $y_r = 0$. From that moment on, the ball and the spring-dashpot move as single system. Thus we have

$$\left. \begin{array}{l} m\dfrac{dv}{dt} = -mg + F \\[2mm] b\dfrac{dy_g}{dt} + ky_g = -F \\[2mm] v_r = 0 \end{array} \right\} \ (y_r \le 0) \tag{6.32}$$

At the moment of impact the ball has a velocity v_0. The acceleration of the ball changes abruptly at that moment to satisfy Eq. (6.32). Throughout the contact phase, the motion the ball-spring-damper is governed by the equation

$$m\frac{d^2 y_g}{dt^2} + b\frac{dy_g}{dt} + ky_g = -mg \qquad (6.33)$$

In the first phase of the motion there is *compression* of the spring until the ball velocity drops to zero (Fig. 6.51a). This is followed by an expansion phase during which the ball starts rebounding. The question is: At what moment the ball detaches from the other body? According to [8] this occurs when the force F drops to zero. At that moment the velocity of the ball and of the spring and damper ends are equal; they continue to move in the same direction gradually separating one from the other (Fig. 6.51a).

Such detachment is difficult to detect numerically. What we need is an *abrupt* change of the motion ball and of the other body at the point of contact. This occurs at zero compression where the velocity is $v_1 > 0$. At that moment by Eq. (6.31) the velocity of the spring-damper end drops abruptly to zero and its motion ceases (Fig. 6.52b). Simultaneously, the force on the ball abruptly changes from the value $-b \cdot v_1$ by Eq. (6.32) to the value 0 by Eq. (6.31). Thus there is a positive impulse, which changes the ball acceleration from value $-g-b \cdot v_1/m$ to $-g$. In effect the body rejects the ball.

We thus assume that the contact with the other body is established when the ball's relative displacement is less or equal to zero, and that the detachment occurs when it is positive again. It is possible that the spring extension never drops to zero in a finite time. This is the case when there is high damping in the system, which corresponds to an inelastic impact.

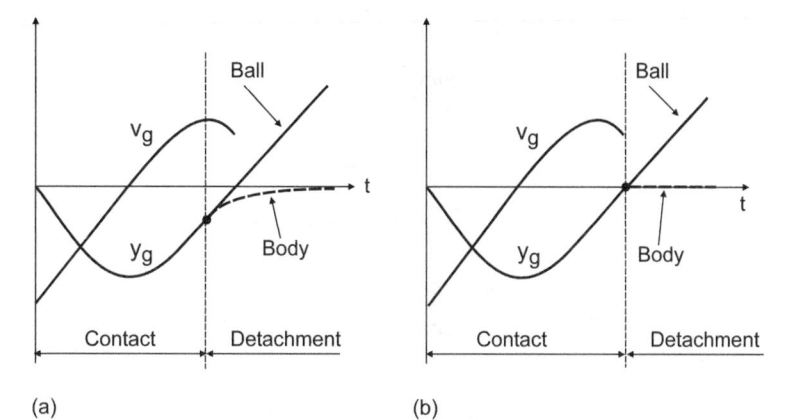

(a)   (b)

Fig. 6.51. Detachment of the ball at (a) zero force, (b) zero compression

To complete the description of this simple model of impact, we calculate the ratio of the ball velocity after and before impact. Solving Eq. (6.33) we get

$$\frac{v_1}{v_0} = -e^{-\frac{\zeta\pi}{\sqrt{1-\zeta^2}}} \qquad (6.34)$$

where ζ is damping ratio of the ball-spring-damper system. In the classical theory of impact, this ratio is known as the coefficient of restitution α. Hence,

$$\alpha = e^{-\frac{\zeta\pi}{\sqrt{1-\zeta^2}}} \qquad (6.35)$$

This is just the *decrement* known from theory of vibrations. For near-elastic impact, i.e. $\zeta \ll 1$, we have

$$\alpha \approx 1 - \zeta\pi \qquad (6.36)$$

Table 6.5 gives values of the coefficient of restitution for various values of damping ratio.

Table 6.5. Restitution coefficient as a function of the damping coefficient

Damping ratio ζ	Coefficient of restitution α
0	1
0.02	0.9391
0.04	0.8818
0.06	0.8279
0.08	0.7771
0.1	0.7292
0.2	0.5266

Now we can formulate a bond graph component that describes the interactions between the two bodies during impact. This is basically a two-port component (Fig. 6.52). The ports are used to connect the bodies experiencing impact.

4
Contact
4

Fig. 6.52. The component representing impact between two bodies

The model of impact is shown in Fig. 6.53. There are two effort junctions e, which represent the velocities of bodies in the normal direction (see Eqs. (6.27) and (6.28)). Summator s evaluates their difference according Eq. (6.29); and the integrator evaluates the difference between the positions of the bodies, as in Eq. (6.30).

The first equation in Eqs. (6.31) and (6.32) describe the motion of the body (the ball); thus, these are not included in the model of contact. This is represented by the capacitive element C and the resistive R elements, which in conjunction with the third effort junction, describes the second equation in Eqs. (6.31) and (6.32).

The switch element Sw changes the condition represented by the third equations depending upon the body relative displacement evaluated by the integrator. The constitutive relation of the switch is simply

$$yr > 0? \quad F : vr = 0 \qquad (6.37)$$

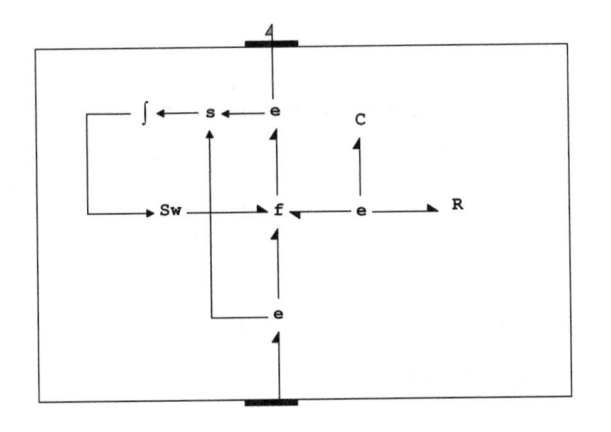

Fig. 6.53. Model of impact

6.4.2 A Ball Bouncing on a Table

We now analyse the motion of a ball, which is dropped from a height h onto a table that is at rest (Fig. 6.54). The ball drops under the action of gravity, hits the table and bounces back. It continues to bounce until it eventually reaches rest.

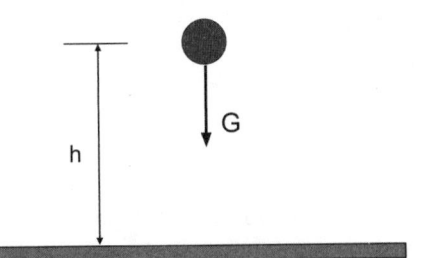

Fig. 6.54. A ball bouncing on a table at rest

We treat the ball as a particle moving in a vertical direction under the action of gravity. The model of the system consists of three components Ball, Ground and Contact (Fig. 6.55). The Ball is represented in the usual way by an inertial element representing the ball inertia in the vertical direction and a source effort describing its weight connected to a common effort junction. The Ground is described by a zero flow source. The last component models the impact of the ball with the table, as described in the last section. We extract several signals such, as the position and velocity of the ball and the position of the table (ground). In this problem this last signal is, of course, zero since the table is at rest. Details of the model can be found in the library under the project name **Bouncing Ball Problem**.

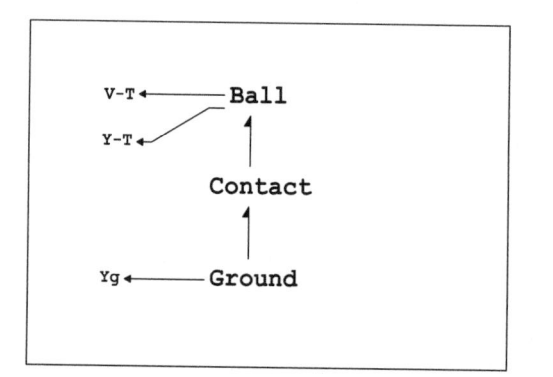

Fig. 6.55. Bond graph model of the ball bouncing on the table

We assume that the table is relatively rigid and that the restitution coefficient is about 0.9. Thus the parameters used for the simulation are:

1. The ball mass m=1 kg
2. Spring stiffness k=1·10⁶ N/m
3. Damper velocity coefficient b= 60 N·s/m
4. Initial ball height above the table h = 1 m

These values correspond to a natural frequency of 1000 rad/s and a damping ratio of 0.03. From Eq. (6.35) we get a coefficient of restitution value α=0.9100. The interval of simulation was chosen as 5 s and the output interval 0.01 s. The complete simulation lasts 0.85 s of processor time. The results are shown in Figs. 6.56 and 6.57.

In Fig. 6.56 we see the characteristic, partially elastic pattern in which the bouncing height gradually diminishes until the ball comes to rest (not shown). Fig. 6.57 shows sudden changes of the ball velocity when it hits the table.

We can estimate the coefficient of restitution by comparing the height of the ball above the table following every rebound. If v is the rebound velocity, the corresponding height of the ball after rebound is $v^2/(2g)$. Thus, the ratio of two successive rebound heights diminishes as α^2. From of the simulation result it is found that the ratio of heights is about 0.828, thus the coefficient of restitution is 0.910 as expected

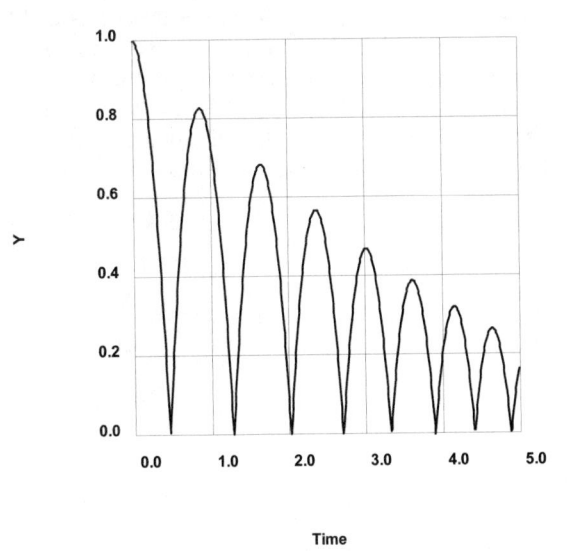

Time

Fig. 6.56. Change of the ball height during the bouncing

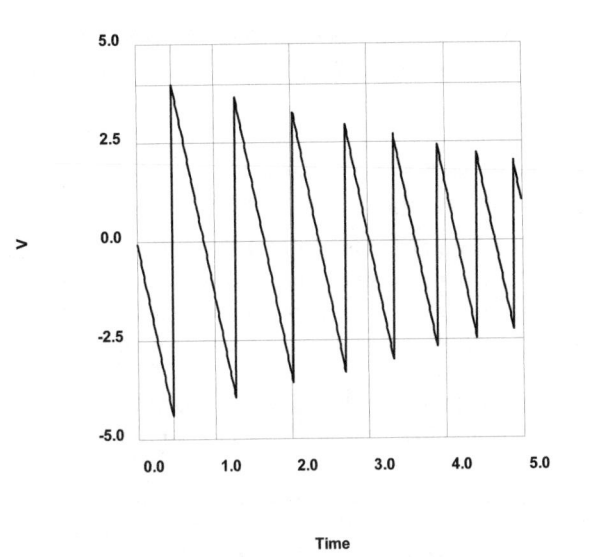

Time

Time: 0.46

V: 3.97745

Fig. 6.57. Change of velocity of the ball

6.4.3 Ball Bouncing on a Vibrating Table

We continue with analysing the motion of a ball bouncing on a table, which is not at rest, but that vibrates with a constant frequency (Fig. 6.58). For low velocities of vibrations, a stable ball motion appears similar to the case of the fixed table analysed in Sect. 6.4.2. On the other hand, if the velocity increases an irregular, chaotic motion occurs. The dynamics of such motions are analysed in [9]. We will not repeat this here, but show by simulation some of characteristic motions that can appear.

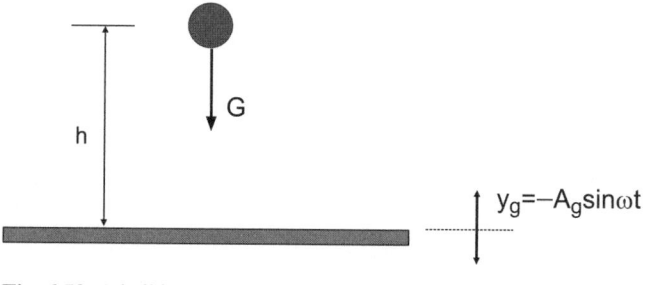

Fig. 6.58. A ball bouncing on a vibrating table

We use the model developed in the last section (Fig. 6.55). We also use the same parameters, but change the velocity generated by the source flow of the component Ground. We fix the amplitude of the table displacement to the value A_g = 0.1m and change its frequency of vibrations. The velocity of the table, as generated by the source, is defined as

$$v_g = -A_g\omega \cdot \cos\omega t \qquad (6.38)$$

That is, it starts to move down. We show two characteristic bouncing motion patterns.

The simulations are run using a simulation interval of 10 s and an output interval of 0.01 s. Fig. 6.59 shows the simulation of the ball motion when the table is vibrating slowly at frequency of 3 rad/s. It can be seen that the ball bouncing follows the table and finally comes to rest with respect to the table.

A quite different ball motion appears at higher frequencies. Fig. 6.60 shows the ball motion during the first 10 s when the table vibrates at 15 rad/s.

To see if the ball comes to rest with respect to the table, Fig. 6.61 illustrates simulation of the ball motion during the first 100 s. It apparently doesn't stop bouncing!

From the figures it can be seen that the interval between two bounces steadily increases until chaotic motions develop. In [9] it was shown that the existence of stable and unstable orbits of various periods can occur. Under certain conditions the ball bouncing from the table at the uppermost position can double its period of motion. As pointed out in [10], doubling the period leads to chaos. This is clearly seen in Fig. 6.60. The reader interested in this and other chaotic systems behaviour can consult [9,10] for a detailed exposition.

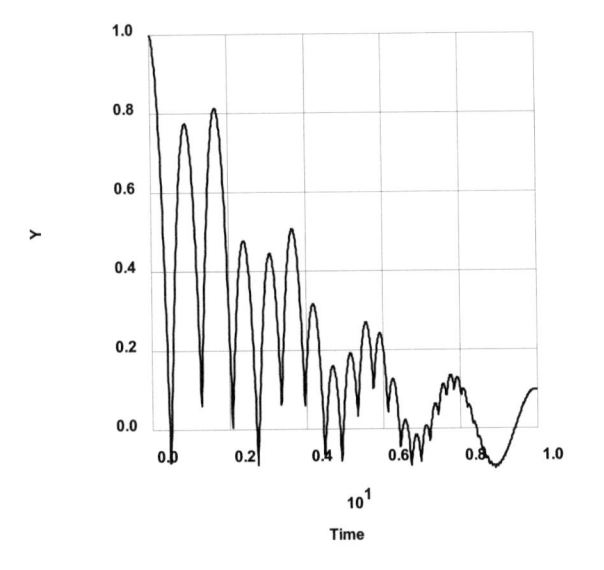

Fig. 6.59. The ball bouncing at a table frequency of 3 rad/s

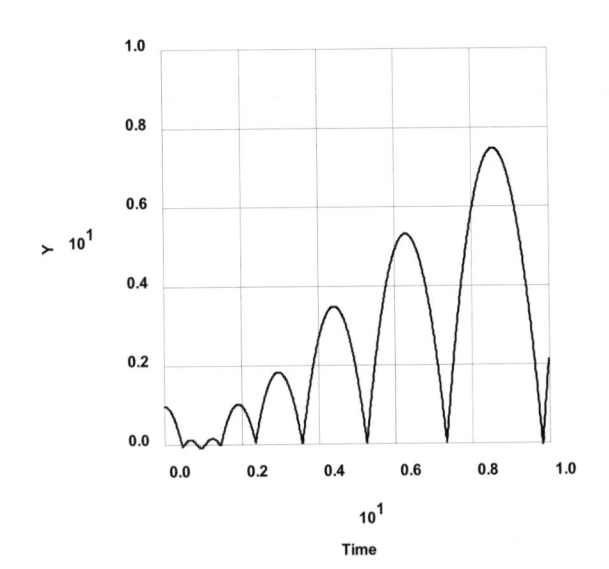

Fig. 6.60. The ball bouncing at a table frequency of 15 rad/s

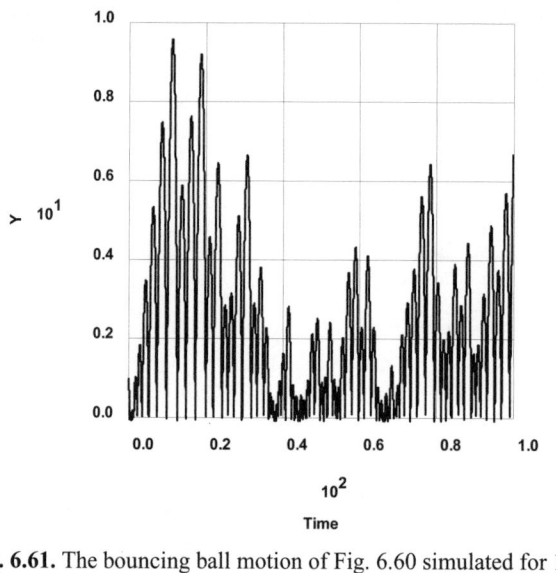

Fig. 6.61. The bouncing ball motion of Fig. 6.60 simulated for 100 s

6.5 The Pendulum Problem

As the last example of simple mechanical systems we return to the see-saw problem of Sect. 2.7.3 (Fig. 6.62). This problem was treated as a simple example of a multibody system, which we will discuss in more detail in Chapt. 9. It was shown that it is possible to develop a bond graph model by systematically decomposing the system into its components.

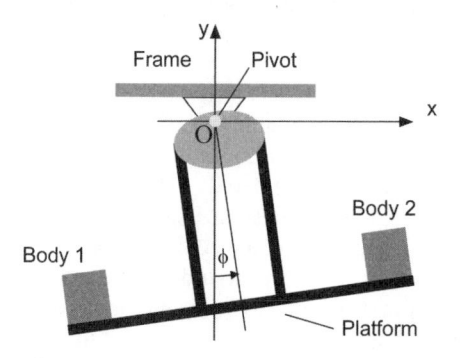

Fig. 6.62. See-saw problem of Sect. 2.7.3

Following the approach described in Sect. 2.7.3 and using BondSim, we can develop a corresponding model. In Fig. 6.63 the basic level of the model is shown in which the main components and the interactions between them are depicted. The model consists of several levels and can be reviewed in the project library under the title See-saw Problem.

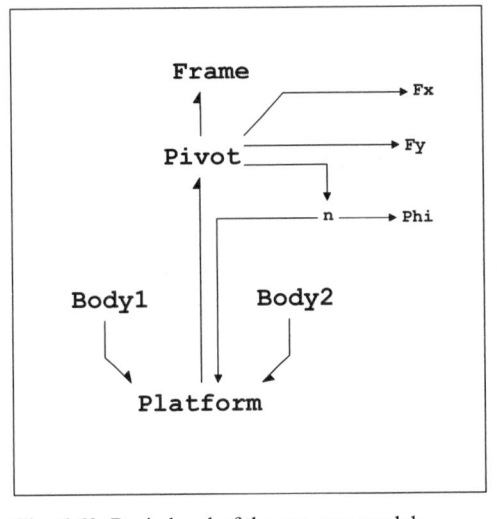

Fig. 6.63. Basic level of the see-saw model

The model developed is based on the direct application of the laws of Dynamics. The corresponding mathematical model consists of a system of differential-algebraic equations (DAEs). It corresponds to multibody models using Lagrange multipliers with constraints on velocities [11,12]. It is well known that such models are of *index 2* type [11,13]. This formulation lies between index 3 formulation when using constraints on the positions of connected bodies and index 1 formulation where all such constraints are eliminated e.g. by differentiation. It should be pointed out that the index 2 formulation appears here quite naturally as a result of the application of the bond graph modelling approach.

The see-saw can be looked at on as a pendulum, as has already been pointed out in Sect. 2.7.3. In [11,13] various formulations of pendulum problem are examined in the context of higher index DAEs. Our formulation is a little different from those in the references citied. Thus, we examine it in some detail.

We use two approaches. The first is simulating the system behaviour by the BondSim program using the default method (Chapt. 5), i.e. the BDF variable coefficient method and the analytical Jacobian matrix. Also we excluded all algebraic variables from the error tests, as is the usual praxis when solving higher index systems by BDF methods. We also scale the algebraic equations, as suggested in [13].

Another approach, which we will use for comparison, consists of reducing the multibody model to a pendulum. The pendulum will be represented in classical state-space form and solved using the implicit Runge-Kutta code RADAU5 of [11].

To simulate the see-saw problem using BondSim, we retrieve the problem from program library (using *Get From* command on *Project* menu) and build the model. We use the following model parameters:

1. Body 1 mass $m_1 = 80$ kg
2. Body 2 mass $m_2 = 20$ kg
3. Platform mass $m_3 = 40$ kg
4. Platform mass moment of inertia (centroidal) $J_3 = 90$ kg·m^2
5. Geometric parameters (Fig. 2.20) a =1.5 m, b = 1.125m, c = 0.375m
6. Initial angle of the platform = 1 rad

The simulation interval was taken equal to 5s and the output interval was 0.05 s, the error tolerances both absolute and relative are equal to $1 \cdot 10^{-6}$ (default). Some results of the simulation are shown in Figs. 6.64 and 6.65. From Fig. 6.64 it can be seen that the see-saw oscillates about the equilibrium position that is at $\varphi = 0.4324$ rad (24.77^0). The period of oscillation is 3.4 s. The amplitude of the force on the frame in the horizontal direction is 349.2 N (Fig. 6.65). The force in the vertical direction (not shown) doesn't change too much, i.e. it oscillates between -1152 N and -1613 N about a value corresponding to the weight of the platform and the bodies on it. Some numerical values also are given in Table 6.6 (the second and fourth columns) and the simulation statistics in Table 6.7 (the second column).

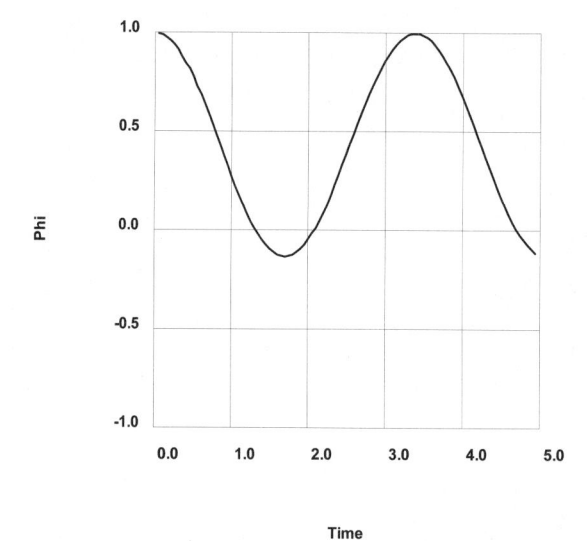

Fig. 6.64. Oscillation of see-saw about its equilibrium position

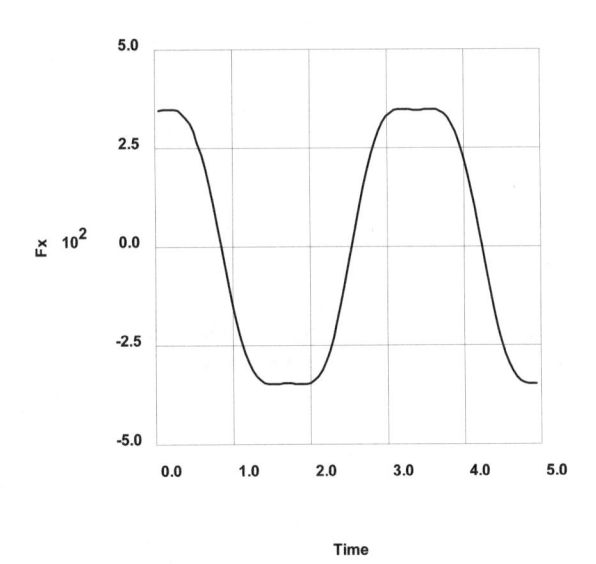

Fig. 6.65. Change of the horizontal force of the see-saw on the frame

Next we develop a simplified model of the see-saw as a pendulum (Fig. 6.66). First we note that the see-saw's centre of mass in not on the body axis. Using the parameters of Fig. 2.20 we can calculate its coordinates in the body fixed frame as

$$\left.\begin{array}{l} x'_C = -(m_1 - m_2)a/m \\ y'_C = -[(m_1 + m_2)(b + c) + m_3 b]/m \end{array}\right\} \tag{6.39}$$

Here m is the total mass of the seew-saw and the bodies on it, i.e.

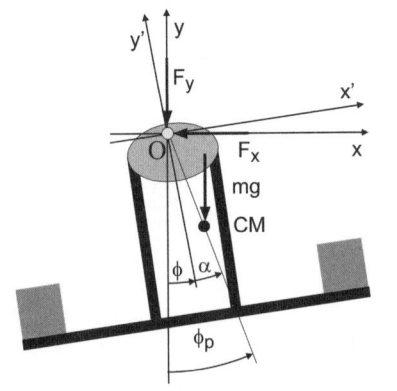

Fig. 6.66. The see-saw as a pendulum

$$m = m_1 + m_2 + m_3 \tag{6.40}$$

The angle made by a line drawn from the origin through the mass centre CM to the y'-axis is given by

$$\alpha = a\tan(x_c' / y_c') \tag{6.41}$$

This line is taken as the pendulum axis. Denoting by φ_p its angle with the vertical axis, we find the angle of the see-saw by

$$\phi = \phi_p - \alpha \tag{6.42}$$

Finally, the length from the origin to the centre of mass is given by

$$l_c = \sqrt{x_c'^2 + y_c'^2} \tag{6.43}$$

Now we can formulate the equation of rotation of the see-saw about the origin (the pin axis) as

$$\left. \begin{aligned} \dot{\phi}_p &= \omega \\ \dot{\omega} &= -\frac{mgl_c}{J_O} \cdot \sin\phi_p \end{aligned} \right\} \tag{6.44}$$

Here the J_O is its mass moment of inertia about the axis of rotation and is given by (Fig. 2.20)

$$J_O = (m_1 + m_2)(a^2 + (b+c)^2) + J_3 + m_3 b^2 \tag{6.45}$$

We also need expressions for the components of the force of the see-saw that acts *on* the frame. We calculate such a force because, in the bond graph model of Fig. 6.63, the efforts associated with the bond connecting the platform and the pin is the force acting *on* the pin, not its reaction.

Hence by applying the law of the mass centre motion we have

$$\left. \begin{aligned} m \cdot \frac{d^2 x_c}{dt^2} &= -F_x \\ m \cdot \frac{d^2 y_c}{dt^2} &= -F_y - mg \end{aligned} \right\} \tag{6.46}$$

Taking also into account that

$$\left. \begin{aligned} x_c &= l_c \cdot \sin\phi_p \\ y_c &= -l_c \cdot \cos\phi_p \end{aligned} \right\} \tag{6.47}$$

we can calculate the force components by

$$\left. \begin{aligned} F_x &= -m \cdot l_c (\dot{\omega}\cos\phi_p - \omega^2 \sin\phi_p) \\ F_y &= -m \cdot (l_c \dot{\omega}\sin\phi_p + l_c \omega^2 \cos\phi_p + g) \end{aligned} \right\} \tag{6.48}$$

Eq. (6.44) is a system of differential equations in state-space form, and Eq. (6.48) are the output equations. These equations can be solved by practically any method. We use the RADAU5 method of [11]. This method is considered by many as one of general-purpose method for solving higher index DAEs.

We use the same parameters as given at the beginning of the section. The other necessary parameters are calculated by Eqs. (6.39)–(6.41), (6.43) and (6.45). The angle of rotation is found from Eq. (6.42). We also use the same simulation parameters as before (the simulation period 5 s, output interval 0.05 s, error tolerances $1 \cdot 10^{-6}$). Results for the angle of rotation and x-component of the force are given in Table 6.6 (third and fifth column).

Table 6.6. Comparison of results obtained by the BondSim simulation and the RADAU5 routine

Time [s]	ϕ [rad]		F_x [N]	
	BondSim	RADAU5	BondSim	RADAU5
0.5	0.7754008987	0.7754021054	293.3272391	293.3285587
1.0	0.2747649236	0.2747643009	-153.4644374	-153.4651285
1.5	-0.09805834991	-0.09805980961	-349.2356995	-349.2361947
2.0	-0.04963962617	-0.04963799260	-345.0137097	-345.0130167
2.5	0.3831236781	0.3831231671E	- 49.42374072	-49.42407961
3.0	0.8559156737	0.8559145601E	330.0201078	330.0194638

Comparing the results obtained by the BondSim simulation and by the RADAU5 solution we see that there are minor differences that are of the order of the error tolerances used. It is interesting to note, in particular, that the accuracy of the force component simulation by the BondSim is in the same range of accuracy as that obtained by direct calculation by Eq. (6.48). This is important because it is an index-2 variable.

The simulation statistics are given in Table 6.7 (third column). The solution of the reduced set of equations is of course much more efficient. But on other hand the performance of BondSim is excellent. We should take into account the fact that the semi-explicit index-2 equations are solved, that the system of equations is relatively large, and that the functions and the partial derivative (Jacobian) matrix are evaluated in the interpreter mode. There is more matrix evaluations and decompositions in the BondSim method, but this is partially a consequence of the variable coefficient form of the BDF solver. On the other hand RADAU5 needs a relatively large number of the function evaluations, which is to be expected from a Runge-Kutta type of solver.

Table 6.7. Simulation statistics for the BondSim and RADAU5 solutions

	BondSim	RADAU5
Equations	52	4
Jacobian matrix elements	129	4
Steps	245	45
Function evaluations	521	288
Jacobian evaluations	276	21
Decompositions	276	35
Elapsed time s	0.250	0.01

References

1. JH Ginsberg (2001) Mechanical and Structural Vibrations, Theory and Applications. John Wiley, New York
2. SW Smith (1999) The Scientific and Engineer's Guide to Signal Processing, 2nd Ed. California Technical Publishing, electronic form http://dspguide.com
3. FP Bowden and D Tabor (1986) The Friction and Lubrication of Solids. Clarendon Press, Oxford
4. D Karnopp (1985) Computer simulation of stick-slip friction in mechanical dynamic systems. Transactions of the ASME, 107:100-103
5. W Borutzky (1995) Represeting discontinuities by sinks of fixed causality. In FE Cellier and JJ Granda (eds) 1995 International Conference on Bond Graph Modeling and Simulation, Las Vegas, Nevada, pp 65-72
6. J m Mera, C Vera (1999) Dry Friction Modelling by Means of The Bond Graph Technique. In JJ Granda and FE Cellier (eds) 1999 International Conference on Bond Graph Modeling and Simulation, San Francisco, California, pp 30-35
7. SS Rao (1995) Mechanical Vibrations, 3rd edn. Addison-Wesley, Reading
8. F Pfeiffer and C Glocker (1996) Multibody Dynamics with Unilateral Contacts. John Wiley, New York
9. J Guckenheimer and P Holmes (1997) Nonlinear Oscillations, Dynamical Systems, and Bifurcations of Vector Fields. Springer-Verlag, New York
10. D Acheson (1997) From Calculus to Chaos, An Introduction to Dynamics. Oxford University Press, Oxford.
11. E Hairer and G Wanner (1996) Solving ordinary Differential Equations II, Stiff and Differential-Algebraic Problems, 2nd Revised edn. Springer-Verlag, Berlin Heidelberg New York
12. D Karnopp (1997) Understanding Multibody Dynamics Using Bond Graph Representations. Journal of The Franklin Institute, Engineering and Applied Mathematics, 334B: 631-642
13. KE Brenan, SL Cambell and LR Petzold (1996) Numerical Solution of Initial-Value Problems in Differential-Algebraic equations, Classics in Applied Mathematics. SIAM, Philadelphia

Chapter 7 Electrical Systems

7.1 Introduction

This section shows that the component model approach developed in Part 1 can be used readily to model electrical and electromechanical components and systems. This is important, as both the mechanical and electrical part of mechatronic and the recently evolved micro-mechanical systems [1] can be modelled and analysed on the same basis, i.e. from the bond graph point of view.

It is well known that electrical systems can be modelled in terms of bond graphs [2,3]. But this approach may appear strange to engineers used to electrical schematics. Models in terms of bonds are usually simplified, e.g. by removing common ground nodes, resistor, capacitor, and other ports. Such models are quite abstract, and often it is not easy to correlate them with the devices of which they are models. In addition, the causality relations discussed in Sect. 2.10 restrict models that could be used for modelling general electrical, as well as of other, engineering systems.

The approach developed here is based on component models, the constitutive relations of which can be freely defined. This enables a systematic approach to the development of electrical components and system models. It is not necessary to represent electrical components in terms of word models only but, as already discussed in Sec. 2.7.2, these can be depicted using graphical electrical symbols. In this way, the bond graph point of view is retained, but on the surface they are represented as electrical schemes. At a deeper level, bond graph elements are employed to model processes in the components.

Models of electronic components are usually described in the form of SPICE models [4]. These are developed for use in SPICE type programs specifically developed for electrical and electronics systems (Sect. 1.7).[1] This type of model can be used in bond graphs, as well. But here it is not necessary to use the SPICE language, and there are no pre-built models of resistors, diodes, transistors, etc. Component models are constructed from the ground up from simpler ones. The same equivalent circuits, constitutive relations, parameter names, and values can be used as in SPICE. Component models can deal easily with macro models that play an important role in developing complex SPICE models. Once developed, component models can be moved into a library and used as needed.

An added advantage of our approach is that component models are transparent, i.e. they can be opened for overview, and modified if required, or maybe a differ-

---

[1] There are various versions of SPICE such as SPICE 2 and 3, HSPICE, PSPICE etc.

ent model of the same component can be developed and tested. The BondSim program is not designed to be a replacement for specially designed programs, such as SPICE, SABER or others, but more as an open environment for developing models and simulating the behaviour of mechatronic and micro-mechanics systems (MMS). The analysis currently supported is essentially in the time-domain.

This section starts with a relatively detailed description of a typical approach to developing bond graph models of electrical systems and their simulation using electrical schematics. This is explained in an example of the simple RLC circuit already described (Sec. 2.7.2). The approach is very similar to that used in Sec. 6.2 for mechanical systems. We then show how SPICE models of electrical circuit elements, such as resistors and capacitors, are developed. We continue with modelling some fundamental semiconductor components. The chapter ends with the analysis of an electro-magnetic system.

7.2 Electrical Circuits

Electrical circuits can be modelled in a similar way to the mechanical systems in Sect. 6.2. Normally, an electrical component palette is used that represents the basic circuit components by electrical circuit symbols. A typical approach to modelling electrical circuits is illustrated by the RLC circuit of Sect. 2.7.2, repeated here in Fig. 7.1.

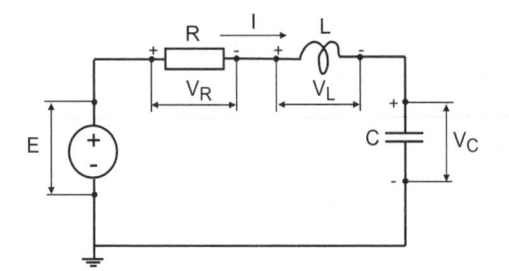

Fig. 7.1. The scheme of the RCL circuit

To begin, we launch BondSim, as described in Sect. 6.2.2, then define a new project, RCL Circuit, using the command *New* on the *Project* menu. When a new empty project document appears, we open the *Electrical component* palette using the command *Use Icon/Electrical* on the *Tools* menu (Fig. 7.2).

The Electrical *component* palette is used to create electrical component objects represented by electrical symbols. The palette contains several buttons used to create basic circuit components, such as *Device, Voltage* and *Current Sources, Ground, Node, Coupled Inductors, Resistor horizontal*, or *vertical*, and others. The particular component created by a button is identified by the tool tip displayed when the cursor is moved over the button.

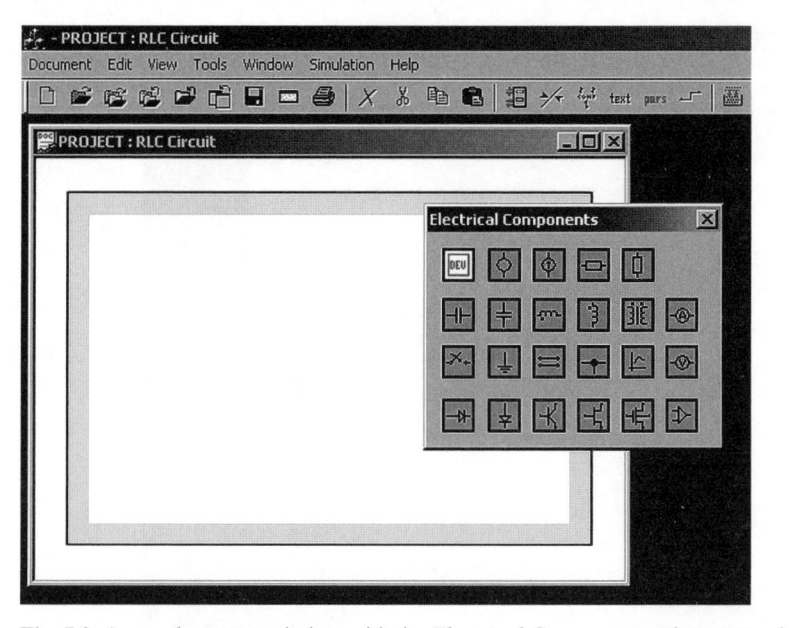

Fig. 7.2. A new document window with the *Electrical Components* palette opened

Model development of the circuit in Fig. 7.1 starts by creating a Voltage source component. Hence, we click the corresponding button in the palete (the second from the left in the first row). We then move the cursor to the place in the document window where we wish to insert the component. As the cursor enters the document it changes its shape to a cross. We create the component near the left document border, but leaving some space. The component created appears in the form of the circuit symbol for a voltage source (Fig. 7.3). Near the symbol a component name VS (Voltage Source) appears; the caret shows that we are in text-editing mode. We can edit the name if we wish, e.g. to E. To end editing, we click outside of the component, as is usual with word component models. After the component has been created, its two power-in and power-out ports are created as well.

We create other circuit components in a similar way. For example, to create a horizontal resistor, click on the *Resistor Horizontal* button. In this case, we retain the default label R, as well as the power flow direction through the component. In a similar way, we create the inductor by clicking on *Inductor Horizontal*. The capacitor is created using the *Capacitor Vertical* button. To complete the circuit, we create the *Ground* component (Fig. 7.3). We also create a *Node* between each of the components. All of these are not always necessary, but we add them anyway to ease the connection of measuring instruments, which can be used for generating display outputs. The left bottom node is used for the connection of the circuit to the ground.

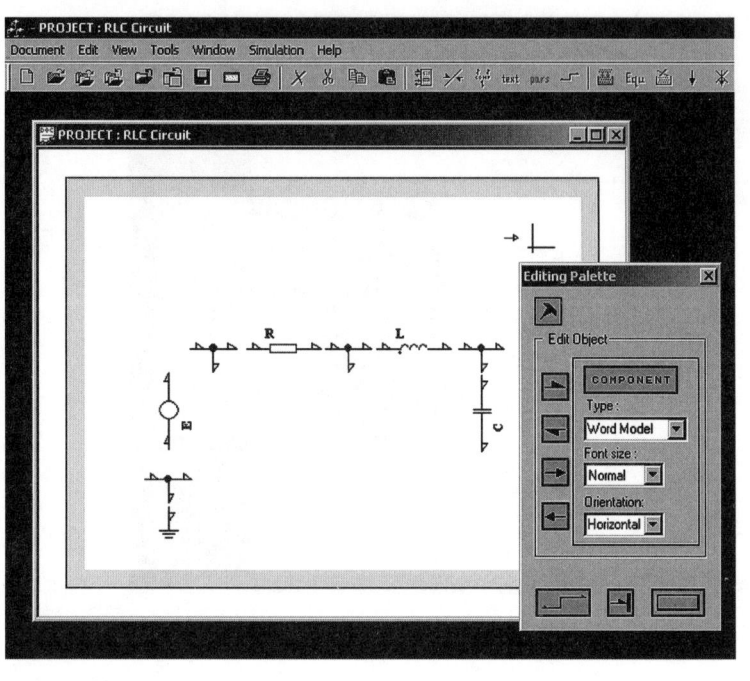

Fig. 7.3. The circuit components corresponding to Fig. 7.1

Before we proceed further, we rearrange the components by dragging them to make space for placing the instruments. We also create a single X-Y Plotter component that will be used to display all variables of interest during simulation. This component is moved to the upper right corner of the document area (Fig. 7.3). We see that the position of some of the node ports, as well as their power directions, are not correct; these must be changed. To do this, and to interconnect the components, we need the standard bond graph *Editing Palette*. We open the *Editing Palette* by clicking its toolbar button, or by using the *Bond Graph Palette* command on the *Tools* menu. The components created so far, together with the *Editing Palette*, are shown in Fig. 7.3.

We assume that power flows from the source through the components to the ground. Thus, we need to change the branch node ports to correspond to that sense of power flow. Also, the positions of some ports are not correct. We generally cannot move node ports because they are, by design, *fixed*. But we can delete them and create a corresponding port at the appropriate place in the node. Thus, for bottom left node, we click the left port to select it and then delete it by clicking the toolbar button *Delete* (or by using the command *Edit/Delete*). Next, we create a port lying opposite to the lower VS port. Because we need a power-out port, we click the corresponding port button on the *Editing Palette*, move the cursor to top part of the Node component, then click again. We also must change the power direction of the right node port. To do this, we select it, then click the toolbar button

Change Port, or use the command *Change Port* on the *Edit* menu. We change all other nodes in a similar way, as required.

Now that all the components are created and their ports have the proper power sense, we interconnect the components by drawing bond lines between corresponding ports. To do this, we click the *Connect Ports* button on the palette. If, for example, we wish to connect the upper voltage source port to the corresponding node port, we click the upper E port. As we move the cursor, a line is drawn. We can move the cursor vertically up to an intermediate point at the level of the port and click the mouse; then move the cursor horizontally to the right until we are over the node port, and then click again to finish the bond line. With this action the ports disappear and the bond line appears with a half-arrow pointing to the node. We similarly connect the ports of all other components. The circuit created in this way is shown in Fig. 7.4.

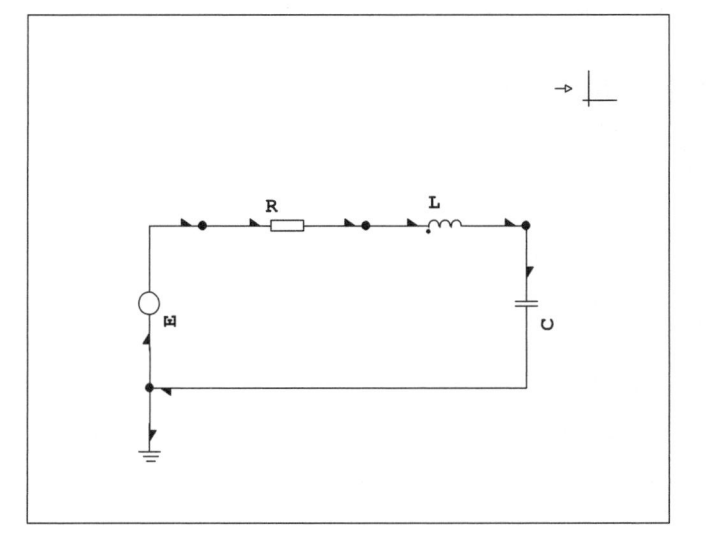

Fig. 7.4. The bond graph model of the RCL circuit

To complete the RCL circuit model, we next define the model of every circuit component (see also Sec. 2.7.2). We start with the voltage source by double-clicking it. This opens the component document, which is empty and contains only the document ports. We create the model as shown in Fig. 7.5. It consists of an effort junction e describing the voltage difference at the voltage source ports, and the source effort SE component that generates the source's electromotive force (*emf*). The constitutive relation describing the voltage generated by the SE is stored in its port. To edit it, we double-click the port. A dialogue appears that is used to assign meaningful names for the port variables, define the model parameters, and edit the constitutive relation that describes the *emf* at the port (Fig. 7.6). We retain e as the name of the effort variable (voltage). The eValue is a parameter defining the constant voltage generated at the port. At port creation it is set to zero; we now change it to 10 (V) by use of the *Parameters* button (see Sect. 4.8.2).

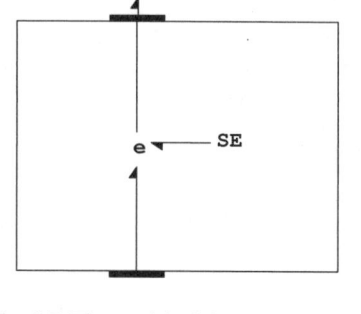

Fig. 7.5. The model of the Voltage Source

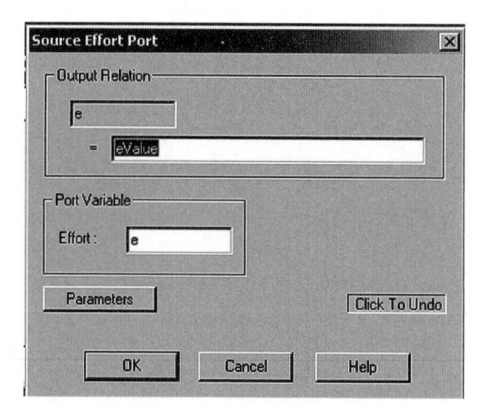

Fig. 7.6. A dialogue for editing the voltage source constitutive relation

Similarly, we create the model of the resistor (Fig. 7.7), which consists of an effort junction e and, this time, a resistive component R. We edit the constitutive relation for the voltage across the resistor by double-clicking the port of the bond graph resistor component R (Fig. 7.8). We retain the default name e for the effort (voltage) and change the flow variable to i. The constitutive relation corresponds to *Ohm's Law*, with the resistance parameter R0 (Sect. 2.5.4). By default, this is set to 1, but we change it to 200 (Ohms) by using the *Parameters* button.

The inductor component is defined similarly (Fig. 7.9). This component uses an inertial component I that defines *Henry's Law* (Sect. 2.5.2). Double-clicking the component I port displays a dialogue used to define the inductor constitutive relation (Fig. 7.10). We change the flow variable to the current i, and the state variable to the coil flux linkage p. In the constitutive relation box we change the inertial parameter I0 to the inductance parameter L. Likewise, we delete the parameter I0 from the parameter list and insert the parameter L with a value of 0.010 (H).

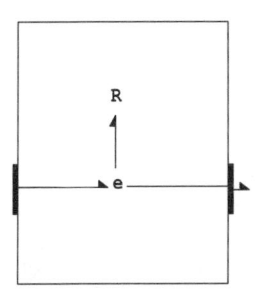

Fig. 7.7. The model of the Resistor

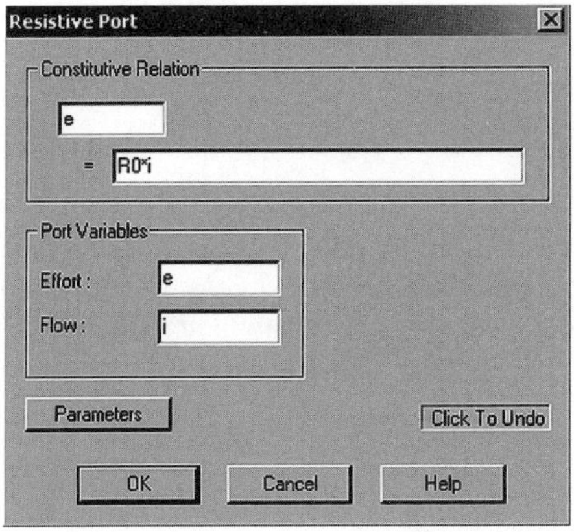

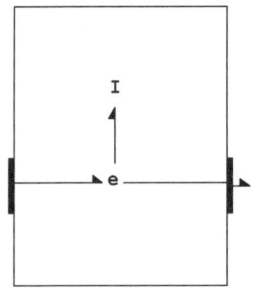

Fig. 7.8. A dialogue for defining the resistor constitutive relation

Fig. 7.9. The inductor component model

Finally, we define the Capacitor model (Fig. 7.11). Again, the effort junction describes the voltage across the capacitor. We use the capacitive component C to describe the processes in the capacitor (Sect. 2.5.3).

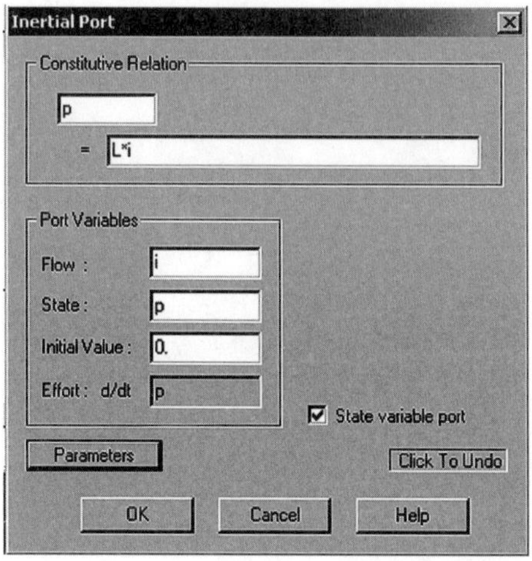

Inertial Port ☒

┌─ Constitutive Relation ─────────────────────────┐

│ p │

│ = │ L·i │

└──┘

┌─ Port Variables ────────────────────────┐

│ Flow :   │ i │

│ State :   │ p │

│ Initial Value : │ 0. │

│ Effort : d/dt │ p │

│   ☑ State variable port

└──┘

│ Parameters │   │ Click To Undo │

│ OK │   │ Cancel │   │ Help │

Fig. 7.10. A dialogue for defining the inductor constitutive relation

The default constitutive relation can be changed by double-clicking component C port (Fig. 7.12). We change the effort variable (the voltage across the capacitor) to e, and the state variable to the capacitor charge q. The parameter C0, the capacitance, is set it to $2.0 \cdot 10^{-6}$ (F).

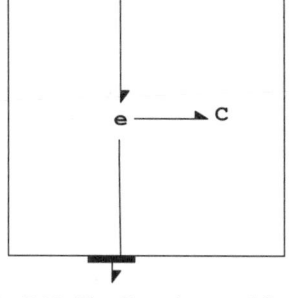

Fig. 7.11. The Capacitor model

To complete the model, it is necessary to define the variables that will be observed during simulation. This is similar to what is encountered when measuring or recording processes in real circuits with instruments. In this example we observe voltages across all components and currents throughout the circuit. This requires the model equivalents of "voltmeters" and "ammeters", respectively.

A voltmeter component is created from the *Electrical component* palette (Fig. 7.2) by clicking the *Voltmeter* button, placing the cursor up to the resistor, then clicking. The voltmeter component appears with two input ports and a single output port (Fig. 7.13). We move the output port to point upward. Using the *Editing Palette*, we create a *control-out* port at the two nodes from both sides of the resistor and connect them to the corresponding Voltmeter ports. The voltmeter, like other components, is created as an empty component that has to be defined.

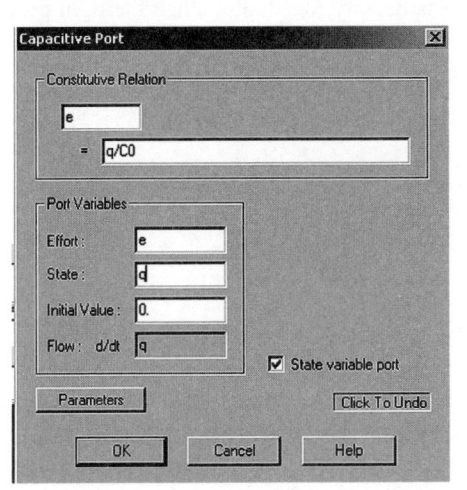

Fig. 7.12. A dialogue for defining the capacitor constitutive relation

We implement a Voltmeter by using the summator component (Fig. 7.14). The summator is created like any other component by selecting it from the list box in the *Editing Palette*. Next, we create two input ports, a single output to the summator, and connect them to the document ports.

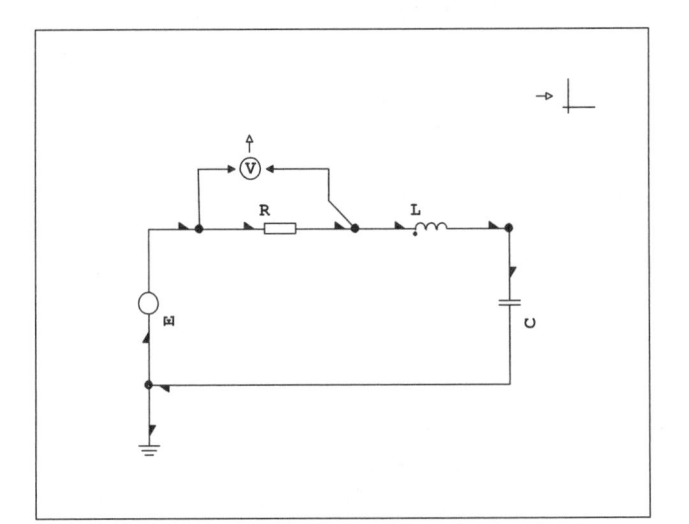

Fig. 7.13. The creation and connection of a Voltmeter

Because we need the voltage difference between the left and right nodes, we double-click the summator left port to confirm that (by default) it has a "+" sign, i.e. it is already in the addition mode. In the same way, we double-click the right port and select the "−" sign from the list.

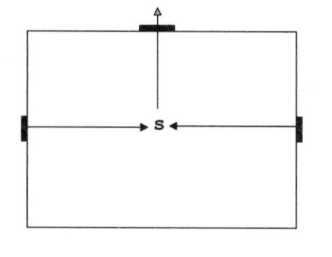

Fig. 7.14. The voltmeter as summator

The current in the circuit can be picked from any of the effort junctions inside the components—the resistor (Fig. 7.7), for example—by adding a control-out port and then connecting it to the outside. But we prefer to do this in a manner similar to that used in actual circuits. Thus, we insert an Ammeter in the return line (Fig. 7.13). To do this, we first must break the line by selecting it, then clicking the *Delete* button. Next, we create an electrical node below the capacitor and set its ports so that power flows from the capacitor to the ground node. We then create an Ammeter component using the *Electrical Components* palette. The component is placed between ground and this new node. We change the direction of the power flow to point leftward. Finally, we connect the ports (Fig. 7.15).

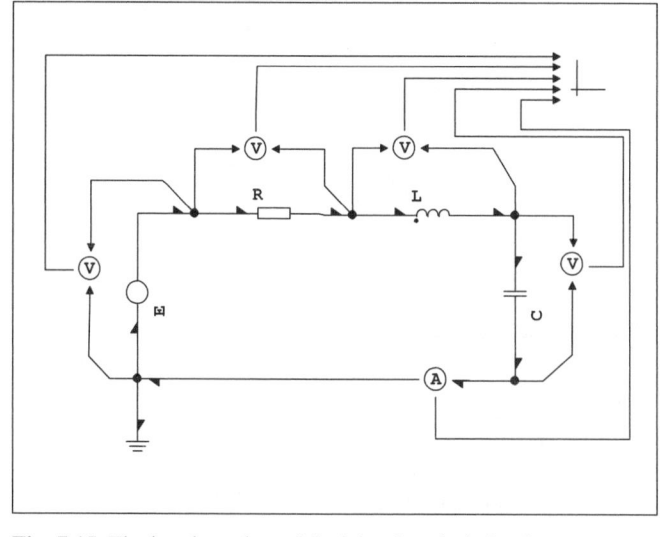

Fig. 7.15. The bond graph model of the electrical circuit

The Ammeter now must be defined. This is very simple: It is just an effort junction, from which information on the current is taken and taken outside (Fig. 7.16).

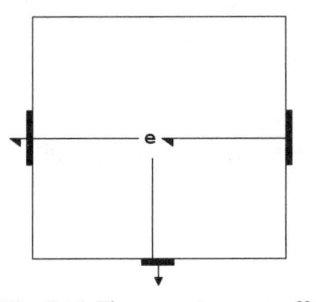

Fig. 7.16. The ammeter as an effort junction

Similarly, we create Voltmeter components for measuring voltages across other components. We do this by copying the Voltmeter component just developed by selecting it, then clicking the *Copy* button on the toolbar. Next, the component is inserted where desired using the *Insert* button (Fig. 7.15). The positions of the ports of the voltmeters lying on the left and right can be easily changed.

We now connect the output ports of all voltmeters and ammeters to the X-Y Plotter component created earlier. Because it only has a single input port, we first create a corresponding number of *control-in* ports at the plotter periphery. It is not important to place them all on the left side of the plotter; they could be placed anywhere around its periphery. At the creation of the plotter ports default names of signals connected to the X-Y Plotter ports are assigned. These are used to label the axes. After the ports are connected, we can change the names of the signals fed to the plotter. All signals are, by default, plotted on the y-axis, the x-axis serving as the *time* axis. This can be changed easily by double-clicking the plotter ports (Fig. 7.17). The dialogue that opens can be used to define names and choose the axis.

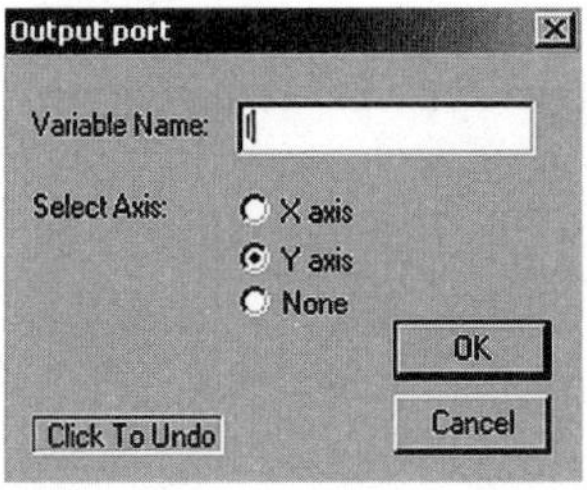

Fig. 7.17. The output variable dialogue

We select only one variable to be used as the x-axis variable. All others must be y-axis variables. We can also choose the *None* option, in which case the variable is not displayed at all. Thus, by double-clicking the current port we define the variable name I for the source current. We choose E for the source voltage, eR for voltage across the resistor, eL for the voltage across the inductor, and eC for the capacitor. Because the expected range of current values (*mA*) is quite different from the range of voltage values (V), we check *None* for the axis, so that the value of the current will not be displayed; only voltages will be displayed. (Later, we can display the time history of the current in the circuit.)

We have thus defined the models of all the circuit components. The models of the nodes are not defined, as they are just flow junctions. The ground is a source effort component and, by default, the voltage at its port is set to 0 (V). It could be changed, if required, as is the case with other source efforts, by double-clicking its port.

To start the simulation we first build the mathematical model (see Sect. 6.2.3) by clicking the *Build* button or selecting the *Build Model* command on the *Simulation* menu. After the model is built, the mathematical model of the circuit can be reviewed, as explained in Sect. 6.2.3.

Before running the simulation, we open the output window by double-clicking X-Y Plotter object to watch how the variables change. Using the *Set Axes* command on the *Simulation* menu, we can also change the default selection of axes, i.e. positive only, or both positive and negative. We select *Positive* for time (the x-axis) and *Positive and Negative* for the y-axis. It is not necessary to select ranges, as the program will do this automatically. It is also possible to change the axis labels. We presently leave these as they are. It should be noted that the computer visual system influences the duration of the simulation markedly. Thus, it is sometimes useful to run the simulation in the background and, when it is finished, to display the results by double-clicking the plotter component.

We start the simulation by clicking the *Run* button on the toolbar, or by choosing the *Run Simulation* command in the *Simulation* menu. This opens a dialogue in which we must enter the simulation interval (Fig. 6.23). The values of the circuit parameters set during model development are:

1. Source voltage E = 10.0 V
2. Resistance R = 200 ohms
3. Inductance L = 0.010 H
4. Capacitance C = $2.0 \cdot 10^{-6}$ F
5. The initial condition for both the inductor and the capacitor was set to zero

Hence, the natural frequency of the system is

$$\omega_n = \frac{1}{\sqrt{L \cdot C}} = 10000 \ 1/s = 1592 \ Hz$$

and the time constant is 0.628 ms. The damping ratio is

$$\zeta = \frac{1}{2} R \sqrt{C/L} = 1.414$$

which implies that the transients in the circuit are highly damped.

We start the simulation by setting the interval to 0.002 s, the output interval to 10 μs, and accept the defaults for all other parameters. The results of the simulations are shown in Fig. 7.18. The diagram shows the transients across the resistor, inductor, and capacitor to a 10 V step in the Source voltage.

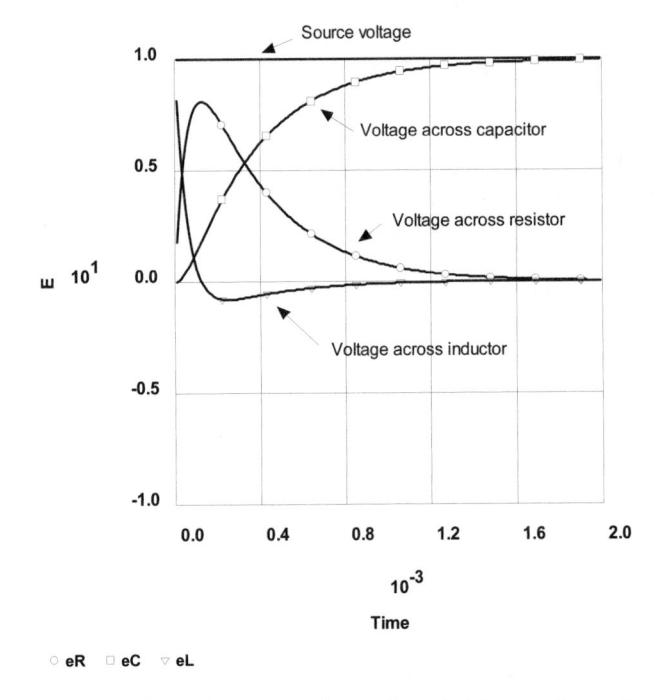

Fig. 7.18. Transient voltages across the resistor, inductor, and capacitor

The legend in Fig. 7.18 is added directly to the plot by right-clicking the mouse and choosing *Insert Legend*. The cursor changes to a cross, and we can select the position where we wish to type the text of a legend. When we are finished with text editing we then click somewhere outside it. A similar technique is used as when editing component object names in a document window. We can move a text object across the output window, or delete it. We can also add arrows using the right mouse button, selecting the *Insert Pointing Arrow* command from the drop-down menu, and then drawing the arrow. The arrow can be moved or stretched in the usual way. We can also remove the markers used to distinguish between plotted curves, in the event that these obscure some important information. This is done by clicking the right mouse button and choosing (reset) *Set Curve Marks*.

We can display a plot of the current in the circuit by means of the X-Y Plotter object. We double-click the input ports, set all voltages to *None*, and set the current to the *y-axis*. The plot can be displayed by double-clicking the plotter object. The resulting diagram, with the y-axis set to *Positive*, is given in Fig. 7.19.

At first, the complete source voltage appears across the inductor coil; then, as the current starts flowing through the circuit and the capacitor accumulates charge, it drops steeply. After the transients die out, there is no current in the circuit and the capacitor is fully charged to a value corresponding to the source voltage.

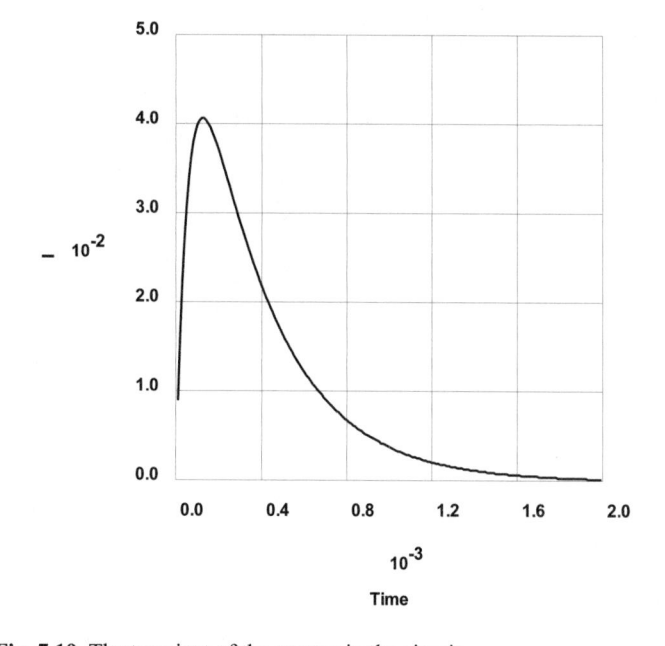

Fig. 7.19. The transient of the current in the circuit

7.3 Models of Circuit Elements

In this section models of basic circuit components, such as Resistors, Capacitors, and Inductors are developed. These can be stored in the library. Formulation of these models follows a SPICE-type description [4–7]. We use the same basic mathematical forms, including similar parameter names, but there are no "SPICE scripts". We treat the electrical components as objects that can be connected to other components by bonds, similar to how real electrical components are wired.

7.3.1 Resistors

The resistor is a two-terminal component usually represented schematically as in Fig. 7.20a. It can be represented as a two-port component in Fig. 7.20b. Its model is very simple and consists of an effort junction and a resistive elementary compo- nent (Fig. 7.21). It thus is described by the following constitutive relations:

$$e_1 - e_2 - e_R = 0$$
$$e = R \cdot i_R$$

(7.1)

Here, e_1 and e_2 are efforts (voltages) at the ports, e_R is the voltage across the resistor, i_R is the current through the resistor, and R its resistance.

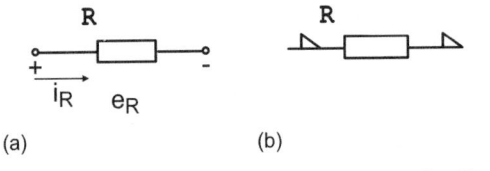

(a)   (b)

Fig. 7.20. The resistor. (a) The circuit symbol, (b) The bond graph representation

The resistance is typically a constant expressed in Ohms. The SPICE software has a built-in model of a resistor that can be exhibit temperature dependence. The bond graph models used here are much more flexible in that respect.

It is a simple matter to describe the dependence of the resistance on temperature as used in SPICE. We can define two temperatures, TNOM and TEMP, as parameters at the resistor document level using the *Parameters* button, or the *Model Parameters* command on the *Edit* menu. The nominal temperature TNOM usually is set to 300 K (27°C), and the value used for TEMP corresponds to the resistor's working temperature. We define (similar to SPICE) two temperature coefficients, TC1 and TC2. SPICE-like temperature dependence can be formulated as

$$R = R_0 * (1 + TC1 * (TEMP - TNOM) + TC2 * (TEMP - TNOM)^2)$$ (7.2)

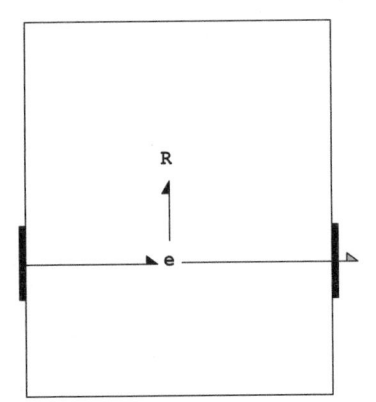

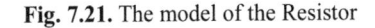

Fig. 7.21. The model of the Resistor

Such an expression can be defined in the resistive component R, or at the resistor document level. We are not restricted to such a linear or quadratic relationship, but other forms can be used as well, e.g. an exponential dependence. We could go a step further and introduce a separate thermal port at the resistor boundary (Sect. 7.4.1). In this way, dependence on temperature can be used instead of the constant temperature parameter in the constitutive relation of Eq. (7.2).

At high frequencies the impedance of resistors drops owing to parasitic capacitance effects [6]. This can be included in the resistor model by adding a parallel capacitive element (Fig. 7.22), i.e. by the use of a flow junction.

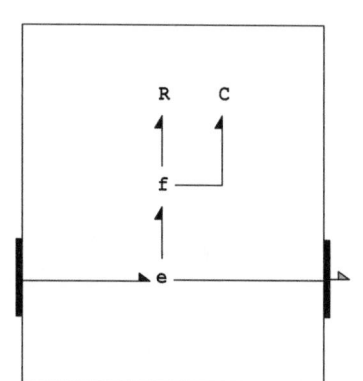

Fig. 7.22. The model of a resistor with a parasitic capacitance

7.3.2 Capacitors

The capacitor is a two-terminal component usually represented schematically as in Fig. 7.23a. We can represent it by the two-port component shown in Fig. 7.23b. Its model consists of an effort junction and a capacitive element (Fig. 7.24).

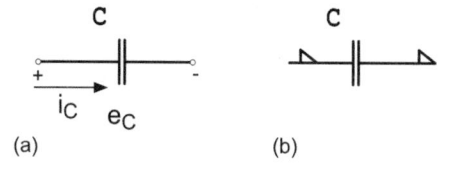

Fig. 7.23. The capacitor. (a) The circuit symbol, (b) The bond graph representation

The constitutive relations for the capacitor can be found easily (Sect. 2.5.3).

$$\left.\begin{aligned} e_1 - e_2 - e_C &= 0 \\ e_C &= q/C \\ i_C &= dq/dt \end{aligned}\right\} \tag{7.3}$$

Here, e_1 and e_2 are efforts (voltages) at the ports, e_C is the voltage across the capacitor, q is the charge, i_R is the current through the capacitor, and C is its capacitance.

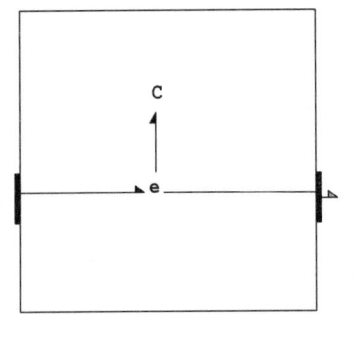

Fig. 7.24. The model of the Capacitor

The dependence of the capacitance on temperature can be described essentially in the same way. A non-linear capacitor can be described using a second equation in Eq. (7.3) that expresses voltage across the capacitor as a function of the charge. In this way, polynomial, exponential, or other non-linear dependence can thus be specified readily.

The capacitor model given above is a reasonably good approximation for frequencies up to about 1kHz [6]. At higher frequencies a series resistor and inductive element gives a better approximation to real capacitors. The model of Fig. 7.24 can be modified to include such effects (Fig.7.25). The parallel resistor models capacitor leakage, which can usually be neglected.

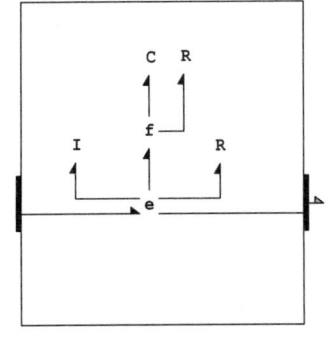

Fig. 7.25. The capacitor model with parasitic effects

The capacitor model of Eq. (7.3) is charge-based; it differs from the SPICE model, in which charge is eliminated. This simplifies the description of the accumulation of charge in circuits and its conservation. There is also some evidence that this type of model behaves better when simulating complex electrical circuits [8].

7.3.3 Inductors

Simple Inductor

The inductor is a two-terminal component usually represented schematically as in Fig. 7.26a. We represent it by the two-port component shown in Fig. 7.26b. Its model consists of an effort junction and an inertial component (Fig. 7.27).

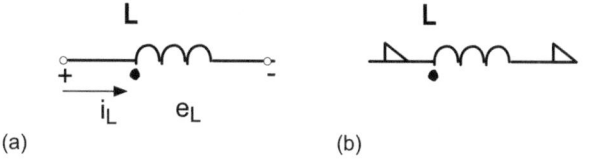

(a)   (b)

Fig. 7.26. The inductor (a) The circuit symbol, (b) The bond graph representation

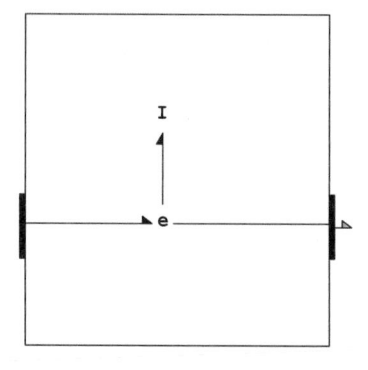

Fig. 7.27. The model of the Inductor

The constitutive relations for the inductor read (Sect. 2.5.3):

$$\left.\begin{array}{l} e_1 - e_2 - e_L = 0 \\ p = L \cdot i_L \\ e_L = dp/dt \end{array}\right\} \tag{7.4}$$

where e_1 and e_2 are efforts (voltages) at the ports, e_L is the voltage across the inductor, p is the flux linkage of the coil, i_L is the inductor current, and L is the inductance parameter.

The inductor's temperature dependence can be described in the same way as for resistors. The model above corresponds to the case in which the inductance is constant. This can be modified to include non-linear dependence on the current. We return to this problem in Sec. 7.5 when analysing electromagnetic systems.

The model of the inductor given above is a reasonably good approximation at medium frequencies [6]. Its behaviour at low frequencies is determined by the resistance of the inductor coils, a process that is not included in this model. At very high frequencies, the winding capacitance comes into effect. The model of Fig. 7.27 can be modified easily to include such effects (Fig.7.28)

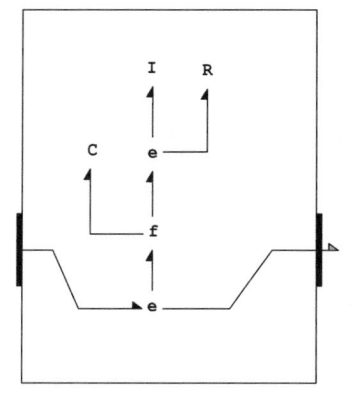

Fig. 7.28. The model of the inductor with parasitic effects

The model of Eq. (7.4) is flux based and differs from the SPICE model in which flux linkage is eliminated from the constitutive relations. It is used here similarly as in the case of the capacitor because it describes the inductor as a dynamic component with explicitly defined internal state variables. There also is evidence that such models are better suited to simulating electrical circuits [8].

Coupled Inductors

The coupled inductor is a four-terminal component usually used for modelling ideal transformers. It is represented schematically in Fig. 7.29a. We represent it by the four-port component of Fig. 7.29b.

The model of the coupled inductor, an extension of the simple inductor of Fig. 7.27, is given in Fig. 7.30. This model consists of two effort junctions at every pair of external (document) ports. These are connected internally to a two-port inertial element. To describe the mathematical model of the component, we use the following variables:

- e_1 and e_2 are efforts (voltages) at the left port, and e_3 and e_4 are those at the right
- e_{L1} is the voltage difference at the left port and e_{L2} is that of the right port
- i_{L1} and i_{L2} are flows (currents) through the left and the right ports, respectively
- p_1 and p_2 are flux linkages of the left and the right inductor coils

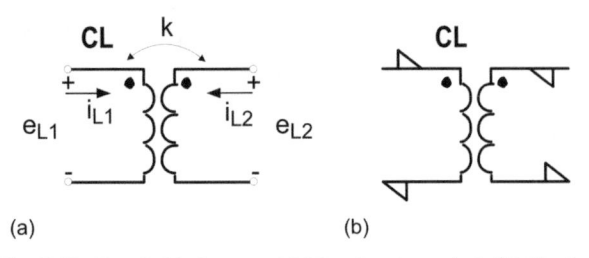

Fig. 7.29. Coupled inductors. (a) The circuit symbol, (b) The bond graph representation

The equations now read:

$$
\left.
\begin{aligned}
e_1 - e_2 - e_{L1} &= 0 \\
p_1 &= L_1 \cdot i_{L1} + M \cdot i_{L2} \\
e_{L1} &= \frac{dp_1}{dt} \\
e_3 - e_4 - e_{L2} &= 0 \\
p_2 &= L_2 \cdot i_{L2} + M \cdot i_{L1} \\
e_{L2} &= \frac{dp_{21}}{dt}
\end{aligned}
\right\}
\tag{7.5}
$$

Here, L_1 and L_2 are coil inductances defined as parameters. Likewise, M is the mutual inductance of the coils. It is usually expressed as

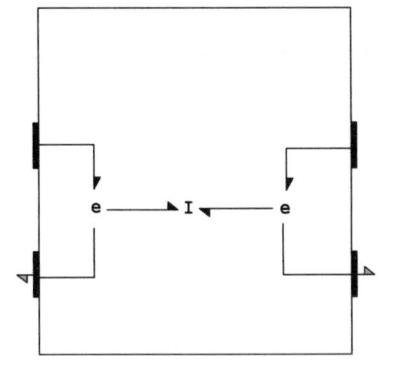

Fig. 7.30. The model of the coupled inductor component

$$
M = k \cdot \sqrt{L_1 L_2}
\tag{7.6}
$$

where k is the coupling coefficient. The relation Eq. (7.6) can be input as a parameter expression at the document level of the component, or inside the inertial component.

The model presented above is linear with constant values of the impedances. It could be modified in a way that is similar to that of the simple inductive component.

7.3.4 Independent Sources

Voltage and Current Sources

The independent voltage source is a two-terminal component that generates a voltage across its port (electromotive force) independently of the current drawn from the source. The circuit symbol used for such a component is shown in Fig. 7.31a. The bond graph component representing the independent voltage source is given in Fig. 7.31b.

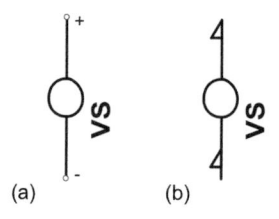

(a)   (b)

Fig. 7.31. The independent voltage source. (a) The circuit symbol, (b) The bond graph representation

It is a two-port component with a half-arrow showing the sense of power delivery. The model used for independent voltage source components is similar to other bond graph models of circuit components and consists of an effort junction and a source effort component (Fig. 7.32).

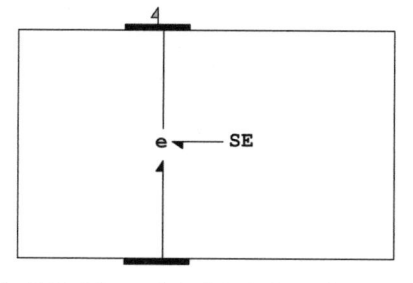

Fig. 7.32. The model of the independent voltage source component

Likewise, an independent current source is a two-terminal component that generates a current that is independent of the voltage across its terminals. The electrical circuit symbol used for such a component is shown in Fig. 7.33a. The corresponding bond graph representation is given in Fig. 7.33b. This is a two-port component with a half-arrow pointing in the sense of power delivery.

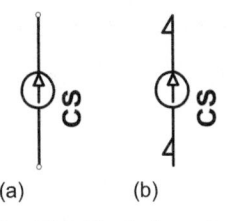

(a)   (b)

Fig. 7.33. The independent current source. (a) The circuit symbol, (b) The bond graph representation

The model of the current source is similar to that of the voltage source and consists of an effort junction and a *source flow* component (Fig. 7.34).

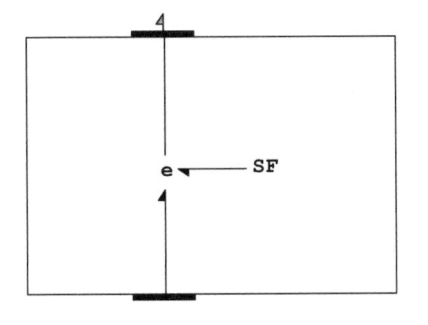

Fig. 7.34. The model of the independent current source

These simple components can be modified to produce better models, real sources, e.g. by adding a resistive element in series or in parallel.

Constitutive Relations

There are no built-in functions for voltages or currents generated by the sources. Voltages or currents, instead, are defined by the constitutive relations of the corresponding SE or SF components (Sec. 2.5.5). We now show how some important relationships can be described for voltage sources, but the same is valid for current sources, as well.

The sinusoidal emf can be described simply as (Fig. 3.35)

$$e = t < TD \ ? \ E1 : E1 + EA * \sin(2 * PI * FREQ * (t - TD)) \qquad (7.7)$$

This function assumes that the voltage has a constant offset of E1 for times less than TD, and then oscillates with a frequency of FREQ (Hz). The constants E1, EA, TD, and FREQ are defined as default parameters at the source document level of the component. The specific values that override them are input in the SE or SF components by using the corresponding dialogues.

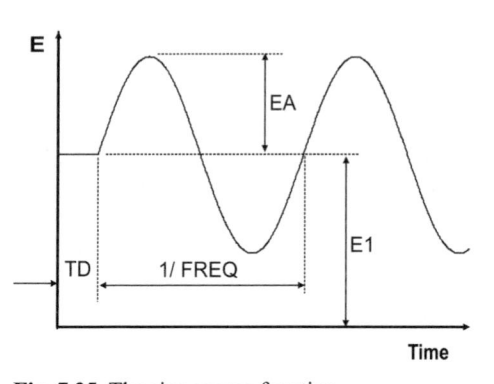

Fig. 7.35. The sine source function

An exponential function can be expressed as (Fig. 7.36)

$$e = t < TD ? E1 : E1 + (E2 - E1) * (1 - exp(-(t - TD)/ TAU)) \qquad (7.8)$$

Here, TD is a time delay and E1 is an offset, as above. E2 is the voltage strength and TAU is a time rise constant that defined by default values, e.g.

E2 = 0
TAU = 1

Specific values that override these default values can be defined in the respective source effort or source flow component.

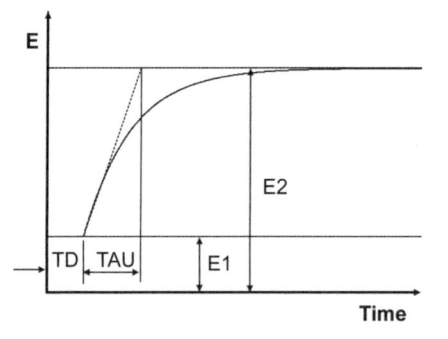

Fig. 7.36. The exponential source function

The law as given by Eq. (7.8) can be extended to include a dropping exponential

$$e = t < TD ? E1 :$$
$$(t < TD2 ?(E1 + (E2 - E1) * (1 - exp(-(t - TD)/ TAU))) : \qquad (7.9)$$
$$(E3 + (E1 - E3) * (1 - exp(-(t - TD2)/ TAU1)))$$

The time parameter TD2 > TD and E3 corresponds to the voltage at the beginning of the fall, i.e.

$$E3 = E1 + (E2 - E1) * (1 - \exp(-(TD2 - TD)/ TAU)) \tag{7.10}$$

The last parameter in Eq. (7.10) is the time constant, defined by a default value, such as TAU = 1. The time constants must not be zero, nor should it be too small; either situation will cause an arithmetic fault and the program will crash.

It is not much more difficult to define a pulse train of arbitrary waveform, such as in Fig. 7.37. To define the pulse, we need the time τ measured from the start of the pulse. This can be found as the *mod* of t - TD with respect to the pulse period PER

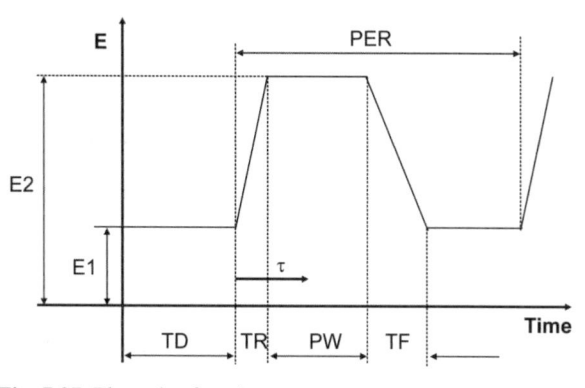

Fig. 7.37. The pulse function

$$\tau = (t - TD)\%PER \tag{7.11}$$

This expression can be used as the argument of a function defining the pulse shape

$$e = t < TD?E1: ((t - TD)\%PER < TR? E1 + KR * ((t - TD)\%PER):$$
$$((t - TD)\%PER < TR + PW ? E2 :$$
$$((t - TD)\%PER < TR + PW + TF ?$$
$$E2 - KF * ((t - TD)\%PER - TR - PW) : E1))) \tag{7.12}$$

E1 and E2 are the pulse offset and pulsed value, respectively, and

$$KR = TR > 0?(E2 - E1)/TR : 0$$
$$KF = TF > 0?(E2 - E1)/TF : 0 \tag{7.13}$$

The effect of such pulses on the circuit of Fig. 7.38 is described next. This circuit model can be found in the library project Filtering of Noise Pulses. The resistor R and capacitor C are used as a high frequency filter of noise transmitted to a load resistor RL [6].

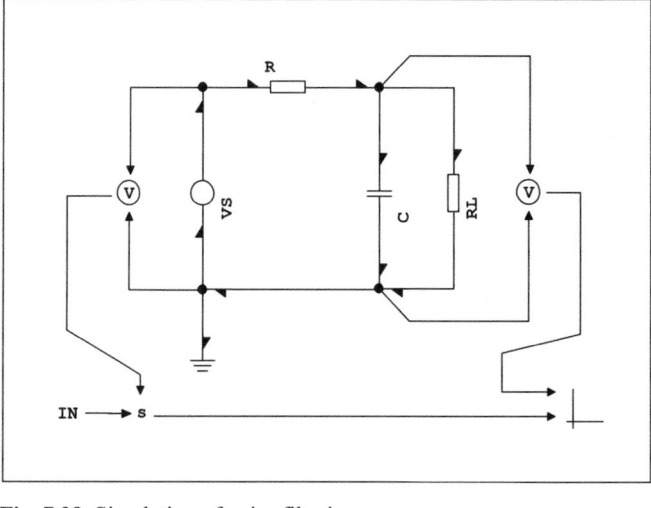

Fig. 7.38. Simulation of noise filtering

Signal noise is simulated by a voltage source that generates a rectangular pulse train with the following parameters: TD = 5 ns, TF = TR = 0, TW = 2 ns, PER = 10 ns, E1 = 5 V and E2 = 6 V. The pulse is defined by

$$e = t < TD\,?\,E1 : ((t - TD)\%PER < TF + PW + TR)\,?\,E2 : E1) \qquad (7.14)$$

The resistors are as in Fig. 7.21, with the resistance of the filter resistor R = 75 Ohm and that of the load RL = 10 kOhm. The capacitor includes parasitic inductance and resistance, and neglects leakage (Fig.7.25). The capacitance C is 1.5 mF. It initially is charged to 7.44417 mF, which corresponds to 4.96278 V. Voltages generated by the source and across the load resistor are measured by the voltmeters and fed to the plotter. Because they are in the same range, the source voltage is displaced by 2.5 V using a summator s and a signal generator IN.

Two simulations were run. In the first, the parasitic resistance and inductance of the capacitor were set to zero. The simulation interval was set to 50 ns and the output interval was a relatively short 10 ps, the better to display transients during the short pulse (2 ns). Results (Fig. 7.39) show that this ideal capacitor efficiently removes the noise.

In another simulation the parasitic parameters of the capacitor were set to L_C=10 pH and R_C=0.001 Ohm. Simulation parameters were as in the previous run. This simulation took 2.7 s of processor time using the default method and error tolerances. Results (Fig. 7.40) show sharp peaks at the rear and the front edges of the pulses. This is to be expected, as the Fourier transform of a pulse contains all frequencies (Fig. 6.32). Hence, they surely will excite higher frequency modes of the capacitor.

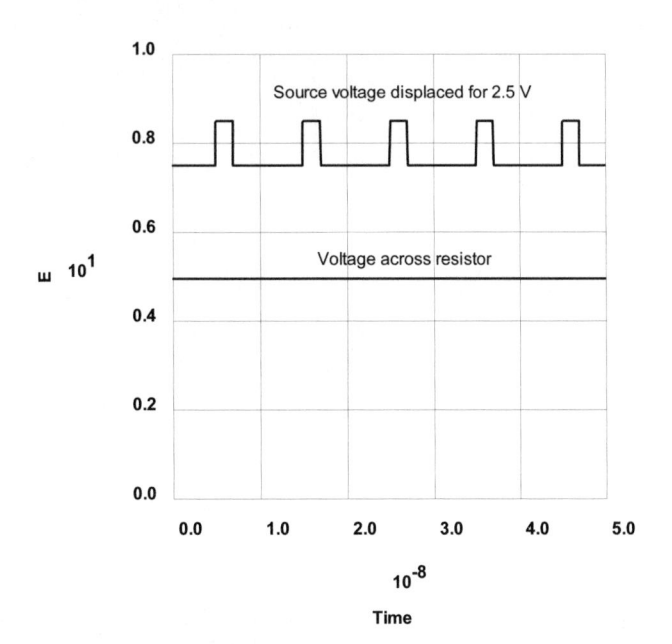

Fig. 7.39. Filtering by the ideal capacitor

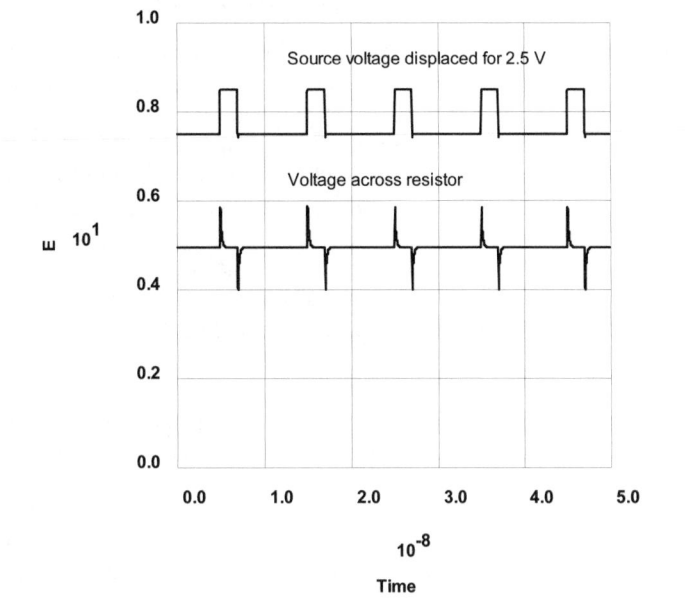

Fig. 7.40. Filtering by a real capacitor

7.3.5 Dependent Sources

Dependent sources are often used to describe the dependence of voltages or currents supplied by sources on other voltages or currents in an electrical circuit. Electric circuit design tools like SPICE define four types of such sources: voltage- and current-controlled voltage sources, and voltage- and current-controlled current sources. We don't need such special devices, because we can use the independent sources of the previous section, to which we add control ports (Fig. 7.41). These are a direct extension of the controlled components discussed in Sec.2.5.8.

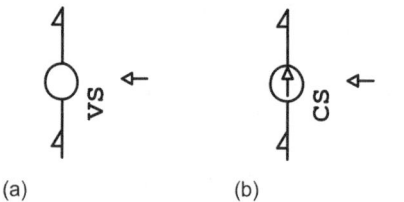

(a)   (b)

Fig. 7.41. The bond graph representation of controlled sources. (a) The voltage source (b) The current source

These models are similar to those of ordinary sources (Fig. 7.42). We need only add a control port to the corresponding source effort or source flow bond graph component and connect it externally. Fig 7.42 shows a model of a controlled voltage source.

Sources can be voltage- or current-controlled, depending upon the type of control variable used. If this variable is a current—taken, for example, from an effort junction—it is a current-controlled source; if the signal is a voltage, perhaps taken from an electrical node, it is a voltage-controlled source. It could, of course, be some other variable type, such as the position of a wheel of a manually controlled source.

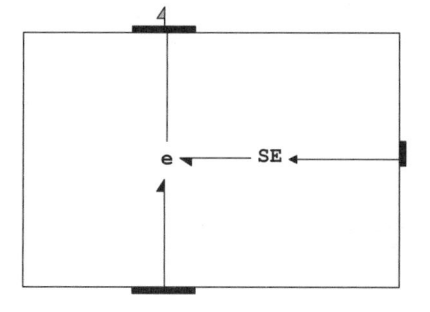

Fig. 7.42. The model of a controlled voltage source

The voltage generated by the source is defined in the **SE** or **SF** component. This can be a linear or non-linear function of the control variable (Sect.2.5.8), e.g.

$$e = k * c \qquad\qquad (7.15)$$

The control c can be the difference of two voltages, or a current flowing somewhere in the circuit.

Using the controlled voltage or current source components has the advantage that the signal serving as the control of the source is clear at first glance.

7.3.6 Switches

Switches are components often found in electrical circuits. A typical circuit symbol used for switches is shown in Fig. 7.43a. We represent electrical switches by the switch component given in Fig. 7.43b.

A switch toggles between an open position, in which there is no electrical connection between its terminals; and a closed position, in which the terminals are short-circuited. In SPICE, the switch is modelled as a controlled resistor that toggles between a high resistance value ROFF and a low resistance RON, depending on the value of the control input. Thus, a controlled resistor can be used to model a switch (Fig. 7.44).

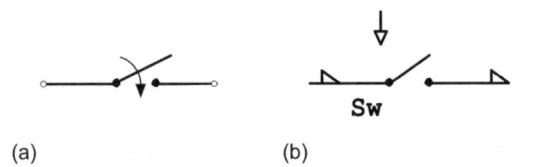

(a)   (b)

Fig. 7.43. The switch. (a) The circuit symbol, (b) The bond graph representation

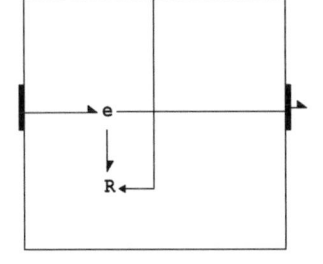

Fig. 7.44. The model of the switch as a variable resistor

The logic of the switch can be described as

$$c < EON?\, e = ROFF*i:RON*i \qquad (7.16)$$

c is the control signal, e is the voltage across the switch, and i is the current flowing through it. EON is the value of the voltage when switching occurs. The constitutive relation can also be defined in such a way that there is a continuous transition from low to high resistance, e.g.

$$c < EON - EW ?\, e = ROFF*i:$$
$$(c < EON ?\, (RON - (ROFF - RON)/EW * (c - EON))*i:RON*i) \qquad (7.17)$$

The parameter EW is the width of the zone where the transition between open and closed switching occurs.

Eq. (7.16) describes a switch that is initially open, i.e. when there is no signal. It is easy to define a switch that is initially closed, e.g.

$$c <= EON?\, e = RON*i:ROFF*i \qquad (7.18)$$

Switches sometimes are used to model logical gates. Fig. 7.45 shows a circuit for simulating an OR logical gate. The circuit model can be found in the library projects under the name OR Gate Test Circuit.

As BondSim currently has no predefined symbol for logical gates, a device symbol can be used. Inputs to the gate are generated by two voltage sources connected to the ground through 1 kOhm resistors. The sources generate pulses of voltages according to Eq. (7.12) (with TD = 0). The parameters of the pulse are as given in Table 7.1. Input signals to the gate are taken from nodes between voltage sources and the resistors. These are fed to the input ports of the gate. The gate is supplied from a separate 5 V source. The IN input components displace the inputs of the gate for better displaying.

Table 7.1. Parameters of the input pulses of Fig. 7.45

Parameters	Input 1	Input 2
PER	50 ns	100 ns
TF	10 ns	10 ns
TR	10 ns	10ns
PW	10 ns	20 ns
E1	0.0 V	0.0 V
E2	3.0 V	3.0 V

Fig 7.46 shows a model of the OR gate implemented using ideal switches. The switches are connected to the ground port by a 1 kOhm resistor. The switching logic of the OR gate switches are defined by Eq. (7.16), with EON = 2.5 V, RON = 1 Ohm and ROFF = 1 MOhm.

When the model is built, a warning appears informing the user that the model is purely algebraic, i.e. without derivatives. It is, however, acceptable to the numerical solver, so we proceed with the simulation. The results of a simulation run for 200 ns with an output interval of 100 ps are shown in Fig. 7.47. (The CPU time of simulation was 1.5 s.).

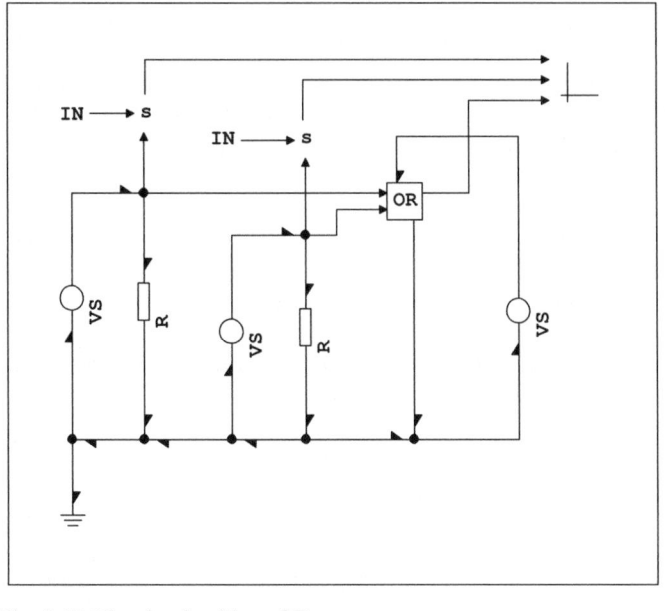

Fig. 7.45. The circuit with an OR gate

Models of logical gates based on switches offer an idealized picture of real devices. They are sometimes useful for representing gates in a simple way; but often they create problems for the simulators. The logical gates are usually built with transistors. Because these introduce some delays, the switching behaviour of real gates is often slightly different from that predicted by idealised models.

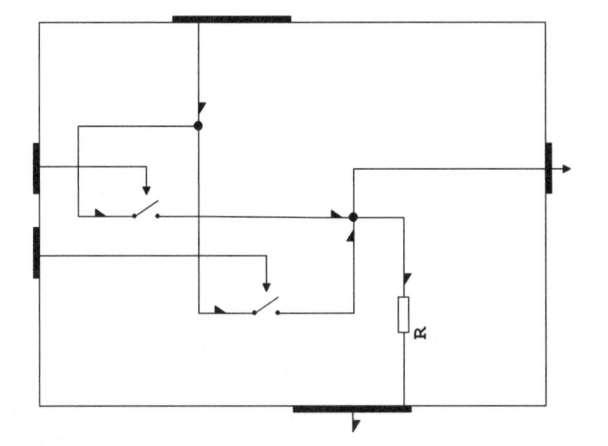

Fig. 7.46. The model of an OR gate using switches

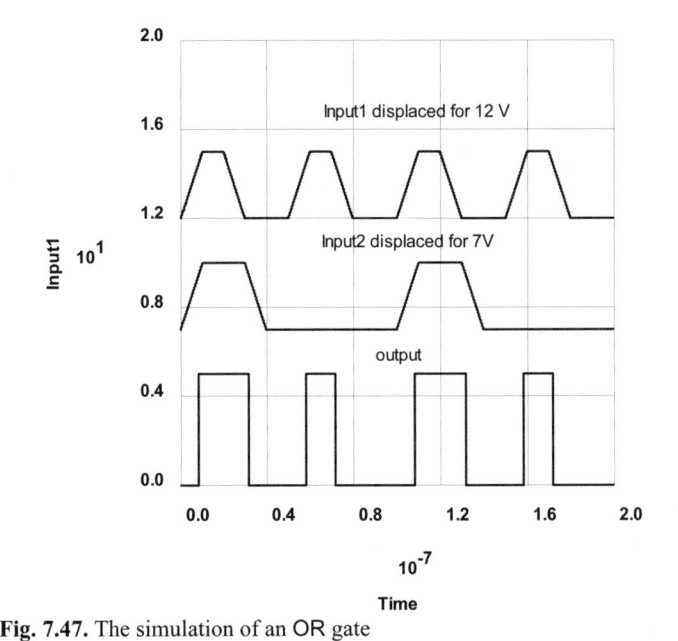

Fig. 7.47. The simulation of an OR gate

7.4 Modelling Semiconductor Components

In this section modelling of semiconductor devices from the point of view of bond graphs and component models is described. Numerous books have been written on the modelling of semiconductors and, in particular, on their SPICE models. Thus, we do not discuss this matter more than necessary, but describe some of the basic bond graph models. The interested reader can consult the literature for details, for example [1, 7, 9, 10].

In the last section it was shown that it is relatively easy to describe SPICE models by bond graphs. The same is true for semiconductors. But it is also possible to use the bond graph to develop models that can take care of effects that are not easy to implement in programs like SPICE, such as thermal effects. The component model approach of this book, supported by the language used for description of the underlying physical relations, offers a good basis for developing such models.

We start by developing suitable models of diodes with their bond graph representation. Then, three main types of transistors are considered: the Bipolar Junction Transistor (BJT), the Junction Field Effect Transistor (JFET), and the Metal Oxide Field Effect Transistor (MOSFET). The models developed correspond to SPICE large signal level 1 models. More complex models can be developed in a similar way. The component model approach is shown to be a powerful method for modelling semiconductors.

7.4.1 Diodes

Static Model

We start with diodes, which are fundamental to the functioning of practically all semiconductor components. The electrical circuit symbol used for the diode is shown in Fig. 7.48a. The bond graph component model of a diode is quite similar in appearance and is shown in Fig. 7.48a.

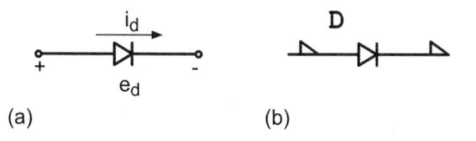

(a)   (b)

Fig. 7.48. The diode. (a) The circuit symbol, (b) The bond graph representation

Diodes are resistive components described by a non-linear constitutive relation

$$i_d = Is \cdot (e^{V_d/V_T} - 1) \tag{7.19}$$

Is is the diode saturation current and v_T is the thermal voltage, defined by

$$v_T = k \cdot T / q \tag{7.20}$$

where $k = 1.3806 \cdot 10^{-23}$ J·K^{-1} is Boltzmann's constant, T is the temperature in degrees Kelvin (K), and $q = 1.6022 \cdot 10^{-19}$ C is the electron charge. Under the nominal working temperature of 300 K (27 °C), the thermal voltage is 0.0258 V. A diode can be represented by a non-linear resistor, as given in Fig.7.49.

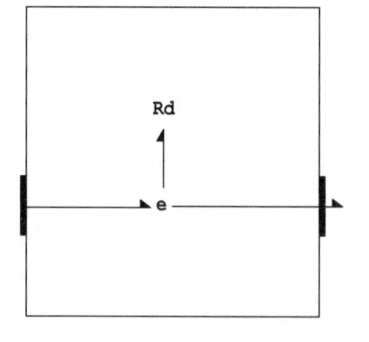

Fig.7. 49. The diode as a non-linear resistor

Models of diodes are sometimes simplified by representing them as internally modulated switches, instead of as non-linear resistive elements (Fig. 7.50). The constitutive relation of such a switch ensures correct switching between the on (forward biased) and off (reverse biased) states, e.g.

$$vd <= 0 \; \& \; id <= 0 \; ? \; id : vd = 0 \tag{7.21}$$

When the voltage across the diode is less than or equal to zero *and* the same is true for the current through the diode, the ports are disconnected and there is no current through the diode. Otherwise, its ports are short-circuited and the voltage across the diode is zero.

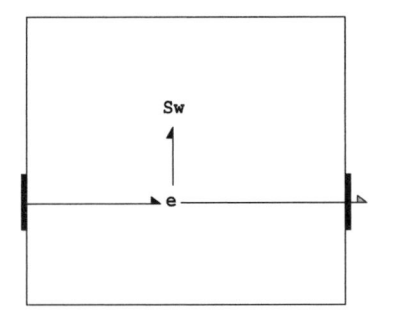

Fig. 7.50. The model of a diode as a switch

Simplified models of diodes as switches are of limited value, as they do not show some of their important features, such as voltage drop. This could be included in the model, of course, but at the cost of increased complexity. From the numerical point of view, they are less convenient than ideal diode models. This is because they model the diode by discontinuous algebraic constraints on the voltages across diodes or currents through them.

Experience shows that there is a departure of the behaviour of real diodes from the ideal law given by Eq. (7.19). This occurs over the greater part of forward and reverse regions, as illustrated in Fig. 7.51 (*see* e.g. [7] for a detailed explanation). To account for such departures, SPICE models diodes as current sources with non-linear diode characteristics and a series resistor. The non-linear characteristic is similar to Eq. (7.20), but introduces the *emission constant* n in the exponent

$$i_d = Is \cdot (e^{v_d' /(nv_T)} - 1) \tag{7.22}$$

The default value of this parameter is 1, which corresponds to the ideal diode, but can be in the range 0.7 to 3 for real diodes. This coefficient describes the departure of the diode at a low forward bias. The series resistor, on the other hand, describes the departure at high forward bias. The voltage appearing in Eq. (7.22) is the *effective* voltage, equal to the voltage across the diode terminals less the voltage drop across the series resistor, i.e.

$$v_d' = v_d - R_s i_d \tag{7.23}$$

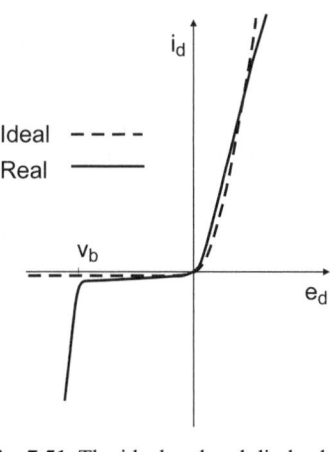

Fig. 7.51. The ideal and real diode characteristics

We follow the same approach but, instead of the current source, a non-linear resistive element Rd is used (Fig. 7.52). The series resistor effect is modelled by a resistive element R added to the resistor effort junction. In this way, the non-linear element effort variable is the effective diode voltage, as given by Eq. (7.23).

The constitutive relation of the non-linear resistive element is given by Eq.(7.22)

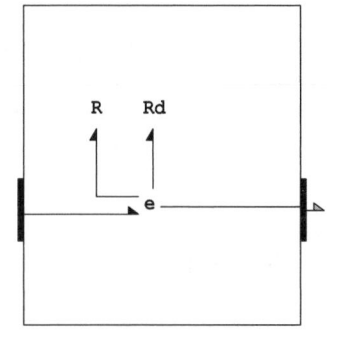

Fig. 7.52. The static model of a real diode

$$id = IS \cdot (\exp(ed/(N * VT)) - 1) \tag{7.24}$$

The saturation current IS, the thermal voltage VT and the emission constants N are parameters of the model. These parameters can be defined at the diode document level by default values similar to that of SPICE. These values can be overridden inside the resistive element.

In the reverse bias region the diode current is practically equal to the saturation current. Some diodes show a stronger dependence on the reverse voltage. This can be taken into account by modelling the diode characteristics separately for forward voltages, as per Eq. (7.24); and using a similar expression, but with a different emission parameter, for the reverse region.

One point that remains to be addressed is reverse diode breakdown. This occurs when the reverse voltage reaches some specific value v_b (Fig. 7.51). The reverse current then suddenly increases to very high values. The process is not necessarily destructive but, owing to the high power dissipated in the diode, it is often critical to determining its useful life. We describe diode breakdown using an approach similar to that in SPICE [4,6,7]. Diode characteristics in the breakdown region can be described by

$$i_d = -IS \cdot e^{-(BV0 + v_d')/v_T} \tag{7.25}$$

When the voltage is less than -BV0, the current starts increasing without bound. This critical value is found from the specification of the breakage voltage value BV and the corresponding current $-IBV$, i.e.

$$-IBV = -IS \cdot e^{-(BV0-BV)/v_T} \tag{7.26}$$

Solving, we get

$$BV0 = BV - VT \cdot \ln(\frac{IBV}{IS}) \tag{7.27}$$

BV0 is found once the breakdown point, defined by BV and IBV, is known. The shape of the function in Eq. (7.26) ensures a smooth transition from characteristics given by Eq. (7.22) at the breakdown.

Taken together, the constitutive relation of the resistive element of Fig. 7.55 can be described as

$$id = ed > -BV0 \; ? \; IS \cdot (\exp(ed/(N*VT)) - 1):$$
$$-IS \cdot \exp(-(BV0 + ed)/VT) \tag{7.28}$$

When applying such a relationship for the simulation of diode behaviour, a few points of caution should be noted. One of these relates to the behaviour of the exponential term at high voltages. If the series resistance is zero, then all of the voltage drop occurs across the diode resistive element. Owing to the exponential character of diode behaviour, the current through the diode could be very high. Such a diode would surely melt down. Thus, the series resistor is an essential guard against this possibility.

Because of the finite precision of floating-point arithmetic, there may be overflow during the numerical calculation. The order of the maximum number that can be represented in double-precision mode typically is 10^{308}. Thus, the maximum value of the exponential term before overflow is about 709, or a voltage across the diode of about 18 V. To prevent this from occurring, it is possible to approximate the exponential function for high values of exponents by a linear function, e.g.

$$ed/(N * VT) > MEXP?$$

$$IS * (exp(MEXP) * (1 + ed/(N * VT) - MEXP) - 1):$$ (7.29)

$$IS \cdot (exp(ed/(N * VT)) - 1)$$

The MEXP parameter is set at 50, but can be changed. It is possible, of course, to limit high currents by using a series resistance [4].

Another important issue is the low conductance at reverse biases, as the expo nential term in Eq. 7.23 very quickly becomes almost zero, and the constant leak-age term remains only. This may cause a problem during simulation because the partial derivative matrix of the system equations can become very badly condi-tioned. To remedy this, it is advisable to use the approach employed in SPICE; that is, to add to the constitutive Eq. (7.29) a constant conduction term of the form GMIN*ed, where is GMIN is very low, e.g. 10^{-12}. This term will not change the diode behaviour appreciably, but can help to solve this problem. With this addi-tion, the constitutive relation of Eq. (7.28) now reads

$$id = ed > -BV0 ? IS \cdot (exp(ed/(N * VT)) - 1) + GMIN * ed:$$

$$-IS \cdot exp(-(BV0 + ed)/VT) + GMIN * ed$$ (7.30)

To illustrate the behaviour of a diode based on the model developed, we ana-lyse the **Rectifier Circuit** project of Fig. 7.53. It consists of a voltage source gener-ating a sine voltage of amplitude 50 V at a frequency of 50 Hz, a diode and a 100 Ohm resistor. The parameters of the diode are given in table 7.2 and correspond to the *Motorola 1N4002* general purpose rectifying diode at 25°C [6]. The break-down voltage is taken at default BG_MAX value, i.e. there is no diode break-down.

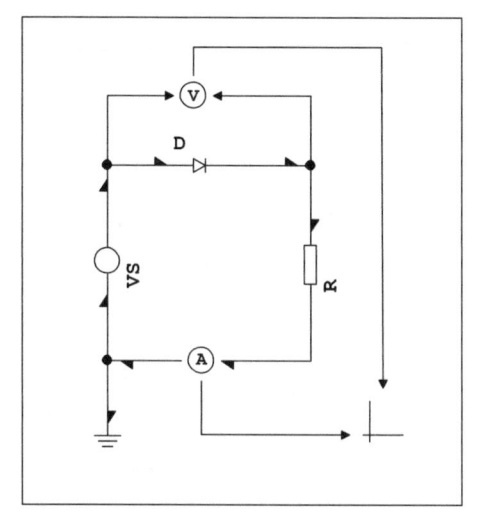

Fig. 7.53. A simple rectifier circuit

Table 7.2. Model parameters of the diode in Fig. 7.53

Parameter	Value
IS	$46.5 \cdot 10^{-12}$ A
RS	0.123 ohm
N	1.35
VT	0.0255 V
BV0	BG_MAX

In the first simulation the interval is set to 10 ms, i.e. to the half period of the voltage generated by the source when the diode is forward biased. The intention is to simulate the diode I-V characteristics. The output interval is chosen as 1 µs. As usual, the voltage across the diode was plotted along the x-axis and the current along the y-axis. The characteristic obtained is shown in Fig. 7.54. It can be seen that the current starts rising at a voltage of about 0.6 V. The reverse part is not simulated because there is little to show.

In the second experiment the simulation interval is set to 0.100 s, i.e. five periods of the voltage. Fig. 7.55 shows the current in the circuit. During the first half-period (0.010 s) the current flows through the resistor, reaching its maximum value of 0.491 A. When the voltage changes its polarity, the diode becomes reverse biased and there is an extremely low current through the circuit.

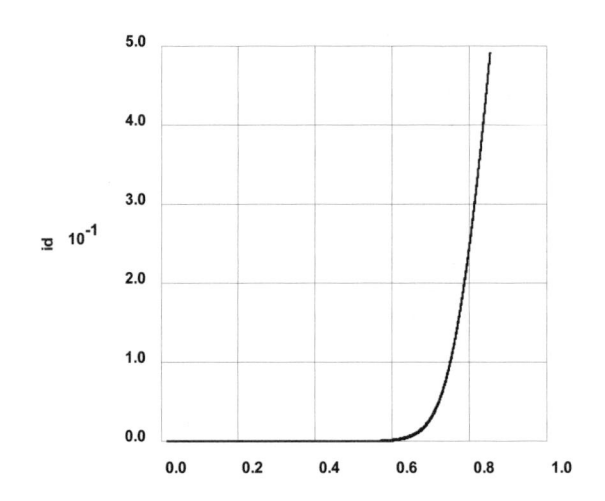

Fig. 7.54. The diode forward voltage vs. current

To analyse the behaviour of the circuit when diode breakdown occurs, the VB0 was set to 40 V and the simulation repeated. Fig. 7.56 shows a plot of the diode current. During the breakdown there is appreciable reverse current through the diode (-0.0944 A).

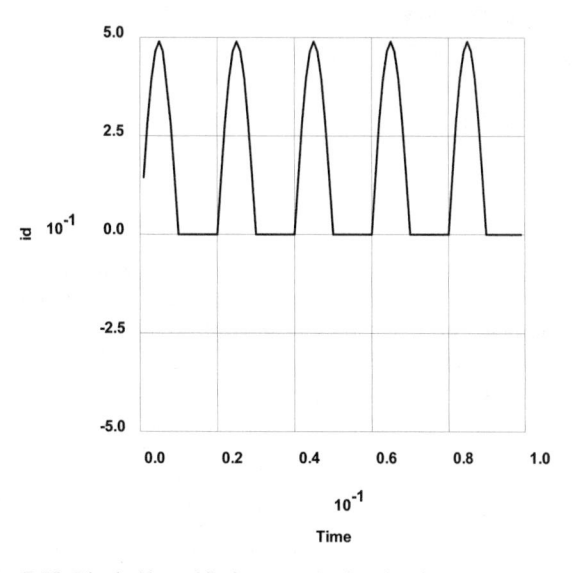

Fig. 7.55. The half-rectified current in the circuit

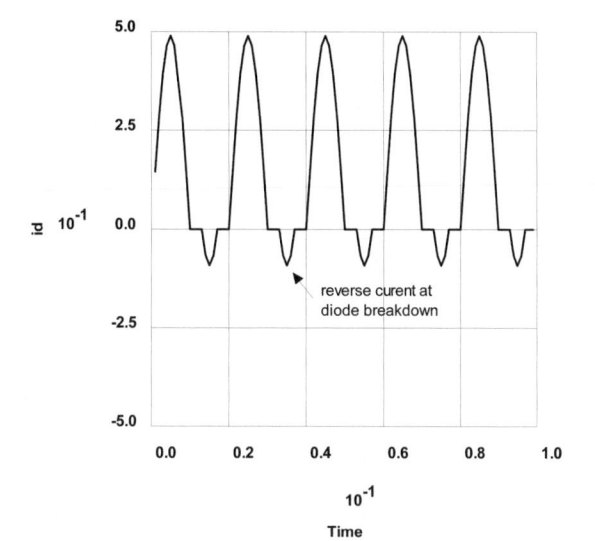

Fig. 7.56. The current in the circuit when there is diode breakdown

It should be noted that the program failed when the diode breakdown characteristics were described according Eq. (7.30). When the breakdown voltage was reached, there was an error message indicating that the program failed to evaluate the partial derivative matrix. To cure the problem, the exponential terms was changed as given by Eq. (7.29).

Dynamical Model

The model developed so far is quasi-static. Such a model, from the perspective of dynamical applications, would be infinitely fast. This is not the case in real diodes, as charge-storage effects limit the velocity of the response. To account for the dynamical effects, we take an approach similar to that used in circuit simulators such as SPICE [4,7]. Our approach is charge-based, however.

There are two mechanisms of charge storage in diodes: charges stored in the depletion layer and charges injected across the layer into neutral regions. Under no bias voltage, the depletion region consists of fixed dopant ions. This layer behaves as a plate capacitor of capacitance C_{j0}. Under negative bias (and under positive bias that is less than the built-in potential), the width of this layer changes, as does the capacitance. We use the charge formulation of such capacitor effects instead of the more common capacitance formulation.

The charge of the depletion region can be expressed as a function of the applied voltage using the formula

$$Q_j = \frac{C_{j0} v_{bi}}{1-m} \left[1 - \left(1 - (v_d / v_{bi}) \right)^{1-m} \right] \tag{7.31}$$

where v_{bi} is the built-potential and m is the grading coefficient. The last coefficient can have a value between 0.5 for abrupt, and 0.33 for linearly graded, junctions. The corresponding capacitance is given by

$$C_d = \partial Q_d / \partial v_d \tag{7.32}$$

From Eq. (7.32) we get

$$C_d = \frac{C_d(0) v_{bi}}{\left(1 - (v_d / v_{bi}) \right)^m} \tag{7.33}$$

The junction charge and capacitance functions, as given by Eqs. (7.31) and (7.33), are only approximate and agree with more exact values at voltages less than about $v_{bi}/2$ [7]. At a voltage equal to the built-in potential, the capacitance is infinite. At that high bias, on the other hand, the charges generated by injection dominate. Thus, it is usually assumed that some error in the charge and the capacitance near the built-in voltage is acceptable. If more accuracy is required, the formula of [11] can be used. We note that, in spite of the fact that we are only interested in charges, capacitances appear naturally when the program calculates the partial derivatives of Eqs. (7.31) by symbolic differentiation. We thus must take care of the behaviour of capacitances, as well.

We follow the approach in [7] that approximates junction charges by Eq. (7.31) up to $FC \cdot v_{bi}$, and employs quadratic interpolation beyond. The corresponding formula reads

$$Q_j = \frac{C_{j0} v_{bi}}{1-m} \left[1 - \left(1 - (v_d / v_{bi}) \right)^{1-m} \right], \, v_d < FC \cdot v_{bi} \tag{7.34}$$

and

$$Q_j = C_{j0}\left(\frac{V_{bi}}{1-m}[1-(1-FC)^{1-m}] + \frac{1}{(1-FC)^m} \cdot (V_d - FC \cdot V_{bi}) + \right.$$

$$\left. \frac{m}{2V_{bi}(1-FC)^{m+1}}(V_d - FC \cdot V_{bi})^2 \right), (V_d \geq FC \cdot V_{bi}) \qquad (7.35)$$

The effect of these charges can be represented by a capacitive element Cj in parallel with the diode's non-linear resistive element (Fig. 7.57). The constitutive relation of the element is defined by Eqs. (7.34) and (7.35), i.e.

$$q = v_d / VJ < FC \; ? \; CJO * VJ * (1 - (1 - v_d / VJ)^\wedge (1 - m)) / (1 - m) :$$

$$CJO * VJ * (F1 + F2 * (v_d / VJ - FC) + F3 * (v_d / VJ - FC)^\wedge 2) \qquad (7.36)$$

where VJ is the built-in potential, CJO the initial junction capacitance value, and

$$F1 = (1 - (1 - FC)^\wedge (1 - m)) / (1 - m)$$
$$F2 = (1 - FC)^\wedge (-m) \qquad (7.37)$$
$$F3 = 0.5 * m * (1 - FC)^\wedge (-1 - m)$$

FC typically has a value of 0.5.

Minority-carrier charges injected into the neutral sections also influence diode dynamics. These are termed diffusion charges [7] and are given by

$$Q_d = TT \cdot i_d \qquad (7.38)$$

where TT is the transit time parameter of the diode. Accumulation of diffusion charges can be represented by a capacitive element Cd in parallel with the junction charges capacitor Cj, as shown in Fig. 7.57.

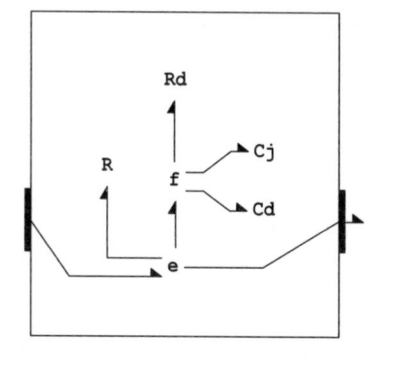

Fig. 7.57. The dynamical model of the diode

The constitutive relation for the charge is given by Eqs. (7.24) and (7.38)

$$q = TT * IS \cdot (\exp(ed / (N * VT)) - 1) \qquad (7.39)$$

To find the time response of the diode we simulate transients in the circuit of Fig. 7.53 when the voltage is pulsed from 0 to 50 V, and then to −50 V. The diode dynamical parameters are given in Table 7.3. Parameters of the generated pulse are (Fig. 7.37): TD = 0 s, TR = 10 ns, PW = 40 µs, and TF = 20 ns. BV0 is set to the maximum value (BG_MAX). Corresponding model can be found in the library under the project name Diode Recovery. The simulation is run for 50 µs with an output interval of 2ns. The results are shown in Fig. 7.58.

Table 7.3. The dynamical parameters of the diode

Parameter	Value
CJ0	$51.5 \cdot 10^{-12}$ C
M	0.333
VJ	0.381
TT	$5.77 \cdot 10^{-6}$ s

When the pulse switches to a negative voltage, the charges injected cause a large reverse current of practically the same value as the forward current (Fig. 7.58). Following the analysis given in [9], the *storage delay time* is approximately equal to 3.91 µs. Only when these charges are removed does the current return to the saturated (or breakdown) value. The necessary transition time depends on the depletion capacitance and circuit resistance. The corresponding time constant can be approximated by $Cj0 \cdot R \cong 0.005$ µs. Thus, the total diode recovery time is dominated by the storage delay time. The value estimated by simulation is 3.96 µs, which agrees well with these figures. It can be checked (by opening the diode voltage plot) that the diode voltage is ≈ 0.8 V until the charges are removed.

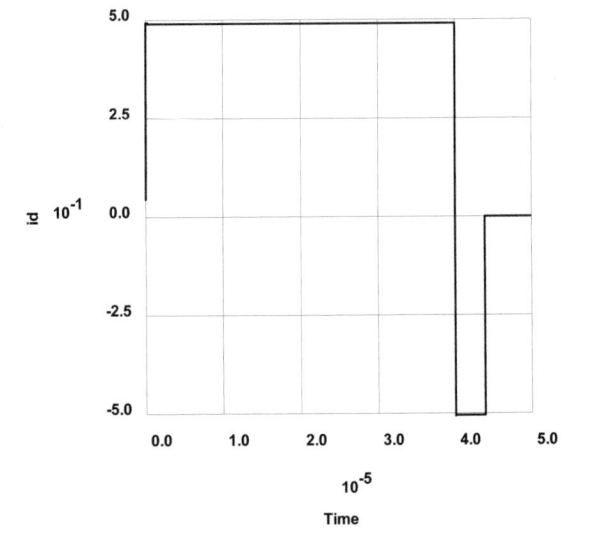

Fig. 7.58. Response of the diode current to a pulse in the supply voltage

Diode Self-Heating

Diodes and other semiconductor devices are known to be notoriously dependent on temperature. It thus is important to model the thermal processes as well as the electrical processes.

Temperature is explicitly contained in the exponent of the diode I-V characteristics through the thermal voltage kT/q. But this is not the only temperature effect because other parameters, such as the saturation current, built-in voltage, and others are strongly influenced by temperature. It is thus of interest to take account of this temperature dependence when modelling the static and dynamic behaviour of diodes.

In SPICE, all circuit parameters are defined at a nominal working temperature of 300 K (27°C). This temperature can be changed to some other *specified* value and all parameters re-evaluated, but the simulations are run at a fixed temperature. Other languages, such as VHDL-AMS [12], are better equipped to deal with the thermal side of device modelling. Bond graphs can also be used to model the thermal effects of devices [2,13]. The approach taken in this book is particularly appropriate for that purpose.

During the diode's operation part of the electrical power is transformed to heat. This heat flows out of the diode to the surroundings; but part also accumulates in the diode, thereby changing its temperature. Depending upon the net heat balance, the diode can heat up or cool down. A change of temperature, on the other hand, influences the electrical characteristics of the diode, both static and dynamic.

To account for such interactions, we add an additional port that serves for the heat transfer with the environment (Fig. 7.59). We label such a diode as DTh (Diode Thermal model).

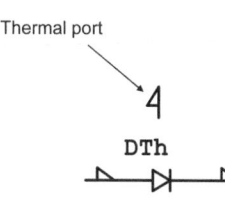

Fig. 7.59. The representation of a diode with self-heating

We develop a thermal model of the diode based on the static model of Fig. 7.52. The proposed model is shown in Fig. 7.60. In addition to the components that model electrical processes, the model also contains a component termed Self-Heating, which models the thermal processes. These components interact through their thermal ports.

Heat generation occurs in elements R and Rd, where electrical power is transformed into heat that flows out. To represent these heat flows, a thermal port is added to every resistive element. The thermal processes are modelled using *pseudo-bond graphs* (Sect. 1.3). In accordance with Table 1.1, the effort-flow pair of variables at these ports are *temperature* temp and *heat flow* fQ at the port.

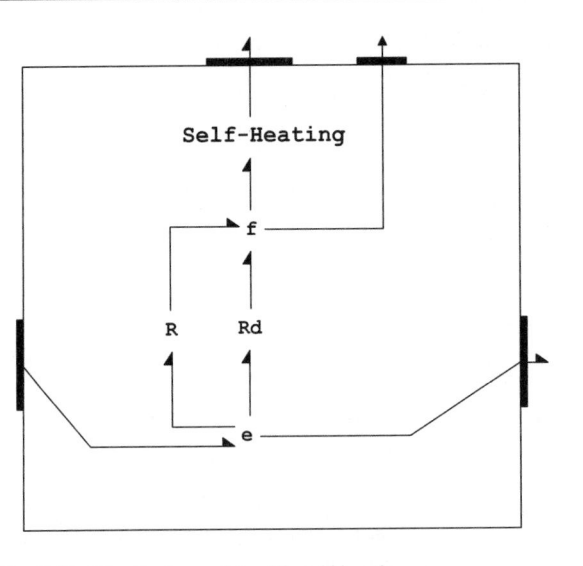

Fig. 7.60. The diode model with self heating

These pseudo bond graphs are convenient for modelling thermal processes. It also is possible to use true bond graphs that are based on *temperature* and *entropy flow* pairs. True bond graphs are more general and also much more complicated than pseudo-bond graphs. They are not really needed here.

The constitutive relations at thermal ports are of the form

$$fQ = e \cdot i \qquad (7.40)$$

where e and i are the voltage and the current, respectively, at the electrical ports. Heat flow is equal to input power. Temperature at the port is defined relative to nominal temperature TNOM. Thus, the absolute (thermodynamic) temperature is

$$T = temp + TNOM \qquad (7.41)$$

The reason for this is that the program is designed to start from the state in which the system is not yet activated; hence, all variables, with the possible exception of certain state variables, are zero. Regarding temperature, we assume that the system is not at absolute zero, but has some predefined value TNOM, which is taken to equal 300 K.

The temperature at a thermal port affects the processes at electrical ports. In the diode resistive element this is through the thermal voltage defined by Eq. (7.20). Thus, we change the constitutive relation Eq. (7.24), replacing the thermal voltage parameter VT by the expression BOLTZMANN·(temp+TNOM)/ECHARGE, where the electron charge and Boltzmann's constants are globally defined.

ECHARGE = $1.6022 \cdot 10^{-19}$ (C)
BOLTZMANN = $1.3806 \cdot 10^{-23}$ (J·K^{-1})

Practically all diode parameters depend on temperature [7]. Temperature dependence of the saturation current is given by

$$is(T) = is(T_0) \cdot (\frac{T}{T_0})^{XTI/n} \cdot e^{-\frac{q E_g(T_0)}{nkT}(1-T/T_0)} \tag{7.42}$$

Where T_0 is the nominal temperature and XTI is a parameter that has a value 3 (for *pn* diodes). $Eg(T_0)$, the energy gap at nominal temperature, is defined (for *Si*) by the parameter EGNOM = 1.115 V.

We describe the right side of the above equation as

IS * (1+ temp/TNOM)^(XTI/N) * exp(ECHARGE * EGNOM *

(temp/TNOM)/(N * BOLTZMANN * (temp + TNOM))) $\tag{7.43}$

and simply substitute it for the saturation current IS in Eq. (7.24).The resulting expression is quite complicated, but this is not a problem for the program. We need only to write it correctly.

We can do it another way, too. The total heat generated is equal to the sum of the heat that flows out of thermal ports, which is represented by a flow junction in Fig.7.60. The junction variable is the temperature temp of the diode junction (relative to TNOM). This can be used to define the saturation current using function components and to return its value through a control input port to the resistive element.

The thermal model of the diode is defined in the component Self-Heating. In general, thermal models of diodes, as well as those of other semiconductor devices, are represented using thermal circuits [14]. That is, continuous thermal processes are discretisized and represented using electrical analogies. We do not need such analogies, for we can deal with this directly by bond graphs or, in particular, pseudo bond graphs [2,13].

Two basic elements are used to build thermal models. The first is the *thermal resistor*, represented by a resistive bond graph element (Fig. 7.61a). The constitutive relation of such a resistor is

$$T_1 - T_2 - T = 0$$
$$fQ = fQ(T) \tag{7.44}$$

i.e. heat flow is a function of temperature. In the linear case, the constitutive relation of the element is simply

$$fQ = T/R \tag{7.45}$$

where R is the thermal resistance expressed in K/W (or °C/W).

The other element represents heat storage (Fig. 7.61b) and is defined by

$$fQ_1 - fQ_2 - fQ = 0$$
$$fQ = \frac{dE}{dt} \tag{7.46}$$

where E is the thermal energy. The accumulated energy is

$$E = C \cdot T \qquad\qquad (7.47)$$

Parameter C is the thermal capacitance and has units of J/K. Strictly speaking, it corresponds to the specific heat at constant volume. But in solids and liquids, including semiconductor materials, work done by expansion of the material is very small compared to the net heat inflow fQ, hence it is usually referred to simply as the material's specific heat. Eq. (7.47) is the constitutive relation of the capacitive element in Fig. 7.61b.

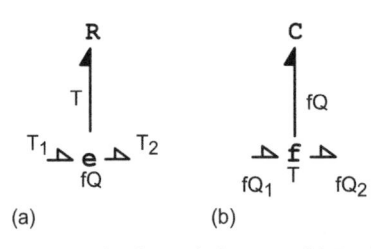

(a)   (b)

Fig. 7.61. The thermal elements. (a) Resistor, (b) Storage

The Self heating component of the diode can be represented by the RC circuit shown in Fig. 7.62. The resistive element represents heat that flows from the junction because of the temperature gradient between the junction and the diode's outside surface. Part of this heat is accumulated in the diode. This is represented by the capacitive element.

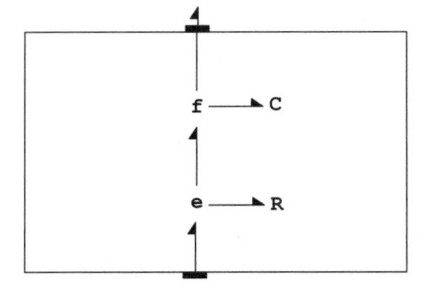

Fig. 7.62. The thermal model of the diode

The diode thermal port is normally connected to components that model the diode environment. This is usually the *heat sink* that removes heat from the diode, ensuring that its temperature is held within acceptable margins (Fig. 7.63). The other side of the sink is usually at ambient temperature, here represented by a SE component. Thermal models of such sinks can be represented in a similar way using a RC circuit.

Heat sinks are usually made of extruded aluminium and for which manufacturers provide thermal data, such as thermal resistance, volume, and material. Complete thermal models also should include the resistance of the heat path to the sink, which depends on the design of the semiconductor components. A component usually is enclosed in a case, and between the case and the heat sink there is some layer of isolation. Thus, the thermal model of the diode and heat sink, as shown in Fig. 7.63, is a simplified one. For proper thermal modelling, details of the design should be taken into account. Detailed models usually consist of several RC segments connected in series [14].

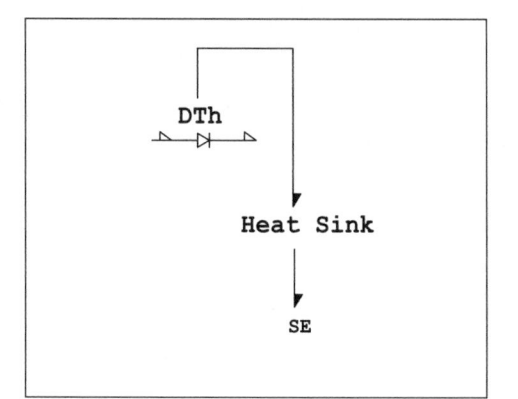

Fig. 7.63. The diode thermally connected to a heat sink

To illustrate the application of models of the diode, including self-heating, we return to the rectifier circuit of Fig. 7.53. We replace the diode model with its thermal model (Fig. 7.64). Information on diode temperature is taken from the heat flow junction (Fig. 7.60) and sent to the display component. The corresponding project can be accessed from the BondSim project library under the title Rectifier circuit with self-heating.

We simulate the temperature rise in the diode following a sudden voltage change from 0 to 50 V. Parameters of the diode thermal model and of the heat sink are given in Table7.4. The other side of the sink is held at nominal temperature. These parameter values have been chose to illustrate the behaviour of the diode under changing temperature conditions; real values could be derived from, for example, cooling tests of real devices. Simulation results are presented in Fig. 7.65.

Table 7.4. Thermal parameter of the circuit

Parameter	Value
Diode thermal resistance	0.1 K/W
Diode thermal capacitance	0.0001 J/K
Heat-sink thermal resistance	10 K/W
Heat-sink thermal capacitance	0.01 J/K

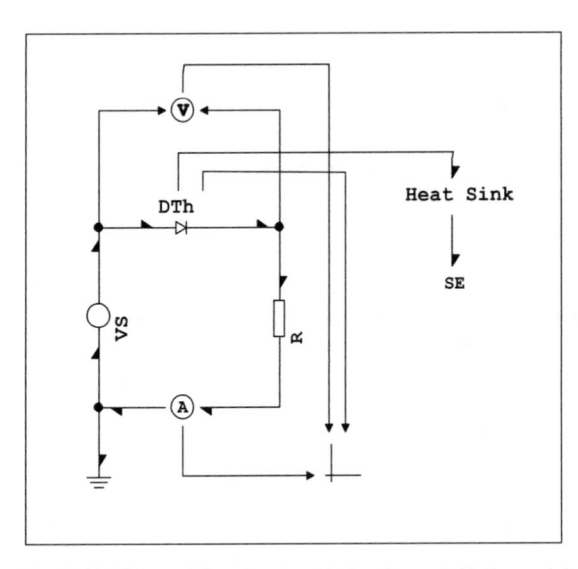

Fig. 7.64. The rectifier circuit with the thermal diode model

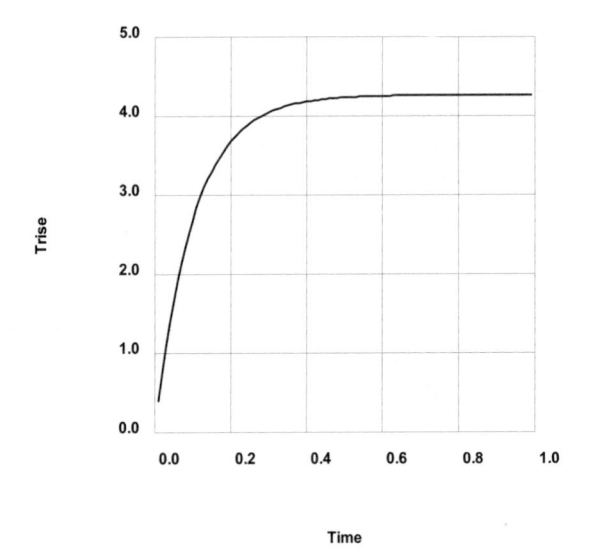

Fig. 7.65. Transient of diode temperature with change in the source voltage

7.4.2 Transistors

The bipolar junction transistor was invented by Bardeen, Brittain, and Schockley in 1947 while working at Bell Laboratories in Murray Hill. Rightly considered to be one of greatest inventions of our time, it earned them the Nobel Prize in physics in 1956. For a long time after its invention the BJT remained one of most important three-terminal devices. It is found in amplifiers and drivers and, even today, serves as one of the most important devices in a wide array of applications.

The field effect transistor (FET) emerged shortly after appearance of the BJT. In fact, many different types of FETs were developed, their difference depending upon how isolation between the gate and channel was implemented. These include the junction FET or JFET, metal semiconductor FET or MESFET, metal-oxide FET or MOSFET, as well as others.

We will not go into the details of the design and functioning of BJTs or FETs, for this is outside the scope of this book. The interested reader can consult specialised books on semiconductors, such as [7,9,10]. We will, however, show how models of some basic types of transistors can be developed as components that can be used to study complex mechatronic or micro-mechanic systems. We follow the modelling approach of SPICE. The models we present are fundamental ones that correspond roughly to *Level 1* models. More advanced component models can be developed in the same way.

We wish to stress that the intention here is *not* to develop special bond graph models that are, perhaps, energetically correct, but often not so easy to comprehend by those who are not experts in bond graphs. Our models will be visually close to the electrical schemas used to describe SPICE models, but they are complete component models without anything hidden .

Bipolar Junction Transistor

The Bipolar Junction Transistor (BJT) is a three-terminal device. The terminals commonly are denoted as the *emitter* E, the *base* B, and the *collector* C. There are two main types of BJTs: the *npn* and the *pnp*. The electrical circuit symbol used for npn transistors is shown in Fig. 7.66a.

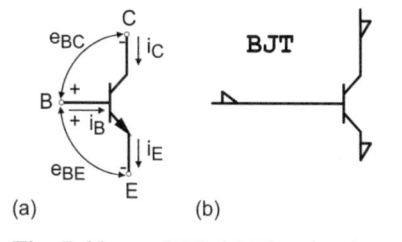

(a)   (b)

Fig. 7.66. npn BJT. (a) The circuit symbol, (b) the bond graph representation

The emitter arrow shows the direction of the current under normal operations. The positive directions of currents at the other terminals are shown. These are slightly different from those in SPICE [4,7]. The corresponding bond graph component representation is shown in Fig. 7.66b. It is assumed that power flows into the component at the base and the collector ports, and flows out at the emitter port. Such a component can be created in BondSim by choosing the *Bipolar Junction Transistor* button from the *Electrical Component* palette (Fig. 7.2). The text BJT is simply a label used for reference and can be changed at component construction time or later.

In the *pnp* type of *BJT*, the positive direction of the currents is just the opposite (Fig. 7.67a). Hence, the *pnp* bond graph assumes positive senses of power-flow at all ports to be the opposite to that of the npn type (Fig. 7.67b). Thus, to create an empty *pnp* BJT component, the *npn* component is created first, then the power direction of each of the three ports is changed. This is accomplished by selecting the component and using the *Change Ports All* command on the *Edit* menu.

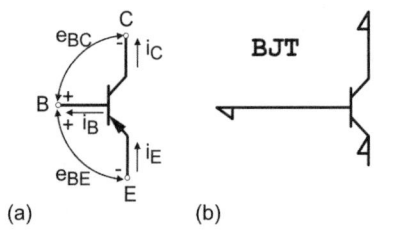

(a)   (b)

Fig. 7.67. pnp BJT. (a) The circuit symbol, (b) The bond graph representation

We develop a model of a BJT based on the Ebers-Moll model, specifically the transport version [7]. The *npn* transistor model is given (Fig. 7.68) consists of two diodes that model two back-to-back pn junctions, and a current source that models the transport of current from the collector to the emitter. Models used for the diodes are slightly modified models of Sect. 7.4.1. They consist of non-linear resistive and capacitive elements, but without the series resistive element (Fig. 7.69). Transistor ohmic resistances are accounted for by the parasitic base Rb, collector Rc, and emitter Re resistors (Fig. 7.68). The constitutive relations for the diode's non-linear resistive elements are

Base-collector diode:

$$i_d = (Is/\beta_R)(e^{e'_{BC}/(n_R VT)} - 1) \qquad (7.48)$$

Base-emitter diode:

$$i_d = (Is/\beta_F)(e^{e'_{BE}/(n_F VT)} - 1) \qquad (7.49)$$

I_s is the transistor saturation current and β_R and β_F are reverse and forward current gains, respectively. The current source represents the transport of charges and is given by

$$I_{CT} = Is(e^{e'_{BE}/(n_F VT)} - e^{e'_{BC}/(n_R VT)}) \qquad (7.50)$$

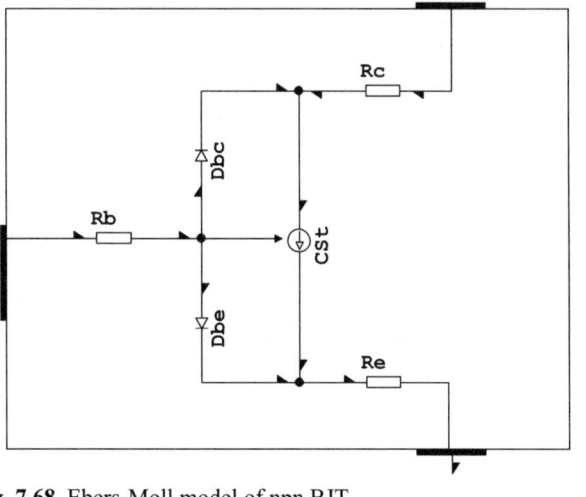

Fig. 7.68. Ebers-Moll model of npn BJT

Voltages v'_{BC} and v'_{BE} are effective base-collector and base-emitter voltages, respectively. This is exclusive of the voltage drops of the parasitic resistances.

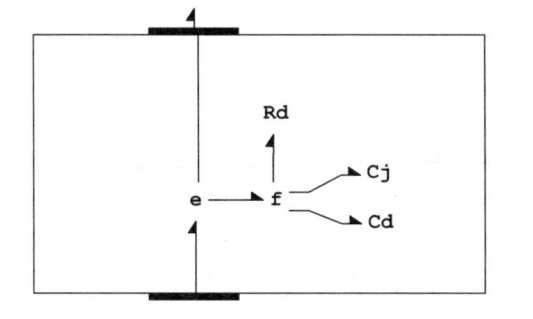

Fig. 7.69. Model of base-collector and base-emitter diode

To generate a current given by Eq. (7.50), it is necessary to supply the current source in Fig. 7.68 with information on the base voltage. The base node is a flow junction; its junction variable is effort, i.e. the base voltage. Thus, we can create a control-out port at the base node and a control-input port at the current source component, then join them by a bond line, as shown in Fig. 7.68. It now is an easy matter to define the current source that implements the constitutive relation of Eq. (7.50). This is shown in Fig. 7.70.

The base-collector and base-emitter voltages are created using two summators. These signals are inputs to the two source flows that generate the current components of Eq. (7.50). Finally, the flow junction is used to add these current components.

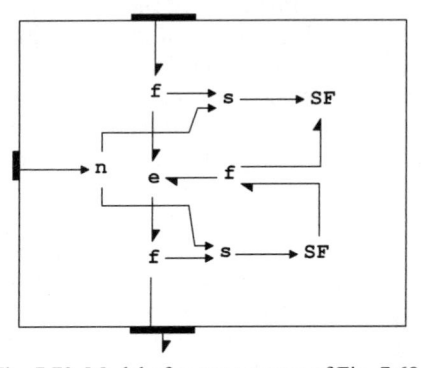

Fig. 7.70. Model of current source of Fig. 7.68

Charge storage is modelled by capacitors that are part of the collector and emitter junction diode models (Fig. 7.69). Constitutive relations of these capacitors are as described in Sect. 7.4.1. The transistor model is a dynamic, large-signal Ebers-Moll model. This doesn't include secondary effects, which could be included by following an approach similar to SPICE [7].

To simulate the transistor characteristics, we create the project npn Transistor Characteristic. The system level model shown in Fig. 7.71 represents the model of a set-up for measurement of *npn* bipolar junction transistor characteristics in the common emitter configuration.

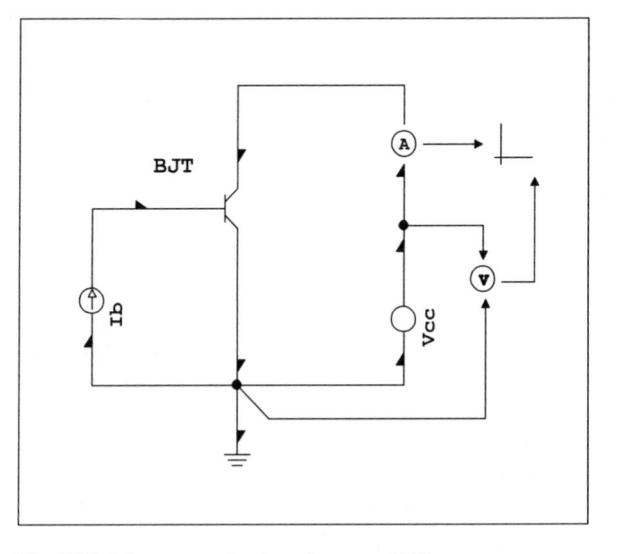

Fig. 7.71. Measurement set-up for a npn BJT

The base terminal of the transistor is fed by a current source and the collector terminal is supplied by a voltage source Vcc that is sinusoidal with an amplitude of 10 V and a frequency of 10 Hz. The voltage of the collector terminal with respect to the emitter terminal is measured with voltmeter V, and the current flowing through the collector terminal is measured by ammeter A. Their outputs are connected to a plotter. We wish to determine a plot of collector current Ic vs. the collector to emitter voltages VCE for the base current in the range of 0 to 50 μA. Parameters of the transistor are given in Tables 7.5 and 7.6.

Default values of parameters of Table 7.6 are defined at the transistor document level; those of the diodes in Table 7.7 are defined at the document level of the respective diodes. The values that override these defaults must be defined on a higher level (counting the system level as 0). Thus, for example, parasitic resistances are defined in the respective resistors. What is a default and what is a specified value is not predefined, as in SPICE; this decision is left to the modeller.

Table 7.5. Parameters of the transitor of Fig. 7.71

Parameter	Value
IS	10^{-16} A
VT	0.0258 V
BR	1
BF	80
RB	5 ohm
RC	1 ohm
RE	0.01 ohm

Table 7.6. Forward and reverse diode parameters of the transistor

Parameter	Value
CJ0	0.0 F
VJ	0.75 V
m	0.333
FC	0.5
TT	0.0 s

During simulation the output normally is stored in, and plotted by, the plotter object. To have a family of curves plotted on the same plot, we must perform several simulation runs and store all resulting values in the same object, then plot them. This can be done relatively simply. We start, for example, with a base current of 0 μA, build the model, and run the simulation for 0.1 s. For better display, we select the output and maximum interval to be 0.0001s. Next, we change the current value to 10 μA. We need not rebuild the model, as it is updated automatically when we click the *OK* button after changing the current. We then rerun the simulation, but in the *Simulation Option* dialogue (Fig. 7.72) we select *Restart* button and uncheck the *Reset plot* check box. The first option forces the simulation to restart from zero; the second prevents release of data from the previous run.

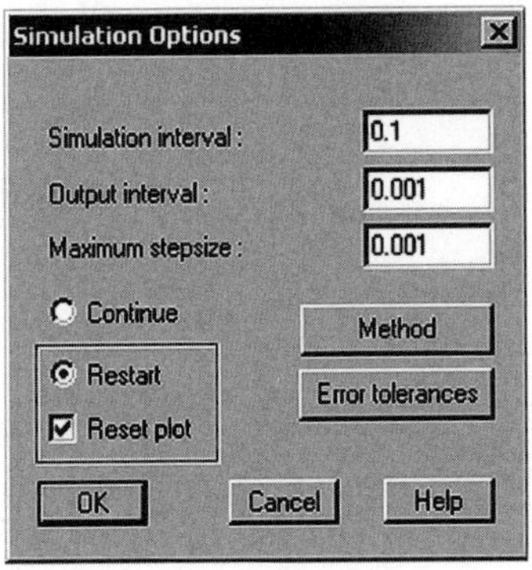

Fig. 7.72. Simulation Options dialogue with Restart and Reset buttons

In the display object (Fig. 7.71) we assign the collector current to the y-axis and label it as Ic, and the voltage to the x-axis and label it as VCA. If the plot window is not already open, we simply double-click the plotter object. Simulations were run with an output interval of 0.1 s (one period of the supply voltage) and a fairly small output interval of 0.0001 s, the better to display the characteristics. In the resulting graph (Fig. 7.73), the curves are labelled by the value of the base current, which are added using the right mouse button as explained in Sect. 7.2.

The diagram shows the familiar emitter characteristics for different values of base current. Because of different current gains for forward and reverse diodes (80 to 1), they are asymmetrical. The collector current changes from 800 µA to 4 mA when the base current changes from 10 to 50 µA. This is in accordance with the forward current gain of 80.

The pnp BJT is similar to the npn type with holes replacing the roles of electrons. Thus, the directions of currents and polarities of the voltages across the emitter, the base, and the collector are opposite to that of the *npn* BJT. The corresponding component representation already has been discussed (Fig. 7.67); its model is developed in a similar way.

We also can create a model of the *pnp* BJT directly by using a copy of the corresponding *npn* model. Starting with the *npn* BJT (Fig. 7.67b), we select the component and change the sense of all ports by the *Change Ports All* toolbar button (or by the corresponding command in the *Edit* menu). This changes the external ports of the component and also the ports of all components inside its document that are connected to them, i.e. ports of resistors Rb, Rc, and Re (Fig. 7.68).

We also need to change the power-flow direction of the ports on the other side of the resistors. Next, we successively select the diodes and the current source in Fig. 7.68 and change all of their ports. Because we must change the direction of the current flow through the current source, we also must change the sense of the effort junction ports connected externally through the flow junctions, but not that of the port connected to the transported current components flow junction (Fig. 7.70). Finally, we must change the summator signs. A model of a *pnp* BJT can be found in BondSim's *Component Library*. The pnp Transistor Characteristics project that parallels the *npn* transistor project discussed above is found in the Bond-Sim projects library.

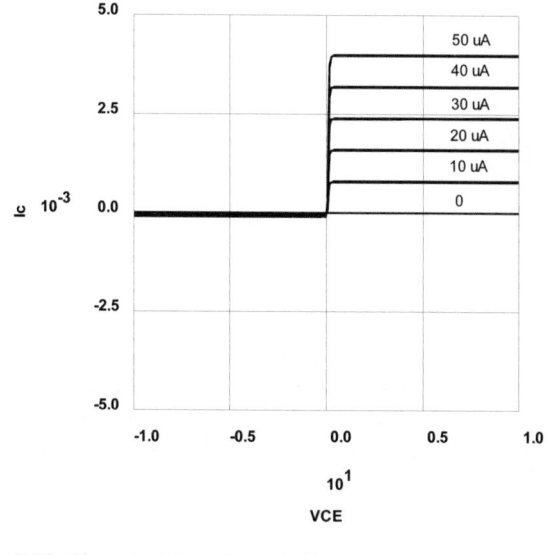

Fig. 7.73. Characteristics of npn BJT

Junction Field Effects Transistor

The Junction Field Effect Transistor (JFET) is a three-terminal voltage-controlled device. Its operation involves an electric field that controls the flow of charge through it. JFET terminals commonly are denoted as *source* S, *drain* D and the *gate* G and are analogous to the emitter, collector, and base terminals of the BJT. The source and drain are ohmic contacts. There is a conductive channel between the drain and the source, through which current flows. The third terminal, the gate, forms a reverse-biased junction with the channel. The conductivity of the channel is modulated by a potential applied to the gate. As the conduction process involves predominantly one type of carrier, JFETs also are called unipolar transistors.

Depending on the type of material involved, a JFET may be referred to as a *n-channel* or a *p-channel* JFET. The electrical circuit symbol used for n-channel JFETs is shown in Fig. 7.74 a. The corresponding bond graph component representation is shown in Fig. 7.74b. It is assumed that power flows into the component at the gate and the drain ports, and flows out at the source port. The component can be created using the *n-channel JFET* button of the *Electrical Component* palette (Fig. 7.2). The text JFET is simply a label used for reference and can be changed when the component constructed or later.

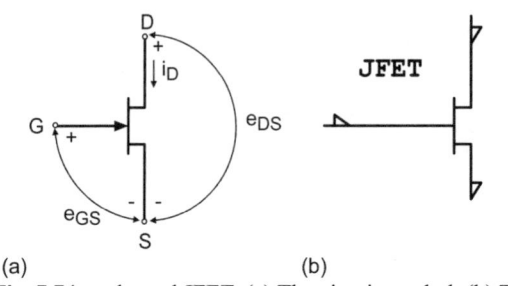

(a)   (b)

Fig. 7.74. n-channel JFET. (a) The circuit symbol, (b) The bond graph representation

The polarities are just the opposite in a *p-channel JFET* (Fig. 7.75). Thus, the p-channel JFET component can be created from an n-channel component by reversing the power flow direction of all ports. This can be done by selecting the component and using the command *Change Ports All*.

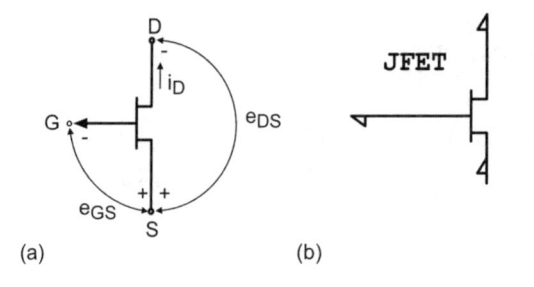

(a)   (b)

Fig. 7.75. p-channel JFET. (a) The circuit symbol, (b) The bond graph representation

It is also possible to change only the gate port. In this case, the drain and the source of the p-channel JFET change places with respect to the n-channel JFET. The n-channel and p-channel JFETs differ in the sense of the gate port power flow. The drain of the n-channel component is a port where power flows *in*; for the p-channel, power flows *out*. For the sources, the opposite is true.

The direction of power flow through JFETs is taken to correspond to the normal mode of operation in which the drain of the n-channel device is at higher potential than that of the source; hence, current flows from the drain to the source. Thus, electrical power flows in at the drain and flows out at the source. For the p-channel device, the source is normally at a higher potential, so current flows from the source to the drain. The power flow direction of the p-channel JFETs is just the opposite.

We develop a model of the n-channel JFET (Fig. 7.76) that consists of two diodes, DS and DD, which model the reverse-biased junctions between the gate and the channel, and a voltage-controlled resistor. Resistors RD and RS represent the ohmic resistances of the drain and the source. This model corresponds to the large-signal SPICE model with the controlled resistor R replacing the dependent current source [4].

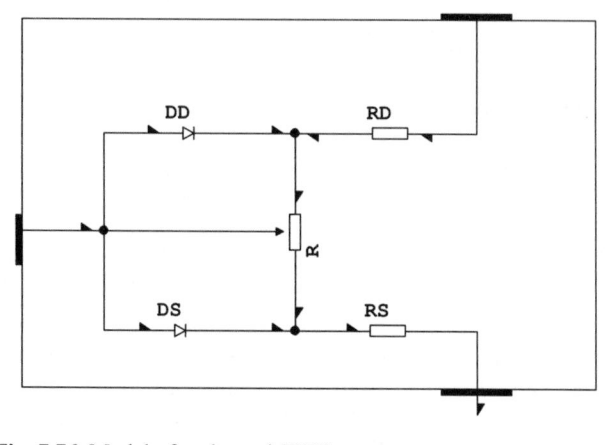

Fig. 7.76. Model of n-channel JFET component

The constitutive relation of the current-controlled resistor is given by [4]

$$id = \begin{cases} 0, & Vgs \leq VT0 \\ BETA \cdot Vds \cdot (2(Vgs - VTO) - Vds)(1 + LAMBDA \cdot Vds), & \\ & 0 < Vds < Vgs - VT0 \\ BETA \cdot (Vgs - VTO)^2 (1 + LAMBDA \cdot Vds), & \\ & Vds > Vgs - VT0 \end{cases} \quad (7.51)$$

The VT0 parameter is the threshold (pinch-off) voltage. This determines the gate bias at which the channel is completely pinched-off and there is effectively no current through the device [4,9]. If VT0 < 0, then at Vgs = 0 the device is in the *on* condition and, under a positive drain to source voltage, current will flow from drain to source. To cut off the device it is necessary to apply a negative gate to the source voltage. Such device operation is known as the *depletion mode*. In the *enhanced mode* VT0 > 0, the device is pinched-off initially and it is necessary to apply a positive voltage larger than the threshold to enable the device to conduct the current.

The diode models consist of a non-linear resistor and a capacitor (Fig. 7.77). Because in JFETs the diodes normally are reverse-biased, there is no diffusion charge. Hence, the charges consist of fixed ions in the depletion region and are represented by a junction capacitor.

Eq. (7.51) shows that at a relatively small drain-to-source voltage Vds the current increases with the voltage until the saturation voltage (equal to Vgs-VT0) is reached. This region is called the linear region of operation. When the drain-to-source voltage increases above the saturation voltage, the drain-source current is practically independent of the voltage and the device is saturated. The parameter BETA is a trans-conductance parameter, and LAMBDA is the output conductance at saturation.

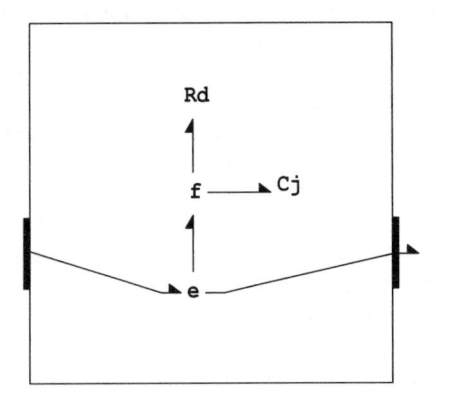

Fig. 7.77. Model of junction diodes

The controlled resistor defined by Eq. (7.51) can be described as shown in Fig. 7.78. Its model is similar to other resistors, i.e. it consists of a resistive element R connected to an effort junction, the ports of which also are connected to the document (external) ports of the resistor. In this way, the effort and flow variables of the resistive element are the drain-to-source voltage Vds and the channel current id. The element also has a control port to collect information on gate bias. Note that the input signal at the document left port is the gate potential (Fig. 7.76). Thus, to create the gate-to-source voltage Vgs a flow junction is inserted, the junction variable of which is the source potential. A summator is used to evaluate the difference of these two potentials, and its output is connected to the resistive element control-input port.

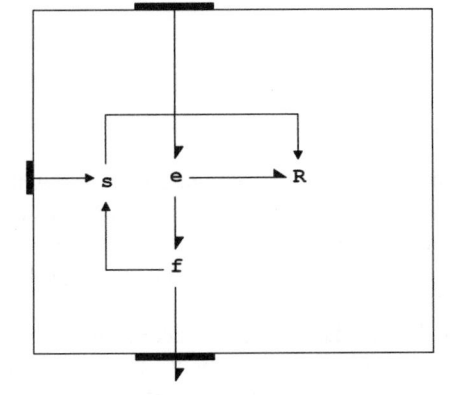

Fig. 7.78. Model of controlled resistor

The constitutive relation of Eq. (7.51) is valid for drain-to-source voltage Vds $\geq$ 0. This corresponds to the normal mode of operation. In the inverted mode (Vds < 0), the drain and source ports switch roles and the current flows in the opposite direction. The corresponding constitutive relation is symmetric [4,7]. Thus, instead of the Vgs voltage, the Vgd voltage is used in Eq. (7.51). This can be achieved by using relationship Vgd = Vgs - Vds. Similarly, Vds is changed to -Vds and the sign of the current is changed.

As was the case with the BJT, we can find the id-Vds characteristics of a JFET by simulation. For this purpose we create a project n-Channel JFET Characteristics, the system level model of which is given in Fig. 7.79. The drain of the JFET is supplied from voltage source VD generating voltage ramp 0–5 V. The gate port is connected to a separate voltage source. The source of the JFET is grounded. We measure the voltage across the drain and source ports by voltmeter V and the current through the component by ammeter A. Instruments outputs are fed to a plotter for display. JFET parameters correspond to the 2N4416 JFET of reference [6] and are given in tables 7.7 and 7.8.

Table 7.7. Basic 2N4416 JFET parameters [6]

Parameter	Value
RD	0.575 ohm
RS	0.575 ohm
IS	$5 \cdot 10^{-12}$ A
VT0	-3.32 V
BETA	0.05 A/V2
LAMBDA	0.00928 1/V

Table 7.8. Junction diode parameters of 2N4416 JFET [6]

Parameter	Value
VJ	0.76 V
CJ0	3.37 pF
FC	0.5
M	0.5
N	1
VT	0.0258 V

To generate id-Vds characteristics of the JFET, several simulations are run with constant gate voltage ranging from 0 to − 3.32V. The simulation interval was set to 1s, which corresponds to increasing the drain-to-source voltage from 0 to 5 V. The output interval was 0.001 s. Output of each simulation is stored in the display component. Results are given in Fig. 7.80.

The transistor is in the depletion mode. As the gate bias becomes negative, conduction of the transistor drops and, at voltage of –3.32 V, is completely pinched-off. The curves also show a linear region in which the source voltage increases with the drain-to-source voltage; and a saturation region in which the current is independent of the drain-to-source voltage.

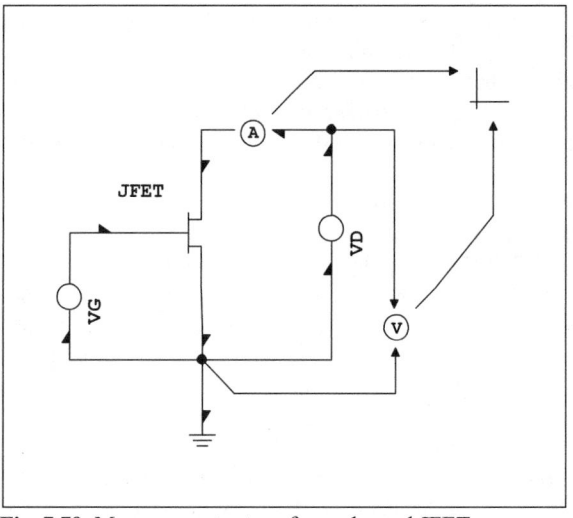

Fig. 7.79. Measurement set-up for n-channel JFET

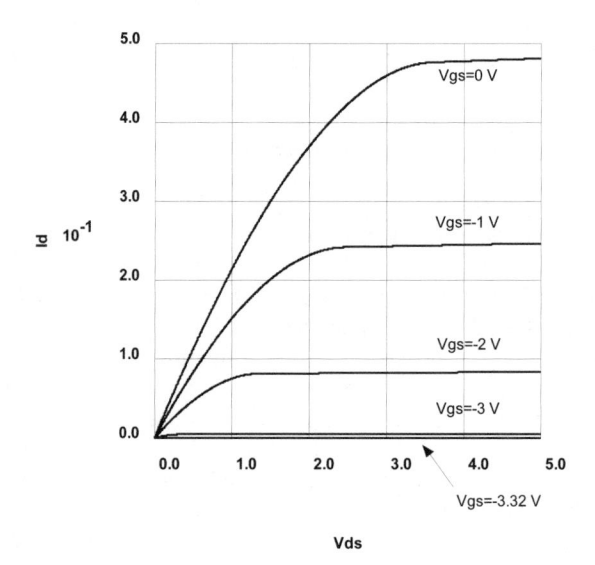

Fig. 7.80. Simulation of characteristics of 2N4416 JFET

The discussion so far has been for n-channel devices, but a similar discussion is valid for p-channel devices. In p-channel devices all voltage and current polarities, including the threshold voltage, and the directions of two gate junctions are reversed [7]. Likewise, greater-than and less-than relations must be reversed. We retain, however, the same constitutive relation as that used for the n-channel device:

Instead of terminal voltages Vgs and Vds, we use reverse voltages Vsg and Vsd. Similarly, VT0 now is the negative of the threshold voltage. Thus, a positive value of VT0 corresponds to a device that is off—in the *enhanced mode*—for both n-channel and p-channel devices. Similarly, the negative value corresponds to a *depletion* mode device, which is initially on, for both n-channel and p-channel devices.

The change in polarity of the voltages across the terminals and of the direction of current is taken care of by a change in the power flow direction through the p-channel device (Figs. 7.74 and 7.75). The model of the p-channel JFET is similar to the n-channel model of Figs. 7.76 to 7.78, but with direction of power flow between external ports reversed. The change of power flow direction should be applied only to the bonds through which power is transferred between external ports. The others are not affected, e.g. the power flow direction to the non-linear resistive element and of the capacitive element of the diode model in Fig. 7.77, as well as that of the controlled resistive element of the resistor in Fig. 7.78. These are the same for both n-channel and p-channel devices. In addition, we need to change the signs of the summator inputs of Fig. 7.78, because these components are used to evaluate Vgs, the voltage used in Eq. 7.51. This way, the p-channel JFET model can be created from the n-channel model by changing the port power flow directions. This is done by using the *Change Ports All* command for all component ports, or the *Change Port* command for a particular port, then changing the summator input signs.

Metal-Oxide Semiconductor Field Effect Transistors

Metal-oxide semiconductor field effect transistors (MOSFET) are so called because their gate is isolated from the channel by an oxide layer [9]. One of the major advantages of MOSFET technology is low cost and the possibility of dense packing. They also enables both p- and n-channel devices to be made on the same substrate, leading to the so-called *Complementary* MOSFETs or CMOSs. Today MOSFETS are one of the most important semiconductor technologies.

Many different models of MOSFET have been developed and are used to describe real devices—ranging from digital to analogue, and of different power capabilities and frequencies—more accurately. We limit our discussion to a basic MOSFET model. More complex models can be developed using a similar approach. Much of the earlier discussion on JFETs applies to MOSFETs and will not be repeated here.

MOSFETS, or MOS, for short, are basically *four*-terminal devices that, in addition to gate, drain, and source terminals, have a *bulk* terminal. Normally, this is connected to a terminal with the most negative potential for n-channel MOSFETS, and to the most positive terminal in case of p-channel MOSFETS. It also can be used for additional control of MOSFETS [9].

The electrical circuit symbol used for n-channel MOS (NMOS) is shown in Fig. 7.81a. Voltage polarities and the current direction correspond to *normal* operation. The corresponding bond graph component is shown in Fig. 7.81b. It is assumed that power flows into the component at the gate, bulk, and the drain ports; and flows out at the source port.

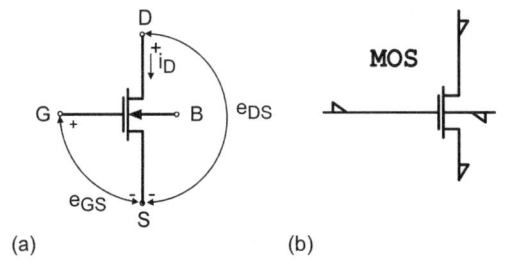

(a)   (b)

Fig. 7.81. N-channel MOSFET. (a) The Circuit symbol, (b) The bond graph representation

Polarities of p-channel MOSFETs (PMOS) are just the opposite (Fig. 7.82a). Similarly, the port power flow senses of the corresponding bond graph components also are reversed (Fig. 7.82b).

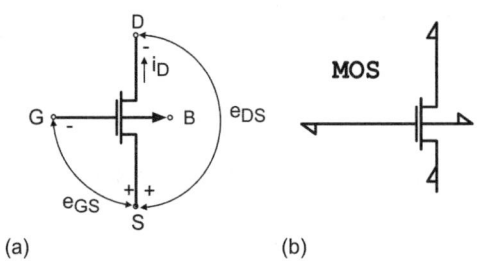

(a)   (b)

Fig. 7.82. P-channel MOSFET. (a) The circuit symbol, (b) The bond graph representation

The NMOS component can be created using the *n-channel MOSFET* button of the *Electrical Component* palette (Fig. 7.2). The text MOS is just a label used for reference to the component and can be changed at this stage or later. The p-channel component can be created from an n-channel component by reversing the power flow direction of all ports. It also is possible to change only the base and bulk ports. In this case, the drain and source of the n-channel MOSFET effectively change places.

The n-channel MOSFET model (Fig. 7.83) is dynamic and corresponds to the Level 1 large-signal SPICE model. We use a controlled resistor to describe the static characteristics of NMOS, as we did in the JFET case.

Static characteristics of NMOS are given by relation [4, 7]

$$id = \begin{cases} 0, & Vgs \leq VTH \\ (KP/2) \cdot Vds \cdot (2(Vgs - VTH) - Vds)(1 + LAMBDA \cdot Vds), & \\ & 0 < Vds < Vgs - VTH \\ (KP/2) \cdot (Vgs - VTH)^2 (1 + LAMBDA \cdot Vds), & \\ & Vds \geq Vgs - VTH \end{cases} \quad (7.52)$$

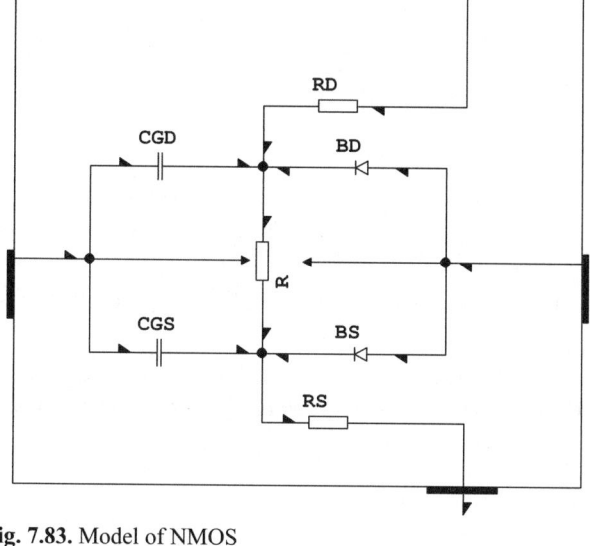

Fig. 7.83. Model of NMOS

where **KP** is effective transconductance. VTH is the threshold voltage that depends on the difference of potential of the body with respect to the source (or drain for PMOS) [7],

$$VTH = VT0 + GAMA \cdot (\sqrt{PHI - V_{BS}} - \sqrt{PHI})$$ (7.53)

GAMA is a body-effect parameter and PHI is the surface potential. The controlled resistor of Fig. 7.83 is represented by the same model as in the JFET model (Fig. 7.78), but using the constitutive relation of the resistive element based on Eq. (7.52) and (7.53). Resistors **RD** and **RS** in Fig. 7.83 are parasitic resistances of drain and source, respectively.

In MOSFET there is no direct resistive current path between the gate and either the drain or source because of the oxide isolation layer. There is charge accumulation, however, that is modelled by capacitors **CGD** and **GCS**, between the gate and the drain and the gate and the source, respectively. In a first-order analysis these can be described by constant capacitances. The capacitance between gate and body is neglected.

There also are junctions between the body, the source, and the drain that are represented by diodes (Fig. 7.83). These junctions are reverse-biased. We use the same model of diodes as in JFET (Fig. 7.77). The capacitances **Cj** in that model describe the accumulation of junction charges.

The discussion regarding direct and reverse operation of JFET applies here, as well. Thus, **Vgs** in Eq. (7.52) is replaced by **Vgd**, **Vbs** is replaced by **Vbd**, **Vds** is changed to **-Vds**, and the sign of the current is changed.

The same applies to PMOSs. The direction of power transfer trough the component is opposite to that of NMOS. Thus, the PMOS model can be created from the NMOS model in the same way as discussed for JFETs. The body port of the PMOS should, however, be connected to the port of higher potential in order that the body junction be reverse-biased, i.e. to the drain for direct operation, and to the source for the reverse.

As a first application of the MOS model, we create a project named **NMOS Characteristics** that simulates the *id-Vds* characteristics of an n-channel MOSFET. The corresponding system-level model is given in Fig. 7.84. The project is similar to that for evaluation of JFET characteristics (Fig. 7.79). The drain junction of the NMOS is set at 0–5 V by the voltage source VD. Its gate port is connected to a separate voltage source. Because ports of the same component cannot be directly interconnected, a separate branch node is inserted that is used to connect the NMOS source and body ports. The node is grounded. There are also instruments for measuring the gate and drain voltages and drain-source current. Outputs of these instruments are fed to a plotter for display. The NMOS parameters used for simulation are given in Tables 7.9 and 7.10.

Several simulations are run with different values of the gate voltage 4–8 V to generate the id-Vds characteristics. The simulation interval was set to 1s, with the output interval 0.001 s. Note that the transistor threshold voltage VT0 = 3.5 V (Table 7.9); hence, this is an enhanced mode NMOS and to conduct current the gate bias must be greater than 3.5 V.

Table 7.9. Static NMOS parameters

Parameter	Value
Rd	10 ohm
Rs	10 ohm
IS	$1 \cdot 10^{-14}$ A
VT0	3.50 V
KP	0.001 A/V^2
LAMBDA	0.0 1/V
GAMA	0.0
PHI	0.6 V

Table 7.10. NMOS capacitances

Parameter	BD	BS
VJ	1 V	1 V
CJ0	5pF	5pF
FC	0.5	0.5
M	0.5	0.5
N	1	1
Cgd	1 pF	
Cgs	1 pF	

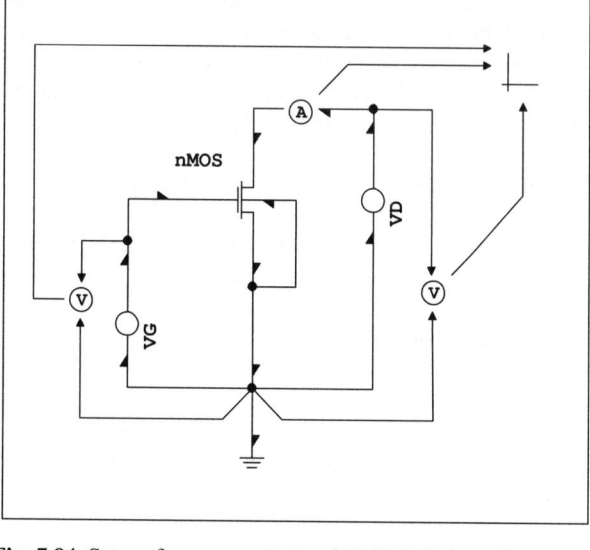

Fig. 7.84. Set-up for measurement of NMOS characteristics

Results (Fig. 7.85) show a linear region in which the source voltage increases with drain-to-source voltage; and a saturation region in which the current is constant.

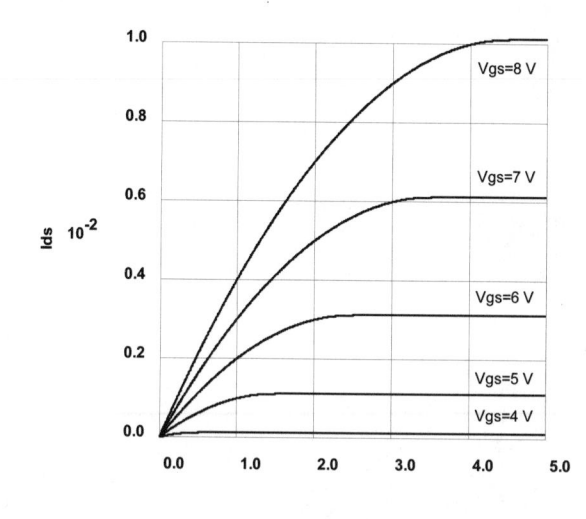

Fig. 7.85. Simulation of NMOS characteristics

The program library also contains a project **PMOS Characteristics** that the reader is invited to analyse on his own.

We conclude this section with analysis of the basic CMOS inverter (Fig. 7.86). It consists of a PMOS and a NMOS transistor, the drains of which are connected through common node. The PMOS is supplied by a 5 V source; the source of the NMOS is grounded. Input voltage is applied to both transistors and the voltage is taken from the common node.

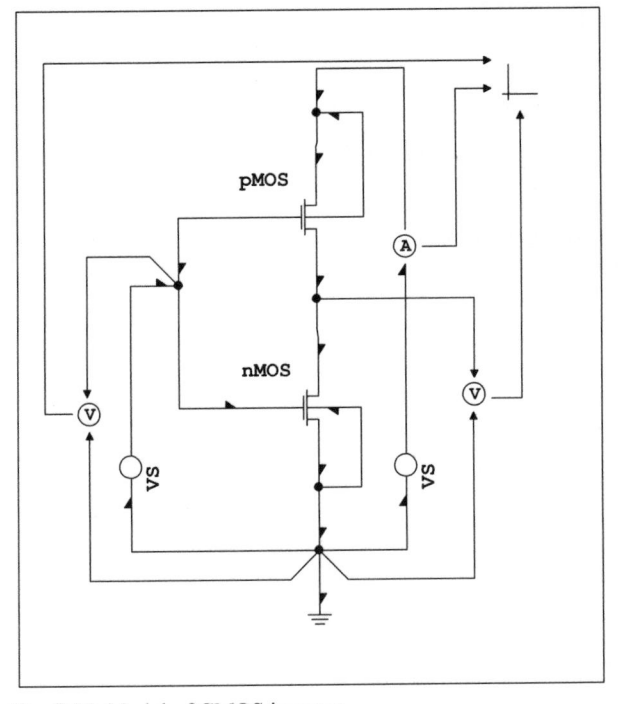

Fig. 7.86. Model of CMOS inverter

Parameters of the inverter are given in Tables 7.11 and 7.12. Note that the threshold voltage of both transistors is 1 V; hence, they are in enhanced mode.

Table 7.11. CMOS static parameters

Parameter	PMOS	NMOS
Rd	1 ohm	1 ohm
Rs	1 ohm	1 ohm
IS	$1 \cdot 10^{-14}$ A	$1 \cdot 10^{-14}$ A
VT0	1 V	1 V
KP	80.0 μ A/V²	80.0 μ A/V²
LAMBDA	0.0 1/V	0.0 1/V
GAMA	0.0	0.0
PHI	0.6 V	0.6 V

Table 7.12. CMOS capacitances (PMOS/NMOS)

Parameter	BD	BS
VJ	0.75 V	0.75 V
CJ0	5fF	5fF
FC	0.5	0.5
M	0.5	0.5
N	1	1
VT	0.0258 V	0.0258 V
Cgd	8 fF / 4 fF	
Cgs	8 fF / 4 fF	

The inverter is driven by voltage

$$V_G = V_0 + V_1 \cdot \sin(2\pi f) \tag{7.54}$$

where $V_0 = V_1 = 2.5$ V and f = 20 MHz.

The simulation was run for 100 ns with an output interval of 100 ps for better resolution. Results are shown in Fig. 7.87.

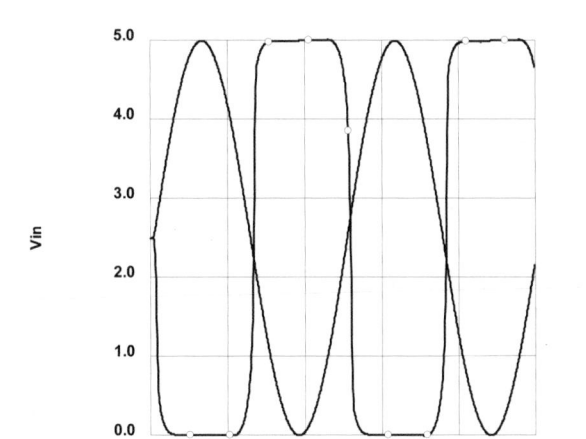

Fig. 7.87. The CMOS inverter simulation

When the input is low, e.g. equal to 0 V, the source-gate voltage of the PMOS is 5 V and is *on* (conducting). The NMOS, however, is *off* because its gate-source voltage is equal to 0 V, below the threshold. Current through both transistors thus is zero. From the static characteristics of the PMOS it follows that the voltage between the source and drain must equal zero. Thus, the inverter output is equal to the PMOS source port voltage, which is 5V. On the other hand, when the gate voltage is high, e.g. equal to 5 V, the PMOS is off and current through the transistors is zero. But now the NMOS is on. This means that its drain-source voltage is 0 V. Because the source port of the NMOS is grounded, the inverter output is also 0 V. Thus, the inverter generates high voltage (5 V) when it's input is low (0 V), and low voltage (0 V) when it's input is high (5 V).

7.4.3 Operational Amplifiers

Operational amplifiers are important integrated circuit components widely used in analogue signal processing. We here consider how they can be represented as components in bond graphs.

Basically, operational amplifiers—op-amps, for short—are represented in electrical circuits by the symbol shown in Fig. 7.88a. It has three terminals, two inputs and one output. The input terminal designated with a "–" is the *inverting* terminal; the other, denoted by a "+", is the *non-inverting* terminal. Bond graph components representing op-amps are shown in Fig. 7.88b. This assumes that power flows outward at the inverting port and inward at the non-inverting port. The op-amp generates power at its output, which is represented by the power-out port.

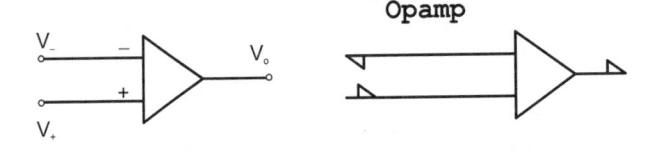

(a)   (b)

Fig. 7.88. Operational amplifier. (a) The circuit symbol, (b) The bond graph representation

The simplest model of the operational amplifier describes it as an ideal voltage amplifier:

$$\left.\begin{array}{l} V_{diff} = V_{+} - V_{-} \\ V_{o} = A_{0} \cdot V_{diff} \\ i_{+} = i_{-} = 0 \end{array}\right\} \tag{7.55}$$

Parameter A_{0} is the so-called *DC open-loop gain*. It assumes that the currents drawn at the input ports are negligible and that there is no limitation of voltage or current at the output port. Thus, the input resistance of the amplifier is infinite and its output resistance is zero. Such a simple model can be represented by the bond graph of Fig. 7.89. It consists basically of a source flow component SF that defines a zero input port current and a source effort SE that generates an output voltage, as given by Eq. (7.55). The last component receives information on the voltage difference (efforts) at the non-inverting and inverting ports. This comes from the flow junction inserted between the effort junction and the source flow component.

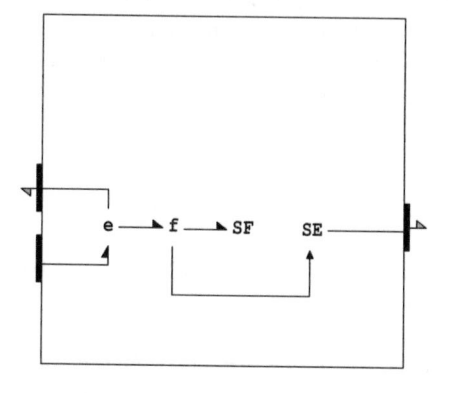

Fig. 7.89. Simple model of op-amp

The power delivered at the operational amplifier output port is much larger than at the input. The excess power comes from voltage sources that supply power to the op-amp through supply terminals. The voltage generated at the output port cannot rise above the voltages at these terminals. The supply terminals are not shown in the simple op-amp model of Figs. 7.88 and 7.89. However, we can take care of the limitation on the output voltage by defining the relationship of the source effort in Fig. 7.89 as

$$\text{eo} = \text{GAIN} * \text{ediff} <= -\text{VLIMIT} ? - \text{VLIMIT} :$$
$$(\text{GAIN} * \text{ediff} < \text{VLIMIT} ? \text{GAIN} * \text{ediff} : \text{VLIMIT}) \tag{7.56}$$

where GAIN is the static gain of the op-amp and VLIMIT is the maximum value of the output voltage. This usually is 1-2 V less than the supply voltage.

We now simulate the characteristics of the operational amplifier defined by the models in Fig. 7.88 and 7.89 and Eq. (7.56). As usual, we create a project named Ideal Op-amp Characteristics and define the model of a set-up for simulating the static characteristics, as shown in Fig. 7.90.

The gain of the op-amp is set to $1 \cdot 10^5$ and the output is limited to ± 4.5 V. The op-amp is driven by the source voltage VS connected to the inverting port by a node. The source generates a sinusoidal voltage. The non-inverting port is grounded. The op-amp output port is connected to ground across a 1 kohm resistor. The output voltage reaches the limiting value at input of $\pm 4.5 \cdot 10^{-5}$ V. To show the characteristics more clearly, we drive the op-amp with a low input voltage with amplitude of $5 \cdot 10^{-4}$ V and frequency of 1 Hz. The frequency is not critical, as the op-amp model is static. To display the transition from the region of linear behaviour to saturation correctly, we must use a fairly small output interval, e.g. 0.001 s or less.

Results of the simulation are shown in Fig. 7.91, which shows that input and output voltages are of opposite sign. This follows from Eq. (7.55) because, for a grounded non-inverting port, the voltage difference Vdiff = -V_ and, hence, the output voltage $V_o = -A_0 \cdot V_-$. The output voltage changes linearly over a short interval around the origin, and is constant elsewhere.

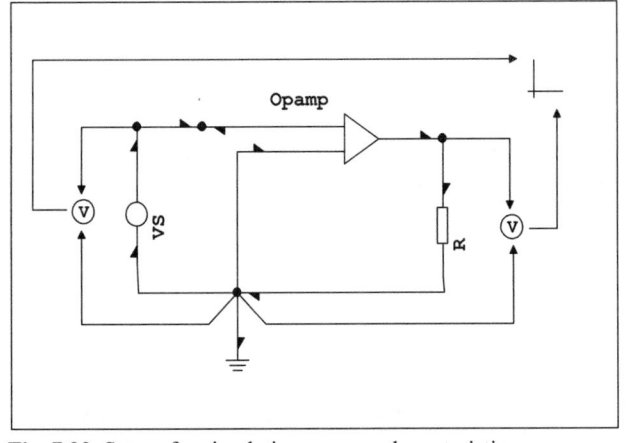

Fig. 7.90. Set-up for simulating op-amp characteristics

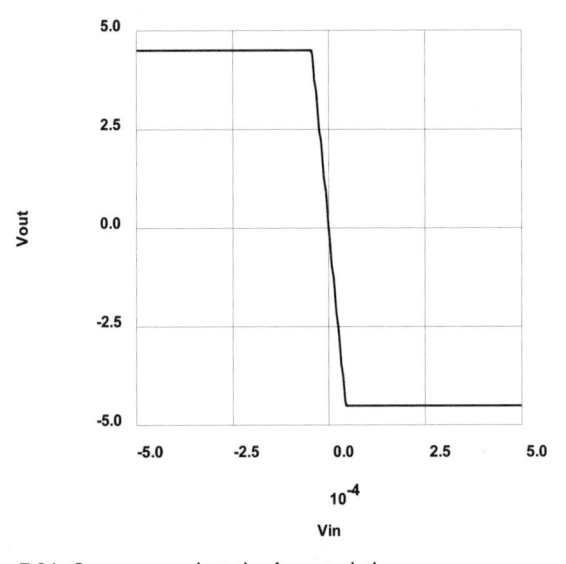

Fig. 7.91. Op-amp quasi-static characteristics

We can improve the model by incorporating finite input and nonzero output resistances, as well as dynamics. The resulting model is given in Fig. 9.92. Compared to Fig. 7.90, the source flow in the input section is replaced by a resistive element that models the input resistance of the op-amp.

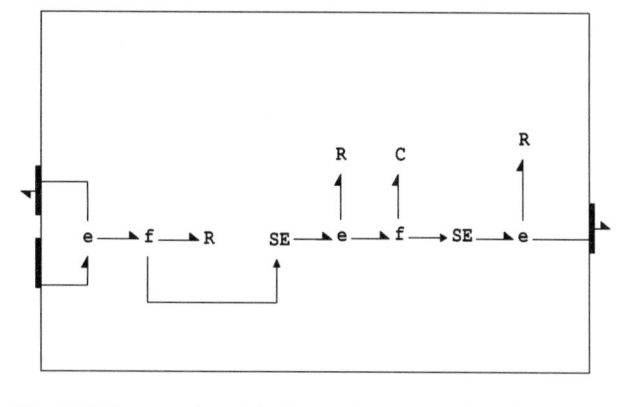

Fig. 7.92. Improved model of operational amplifier

Similarly, a resistive element is added in series between the source effort and output port. This models the output resistance. Finally, an intermediate section is added that models the dynamics of the operational amplifier. Real amplifiers are built of many bipolar junction transistors, JFETs, or MOSFETs, so the dynamics are quite complex. A capacitance usually is added to the op-amps high-gain node to control its dynamics over the useful range of frequencies. In simplified models amplifier dynamics often are approximated by first-order dynamics [1, 4, 5], as represented in Fig. 7.92 by the R and C elements.

Denoting by e_c the effort (voltage) on the capacitive element, the current is

$$i = \frac{d}{dt}(C \cdot e_C) = C \frac{de_C}{dt} \tag{7.57}$$

The same current flows through the resistor. Thus, the dynamics of the operational amplifier of Fig. 7.92 is given by

$$R \cdot C \cdot \frac{de_C}{dt} + e_C = A_0 \cdot e_{diff} \tag{7.58}$$

The natural frequency of this simple first-order system is

$$\omega_n = \frac{1}{RC} \tag{7.59}$$

and the dynamics of the op-amp can be described by

$$\frac{de_C}{dt} + \omega_n e_C = A_0 \cdot \omega_n \cdot e_{diff} \tag{7.60}$$

The resistance R and capacitance C are chosen to approximate to the dominant low-order natural frequency of the op-amp. This is taken as the bandwidth of the op-amp, i.e. the frequency at which the output amplitude drops to $1/\sqrt{2}$, or $-3dB$ of the DC value. Thus, for an op-amp with a bandwidth of 10 Hz = 62.83 rad/s, if we choose Rdin = 1 kOhm the capacitance is Cdin = $15.92 \cdot 10^{-6}$ F.

The product $A_0\omega_n$ on the right of Eq. (7.60), the *gain-bandwidth product*, is an important characteristic of the op-amp. It usually is expressed in Hz—as A_0f_n , where $f_n = 2\pi\omega_n$–and typically is of order of MHz. In the example above, with the DC gain $1\cdot10^5$ and the bandwidth 10 Hz, it is exactly 1 MHz.

In the model of Fig. 7.92 we limit the output voltage at the source effort in the output section by Eq. (7.56) using unity gain. The amplifier DC gain is defined in the source effort of the intermediate section. This is, in effect, similar to the diodes used in the output stage of the op-amp models of [4].

Operational amplifiers often are used in circuits for analog operations on signals, such as amplifying, integrating, differentiating, and filtering. We analyse here only one such common circuit, the *Inverting amplifier* (Fig. 7.93).

The project Inverting amplifier in Fig. 7.93 is a slight modification of the operational amplifier circuit of Fig. 7.90; it has an added input resistor R and also a feedback resistor Rf. This connects the output port to the input node, which in turn is connected to the inverting input port of the op-amp. Loading the operational amplifier output port is not needed here.

We estimate the close loop gain of the amplifier by summing the currents at the input node under the hypothesis of an ideal operational amplifier of very high DC gain.

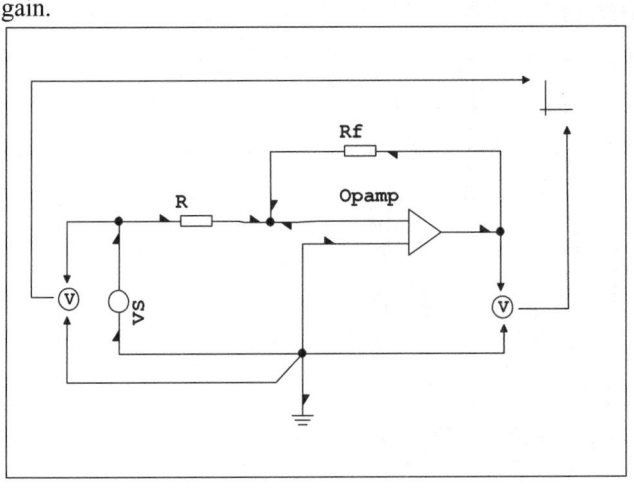

Fig. 7.93. Inverting amplifier set-up

Under this hypothesis the potential of the node is equal to the ground potential, and the current drawn by the op-amp is zero. Hence we have

$$\frac{V_{in}}{R} + \frac{V_0}{R_f} = 0 \qquad (7.61)$$

or

$$V_0 = -\frac{R_f}{R} V_{in} \qquad (7.62)$$

The voltage gain of the inverting amplifier is, for sufficiently high open-loop gain, given by the ratio of the feedback and input resistors. If we set R = 1 kOhm and Rf = 25 kOhm, the gain is 25. This can be verified by simulation. We use a sufficiently low input amplitude of 0.002 V, such that the amplifier is not saturated (set to VLIMIT = 5 V). The frequency chosen for the input voltage, 1 Hz, is much less than the bandwidth frequency of 10 Hz. The parameters of the operational amplifier are given in Table 7.13. The simulation was run for 5 s corresponding, to 5 periods of the input voltage, using an output interval of 0.01 s. The results (Fig. 7.94) show that the output amplitude is 0.05 V, as expected (that is: 0.002 V·25 = 0.05 V). Also, the output phase is opposite to that of the input.

Table 7.13. Parameters of op-amp model

Rin	1 Mohm
Rout	100 ohm
GAIN	$1 \cdot 10^5$
VLIMIT	5 V
Rdin	1 kohm
Cdin	15.92 μF

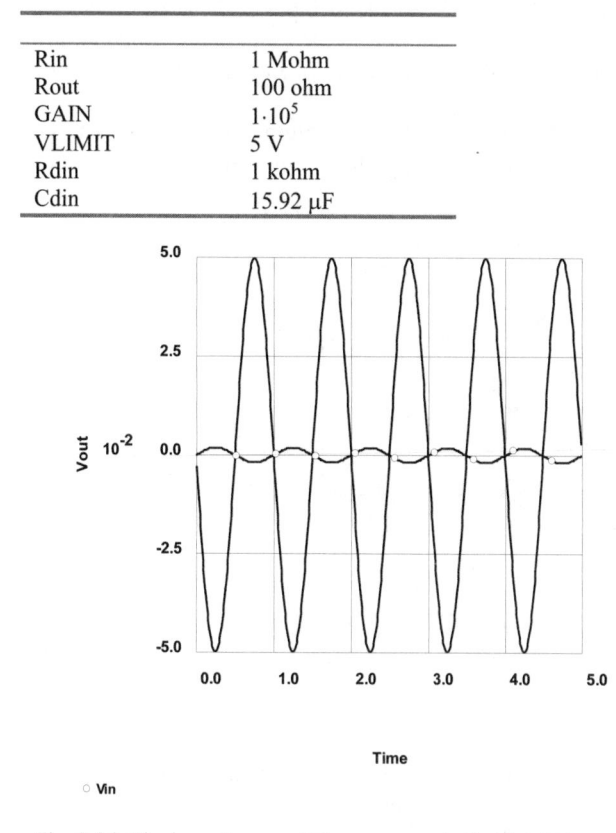

Fig. 7.94. The inverting amplifier response to 2mV at 1 Hz

We expect that the output amplitude will roll off when the frequency reaches the amplifier bandwidth. Because of feedback around the operational amplifier, this frequency generally will be much higher than that of the open-loop operational amplifier, which is set at 10 Hz.

It can be shown ([1]) that the close loop bandwidth of the inverting amplifier can be approximated by

$$(f_n)_{close-loop} = \frac{A_0 f_n}{1 + \frac{R_f}{R}} \qquad (7.63)$$

This relationship is valid if the open loop gain A_0 is much larger than $1 + R_f/R$, a criterion that normally is satisfied. In our example, the gain-bandwidth product $A_0 f_n$ equals 1 MHz. Hence, for the close-loop gain $R_f/R = 25$, we get, by Eq. (7.63), the close-loop bandwidth of about 38 kHz. The expected amplitude is 0.05 $/\sqrt{2} = 0.03536$ V.

We repeat the simulation with an input voltage of the same amplitude, but with a frequency of 38 kHz. The simulation interval was set to 0.0002 s and the output interval to $2 \cdot 10^{-7}$ s. Results (Fig. 7.94) show that the output amplitude has dropped to 0.03547 V. There also is a time lag of the output of about 16.5 μs that corresponds to the phase lag of 3.94 rad (1.254 π), where π radians is due to the signal inversion and the other $0.254\pi \approx \pi/4$ rad is true dynamic phase lag.

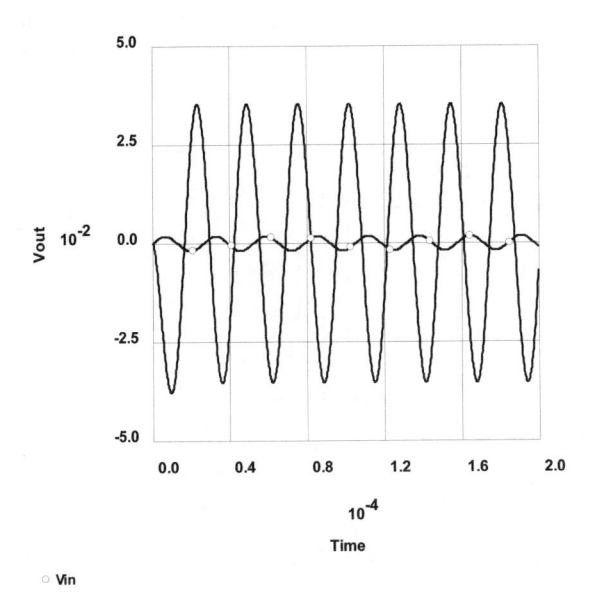

Fig. 7.95. Response of the amplifier to 38 kHz input

7.5 Electromagnetic Systems

This section analyses application of the bond graph approach to modelling electromechanical systems. The example chosen is an electromagnetic actuator. Permanent magnet data for the actuator were calculated by a magnetic field analysis program. These data are used as input to the BondSim program. More information on modelling magnetic systems by bond graphs can be found in [2].

7.5.1 Electromagnetic Actuator Problem

The schematic of an electromagnetic actuator is shown in Fig. 7.96. The actuator consists of a plunger, switching coils, and a permanent magnet for generating the holding force. The electrical driver circuit is represented by a capacitor, a resistor, and an ideal (lossless) switch. The plunger moves in the actuator assembly between hard stops from a value $x = x_{min}$ to a value $x = x_{max}$, where the displacement x is measured from the lower pole end. The electromagnet's behaviour is described by tabulated values of flux in one turn (electrical side), and force on the plunger (mechanical side), for a range of ampere-windings and plunger positions between x_{min} and x_{max}. These values were calculated using the magnetic field analysis program.

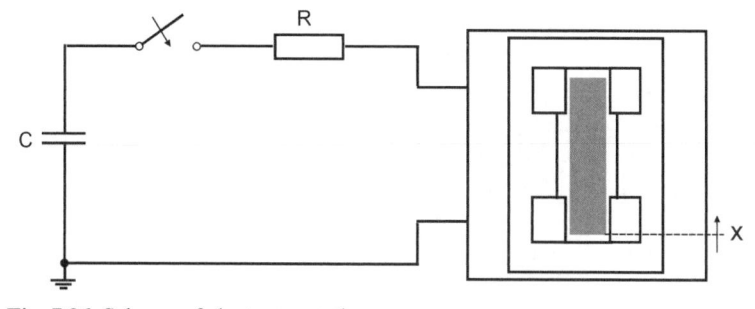

Fig. 7.96. Scheme of electromagnetic actuator

System parameters are:

1. Capacitor $C = 68$ mF, initially charged to 50 V (3.4 C)
2. Resistor $R = 0.2$ Ohms
3. The switch starts closing at 10μs, and closing takes 1μs. The resistance changes linearly from $1 \cdot 10^9$ (open) to $1 \cdot 10^{-9}$ (closed) Ohms
4. The plunger has mass 5.575 kg, its weight is neglected
5. Hard stop positions are at $x_{max} = 0.0199$ m and $x_{min} = 0.0001$ m. The material posses high stiffness coefficient ($1 \cdot 10^9$ N/m) and a high damping factor ($1 \cdot 10^6$ N·s/m), corresponding to impact without rebound.

7.5.2 System Bond Graph Model

We start by defining the basic structure of the Electro-magnetic Actuator system, as shown in Fig. 7.97. The model consists of an electrical part, resistor R, capacitor C, the switch, and the Magnetic Actuator component. During simulation the variables that are displayed are the current through electric circuit, the position and velocity of the plunger, and the total force on the plunger.

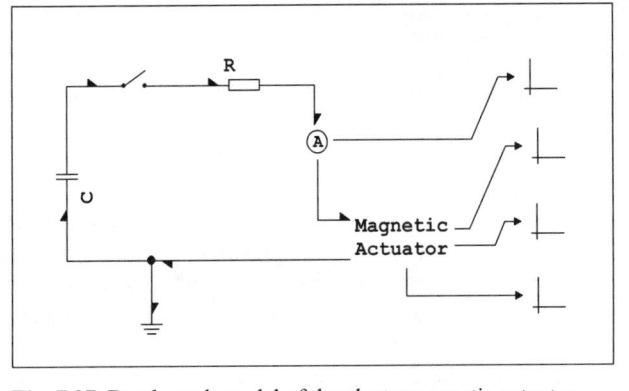

Fig. 7.97. Bond graph model of the electromagnetic actuator

7.5.3 Electromagnetic Flux and Force Expressions

Before proceeding with decomposition of the Magnetic Actuator component, we must define the Flux(x,I) and EMForce(x,I) functions. These interpolate data from tables of flux and force, respectively. Each is a function of plunger position x and ampere-winding i. In BondSim, one- and two-variate cubic B-spline interpolations are used for functions defined by one- and two-dimensional tables, respectively [15].

To define, for example, the Flux function, we select the *Edit Tabular Function* command on the *Edit* menu, then *New* from the menu that drops down. In the dialogue that then opens, we chose between *One* or *Two Dimensional* functions. Flux, as well as EMForce, has two arguments; thus, we check the *Two Dimensional* functions box and click the OK button. This opens the *Create New Tabular Function* dialogue, in which we define the function name, Flux. This is the name that will be used when referring to the function in an algebraic expression. We accept this name by clicking the OK button. A dialogue again opens, this time for defining the names and the values of the variables corresponding to the rows and columns of the function table. The row variable represents the first argument, and the column variable the second argument of the function we are defining (Fig. 7.98). Thus, in Flux(x,I) the row variable is *Position* (x) and column variable is *Ampere-Winding* (i).

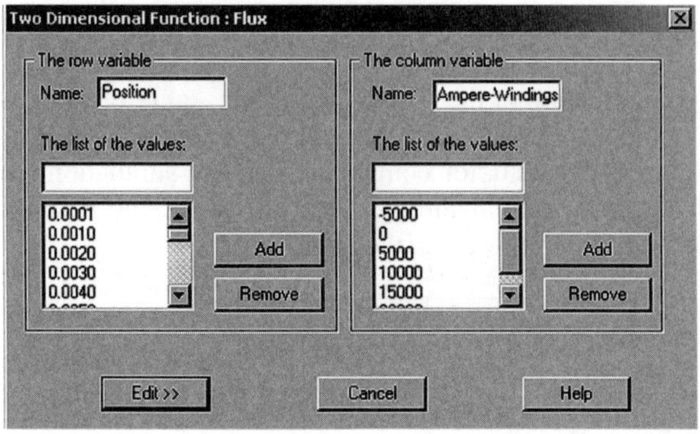

Fig. 7.98. Dialogue for defining values of variables

When we have added all values for both variables, we edit the values of the function by clicking the corresponding button. A grid appears that, in its leftmost column and topmost row, contains the values we have just defined (Fig. 7.99). We can type the values of the function in the cells of the grid, i.e. the flux value corresponding to a *Position* and an *Ampere-winding*.

The function can be viewed using the *Show Plot* button. A *Plot Options* dialogue appears that offers several possibilities for the plotting. The Flux function appears as in Fig. 7.100. Similarly, we can define the EMForce function (Fig. 7.101).

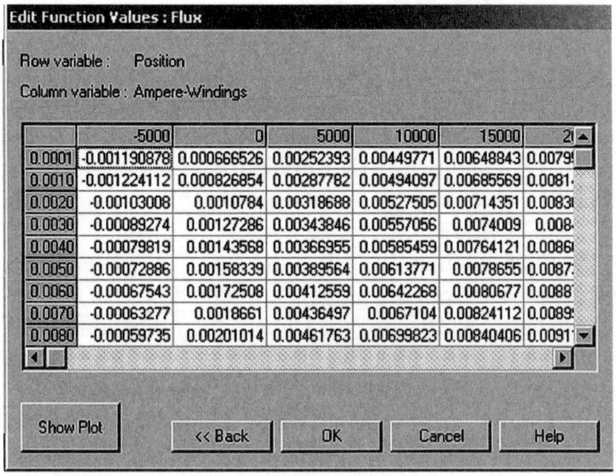

Fig. 7.99. Editing of the function values

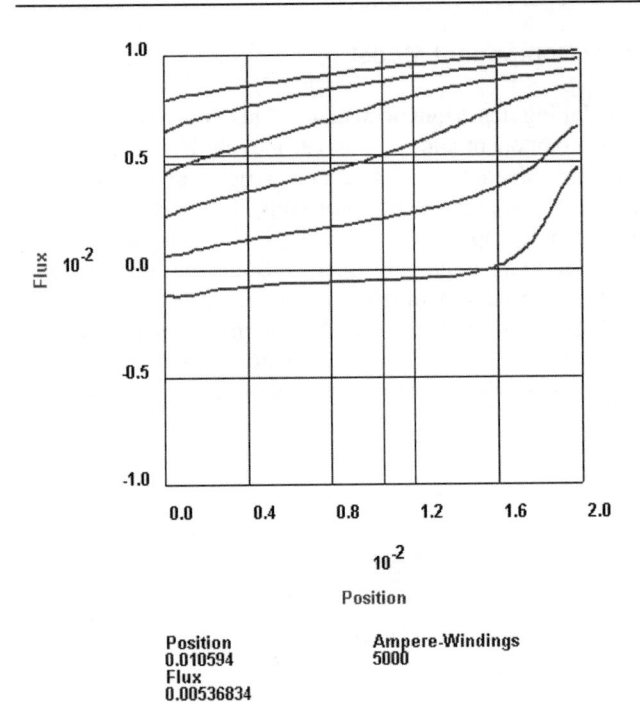

Position
0.010594
Flux
0.00536834

Ampere-Windings
5000

Fig. 7.100. Plot of Flux function

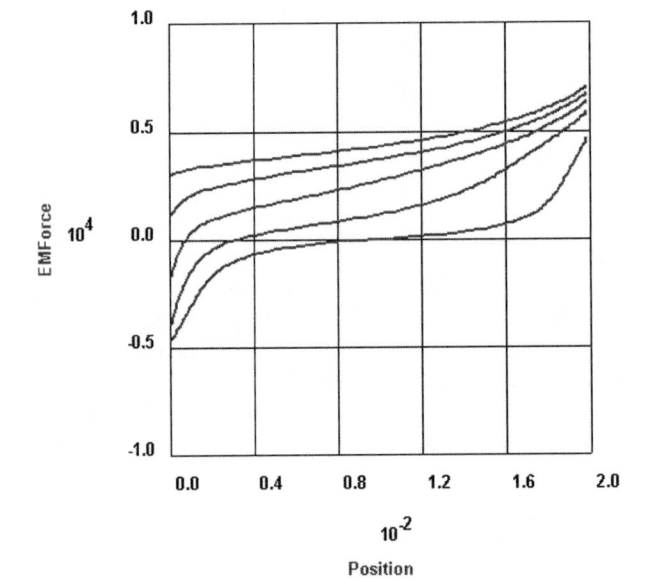

Fig. 7.101. Plot of EMForce function

7.5.4 Magnetic Actuator Component Model

We now continue with modelling the magnetic actuator. This consists of an Electro-Mechanical Conversion component and the Plunger moving between the hard stops (Fig. 7.102). The Electro-Mechanical Conversion component has two ports on the left where current flows through the actuator coils. The port on the right corresponds to the magnetic pole gaps where interaction with the plunger takes place. The plunger is modelled in a similar way as in the Bouncing Ball problems of Sect. 6.4. The plunger is represented simply as a particle (its weight is neglected). It interacts with hard stops, at the top and the bottom, through a Contact component taken from the library. This describes impact without rebounding (Figs. 6.52 and 6.53). The mass, stiffness coefficients, and the damping coefficient are given in Sect. 7.5.1.

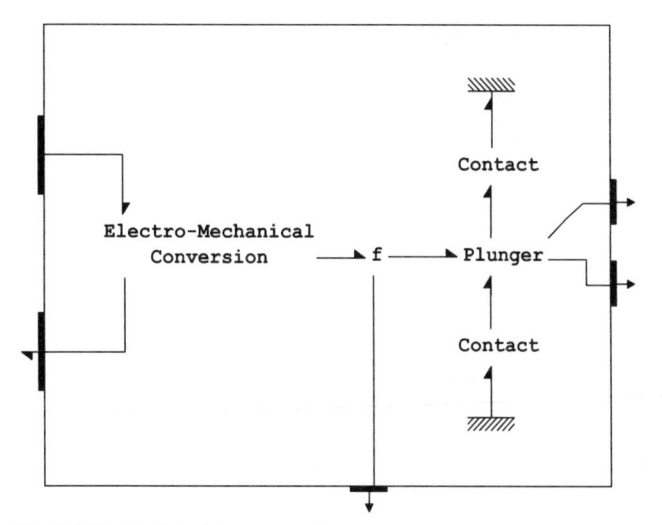

Fig. 7.102. Model of the magnetic actuator

Electro-mechanical conversion is of fundamental importance for the functioning of the complete system. Using bond graphs, it can be represented very compactly. The electromechanical conversion is described by two fundamental components: a capacitor and a gyrator (Fig. 7.103).

The capacitor has two ports, magnetic and mechanical, and describes the storage of magnetic and mechanical energy in the actuator. The constitutive relations for these two ports are:

Magnetic port:

$$\left. \begin{array}{l} \text{phi} = \text{Flux}(x, M) \\ e = \dfrac{\text{dphi}}{\text{dt}} \end{array} \right\} \tag{7.64}$$

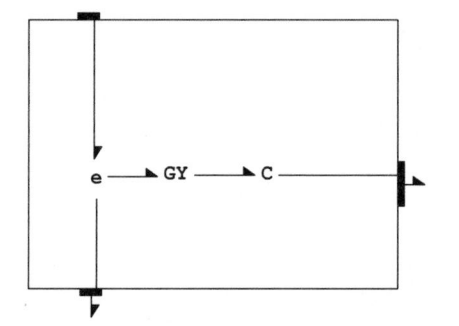

Fig. 7.103. Modelling of electromechanical conversion

Mechanical port:

$$\left.\begin{array}{l} F = \mathsf{EMForce}(x,M) \\ v = \dfrac{dx}{dt} \end{array}\right\} \qquad (7.65)$$

The gyrator describes relations between voltage V_d across coil ports and e.m.f. e, and between the magneto-motive force M and the current i through the coils:

Gyrator:

$$\left.\begin{array}{l} V_d = N \cdot e \\ M = N \cdot i \end{array}\right\} \qquad (7.66)$$

where N is the number of coil turns. Eqs. (7.64)–(7.66) are familiar conversion relations from electromagnetics.

7.5.5 Simulation of Magnetic Actuator Behaviour

The simulation of the electromagnetic actuator was done to determine how the system variables change during switching on the actuator. Simulations were made using a simulation interval of 0.050 s. The output interval of 0.00001 s was chosen to determine characteristic switching points accurately. The processing time was 5.2 s. Some of the results are shown in Figs. 7.104 to 7.106.

Figs. 7.104 and 7.105 show that the force on the plunger steadily increases until it crosses the zero value at 0.02253 s. Because the plunger is depressed to the hard stop, the real motion starts a little later, at 0.02358 s. The plunger hits the upper hard stop after 0.03753 s. The final value of the force at 0.04999 s is 5575 N.

Changes of current through the coils are shown in Fig. 7.106. The current first rises steeply, then reaches its maximum, 90.05 A, at 0.0258 s. It then drops down and reaches its minimum, 0.7481 A, when the plunger hits the hard stop. After that it rises somewhat to a final value of 20.45 A. This is the current that holds the plunger pressed to the upper hard stop.

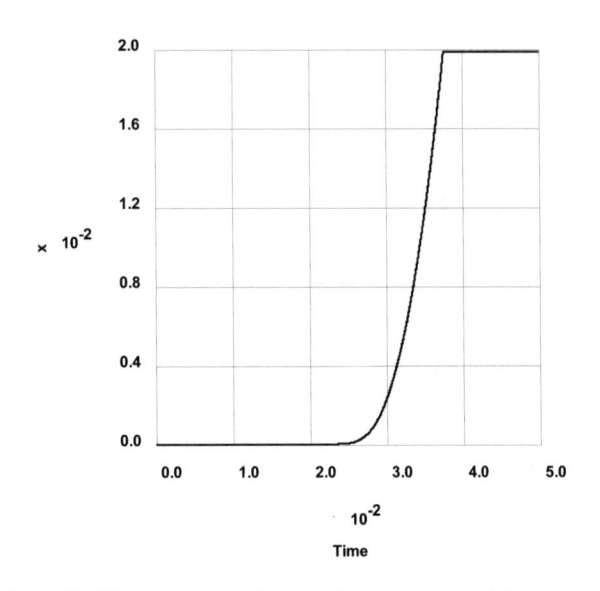

Fig. 7.104. The plunger position motion during switching on

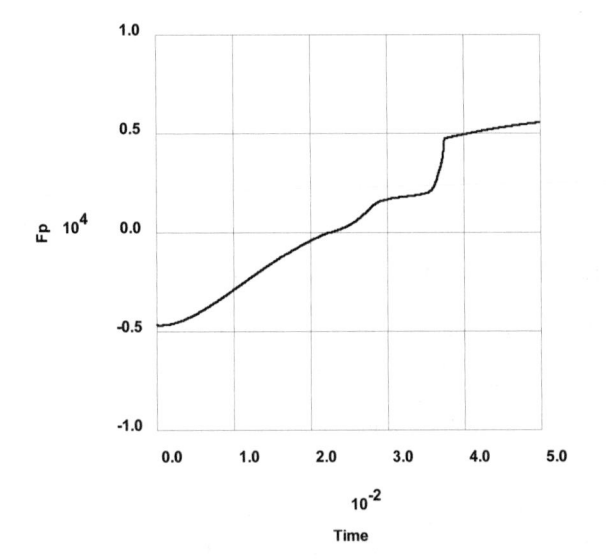

Fig. 7.105. Total force on the plunger during switch-on

This section, as well as the previous sections of this chapter, shows the effectiveness of the bond graph approach in solving complex electronic and electromechanic systems.

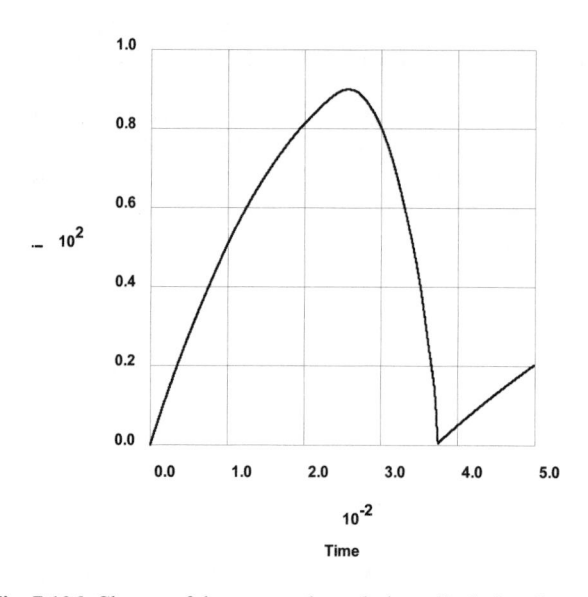

Fig. 7.106. Change of the current through the coils during the switch-on

References

1. SD Senturia (2001) Microsystems Design, Kluwer Academic Publishers, Boston
2. Dean C Karnopp, Donal L Margolis and Ronald C Rosenberg (2000) System Dynamics: Modeling and Simulation of Mechatronic Systems, 3rd edn. John Wiley, New York
3. Peter Gawthrop and Lorcan Smith (1996) Metamodelling: Bond graphs and dynamic systems. Prentice Hall, Hemel
4. A Vladimirescu (1994) The Spice Book (1994), John Wiley & Sons, New York
5. J Keown (2001) OrCad Pspice and Circuit Analysis. Prentice Hall, Upper Saddle River
6. R Kielkowski (1995) Spice Practical Device Modeling, 2nd edn. McGraw-Hill, New York
7. G Massobrio and P Antognetti (1993) Semiconductor Device Modeling With Spice, 2nd edn. McGraw-Hill, New York
8. R Märtz and C Tischendorf (1997) Recent Results in Solving Index-2 Differential-Algebraic Equations in Circuit Simulation. Siam J. Sci. Comput. 18:139-159
9. J Singh (2001) Semiconductor Devices, Basic Principles. John Wiley & Sons, New York
10. T Tieze and C Schenk (1999) Halbleiter-Schaltungstechnik, Springer-Verlag, Berlin-Heidelberg
11. P Van Halen (1994) A Physical Charged-Based Model for Space Charge Region of Abrupt and Linear Semiconductor Junction. In Proc. of the 1994 International Symposium on Circuits and Systems, London, pp. 1.403-1.406

12. E Christen, K Bakalar, AM Dewey and E Moser (1999) Analog and Mixed Signal Modeling Using VHDL-AMS Language (tutorial). The 36th Design Automation Conference, New Orleans

13. J Thoma and B O Bousmsma (2000) Modelling and Simulation in Thermal and Chemical Engineering, A Bond Graph Approach. Springer/Verlag, Berlin-Heidelberg

14. Allen R Hafner and David L Blackburn (1993) Simulating the Dynamic Electrothermal Behavior of Power Electronic Circuits and Systems. IEEE Transactions on Power Electronics 8: 376-385.

15. Carl de Boor (1978), A Practical Guide to Splines, Springer-Verlag, New York

Chapter 8 Control Systems

8.1 Introduction

Control system theory and praxis play important roles in mechatronics. Their fundamental role is in the control of mechanical motion. Control systems are commonly described in terms of block diagrams, where input-output relations are represented by transfer functions. This offers a simplified picture of system processes from a control point of view. The theory of control systems is well documented in numerous books, so we do not cover it here. What we wish to demonstrate is how control actions taking place in mechatronic systems can be modelled using the bond graph approach developed here.

The bond graph elements required to deal with the control of mechatronic signal interfaces are defined in Sect. 2.5.8. In Sec. 2.6 basic block diagram components are defined that can be used to synthesize the signal processing of control loops. Control actions in real components can thus be simplified and represented by block diagram components. These component models control actions in the time domain. Modelling is not restricted to linear relations, but is applicable to non-linear systems and even those that feature discontinuities. Creation of such components, or development of more complex components, follows the philosophy of the component modelling approach discussed in previous chapters.

The next section explains the basic techniques of block diagram components as applied to a simple control system. Some modelling details specific to block diagram components are given. It will be seen that there is not much difference from other power port components.

After that, a short overview of the modelling approach to control systems is given. This focuses mainly on modelling PID controllers in servo loops. Finally, modelling and simulation of a DC servomotor is presented.

8.2 A Simple Control System

We start by modelling a typical feedback control system (Fig. 8.1). This system consists of a controller that regulates an object's motion. The motion in question is defined by a reference input and output information collected by a sensor. There are also disturbances acting on the object. This system can be implemented either as a position- or velocity-control system. In mechatronics, the object typically is a mechanical linkage actuated by a motor that executes the control action.

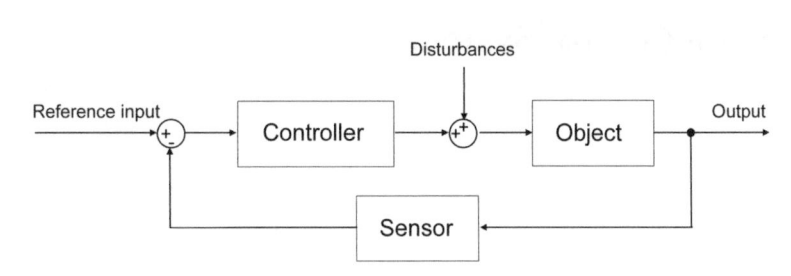

Fig. 8.1. A simple control system

We begin the modelling process by creating a new project named **Simple control system**. Starting from the left in the diagram (Fig. 8.1), we first create the input component that generates the reference signal. This component is created by selecting the *Input* elementary component from the component type list in the *Editing Palette* (Fig. 8.2), then creating it at the left in the project document window.

Input components are signal generators and can have only a single control-out ports. Such a port is created in the usual way, by selecting it from the palette and inserting it at the right border of the **Input** component inside its boundary. Double-clicking the port and using the *Parameter* button confirms that, when created, the **Input** generates constant signal preset to zero. This value can be changed later.

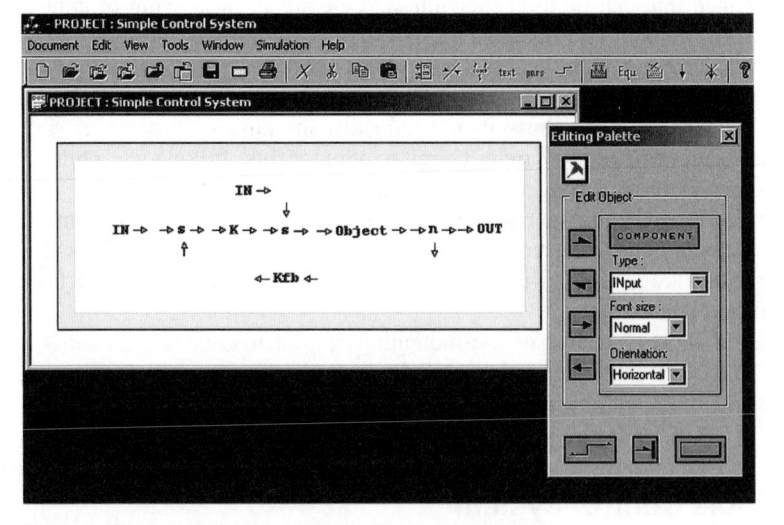

Fig. 8.2. The creation of the control system components

We next create the component corresponding to the summation of the reference and feedback signals: Select *Summator* in the component type list and create it to the right of the **IN** component. By default, the newly created component has a predefined name, **s**, which of course can be edited to something more descriptive. (We accept the default name in this example.)

Next, insert two control-input ports and a single control-output port into the summator component. By default, all input ports implement addition; that is the summator output is defined as cout = cin1 + cin2

In the present case, the feedback signal is subtracted from the input reference (Fig. 8.1); we must therefore change the default addition operation of the summator feedback port to implement this subtraction. First, double-click the port to open the *Summator Input* dialogue (Fig. 8.3). Then select the subtraction symbol, "-", in the *Summator Sign* combo box. We also can change the name of the feedback signal to, for example, cfb—short for "feedback control signal", as well as the names of all other control port signals.

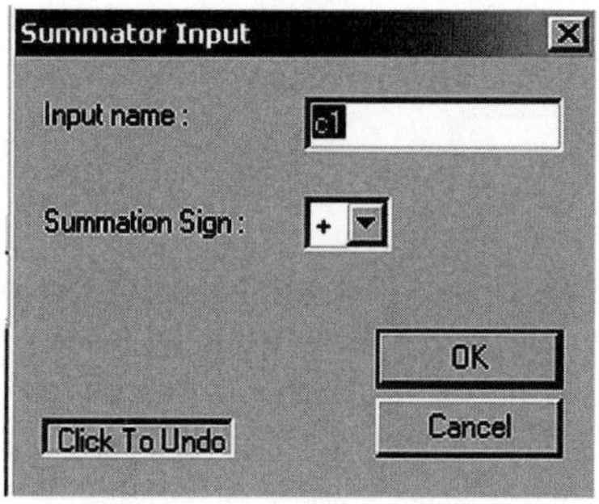

Fig. 8.3. Selecting the sign of the summator input

The controller in this example is implemented as a simple proportional (linear) control law, given by

$$cout = K \cdot cin \tag{8.1}$$

where K is the controller gain. This is accomplished by first selecting *Function* from the component type list box, then changing the name of the component to K.

Next, insert a single control-input port on the left of the component and a control-out port on its right. By default, the constitutive relation of the component is linear, which corresponds to Eq. 8.1. We can confirm this by double-clicking the output port. The value of the gain defaults to 1.

The next step is to insert a summator that adds the disturbances, and an IN component above it to input disturbances. The object of the control (the object) is created as a word model component named *Object*. This has a control input port and a control out port. Its model is defined later.

Next, we create the node that picks up the feedback signal (Fig. 8.1): Choose *Node* from the component type list and create it at the right of the Object. A node component can have a single input and several output ports. The control-input port must be created before the two control-out ports that regulate, respectively, the system output and feedback connections (Fig. 8.2).

To close the servo loop, the sensor component must be created. This is represented by a simple linear function component denoted Kfb.

The final component, Out(put), displays the control system's output. Note that every signal has to be fed to some destination; there can be no 'dangling' signals. By connecting all ports with bond (signal) lines, as in Fig. 8.1, we finish the first (system) level of the control model (Fig. 8.4).

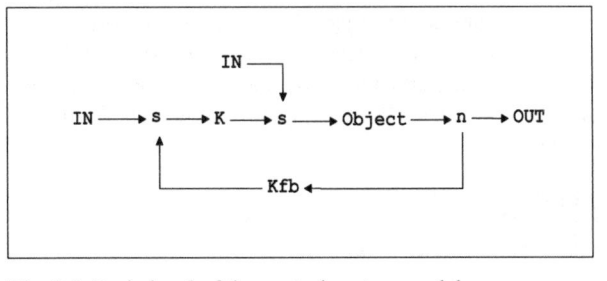

Fig. 8.4. Basic level of the control system model

At this point we introduce a slight modification to the model to permit display of the reference signal. To accomplish this we disconnect the left Input, insert a node, and connect it to the output component. It is not necessary to redraw the complete block diagram: Use the mouse to select the rectangular region around it, then move the selected region to the right. Eventually, we need to enlarge the document window by dragging its right edge (with help of the *Size document* button in *Editing Palette*). The final form of the block diagram is as illustrated in Fig. 8.5.

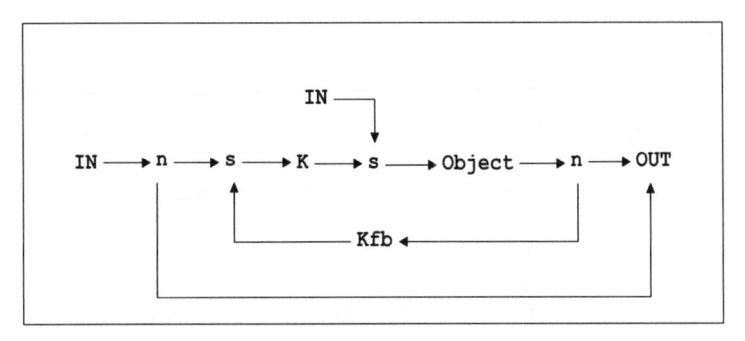

Fig. 8.5. The modified control system model

The object of control in mechatronic applications is a mechanical object, such as a robot link or a headlight assembly in an automobile driven by an actuator. The object's dynamics are commonly described by a transfer function relating input and output, such as in the following formula:

$$Cout(s) = G_{ob}(s)Cin(s) \qquad (8.2)$$

Cout and Cin are Laplace transforms of the object input and output, respectively, and s is the complex variable. We use the following transfer function as a simple model of this object:

$$G_{ob}(s) = \frac{K_{ob}}{(T_{ob}^2 s^2 + 2\varsigma T_{ob} s + 1)s} \tag{8.3}$$

Here K_{ob} is the object's static gain, T_{ob} is the time constant of the object, and ς_{ob} is the damping ratio. We cannot, however, model this transfer function directly; we must transfer it back to the time domain. The dynamics corresponding to Eqs. (8.2) and (8.3) now read:

$$T_{ob}^2 \frac{d^3 cout}{dt^3} + 2\varsigma_{ob} T_{ob} \frac{d^2 cout}{dt^2} + \frac{dcout}{dt} = K_{ob} cin \tag{8.4}$$

To describe this equation in block diagram form using only elementary components—i.e. integrators, functions, summators, etc.—Eq. (8.4) can be written as

$$\frac{d^2 cout}{dt^2} = \left(\frac{d^2 cout}{dt^2}\right)_0 +$$

$$\int_0^t \frac{1}{T_{ob}^2}(K_{ob} cin - 2\varsigma_{ob} T_{ob} \frac{d^2 cout}{dt^2} - \frac{dcout}{dt})dt \tag{8.5}$$

with

$$\frac{dcout}{dt} = \left(\frac{dcout}{dt}\right)_0 + \int_0^t \frac{d^2 cout}{dt^2} dt \tag{8.6}$$

and

$$cout = (cout)_0 + \int_0^t \frac{dcout}{dt} dt \tag{8.7}$$

The first term on the right of each equation is the initial value of the expression. Written in this fashion, these equations can be described directly by elementary block diagram components.

The input and output variables are the variables transferred to and from the control ports of the Object (Fig. 8.4). To describe its model according to Eqs. (8.5) – (8.7), we create a new model document by double-clicking it. In the document window that opens, we create a model of the control system object in a way that is similar to the way in which we developed the basic model. The resulting model is shown in Fig. 8.6.

Note that the input port of the gain (function) component Kob and the output of the last integrator are connected to the document ports that serve as the internal connectors of the outside component ports.

The function component FUN1 describes multiplication by $1/T_{ob}^2$ in Eq. (8.5), and the first integrator describes the corresponding integration. The Fun2 component of the inside feedback loop describes multiplication by $2\varsigma_{ob} T_{ob}$. The other two integrators describe the integrations represented in Eqs. (8.6) and (8.7). Finally, the summator describes the additions inside the parenthesis in the integrand of Eq. (8.5).

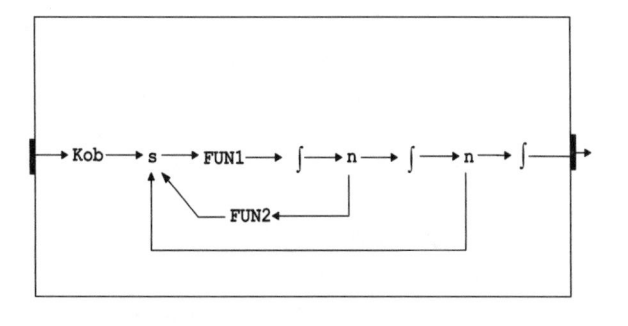

Fig. 8.6. The dynamics of the Object

The initial conditions that appear in Eqs. (8.5) – (8.7) can be set by double-clicking the integrator's output ports. In the dialogue that appears (Fig. 8.7), the initial values of the output variables can be set in the *Initial Value* edit box, the default of which is zero.

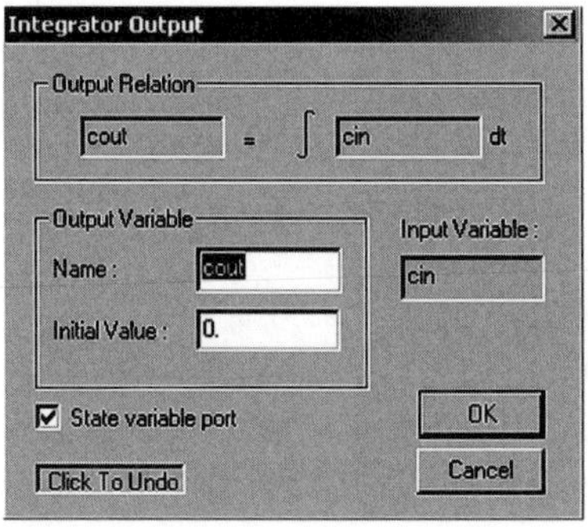

Fig. 8.7. The integrator Output dialogue

The control system model is now complete, but before simulating the system several parameters must be set and the form of the input function must be defined.
We use the following parameter values:

Controller gain	$K = 30$
Object gain	$K_{ob} = 2.77$
Object time constant	$T_{ob} = 5.41 \cdot 10^{-3}$ s
Object damping ratio	$\zeta_{ob} = 1.35$
Feedback gain	$K_{fb} = 1$

The default values are accepted for all other model parameters.

We first analyse the system response to a unit step input with no disturbance. We set the output of the left INput object to 1, that of the upper INput object to 0, and then build the model. The simulation interval is set to 0.2 s, the output interval and maximum step-size to 0.001 s, and defaults are accepted for all other settings (error constants, method etc.).

The simulation results (Fig. 8.8) show that the settling time is about 0.12 s; and that there is an overshoot of about 26%.

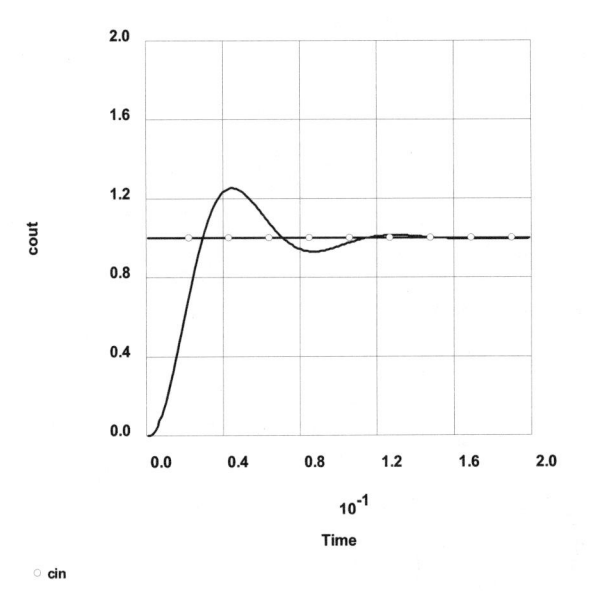

Fig. 8.8. The response of the system to a unit step input

We next simulate the response to a unit disturbance with the input equal to 0. As can be seen in the next graph (Fig. 8.9), the output does not return to zero, but settles at a value of 0.0333.

Better response behaviour can be achieved using a better controller, e.g. a controller with proportional-integration and derivative action (PID) described in the next section.

We end the discussion of this simple control system with the analysis of the response to a sinusoidal input. The reference signal generated by the IN component of Fig. 8.5 is now defined as a sine function of amplitude cValue and frequency FREQ Hz. The formula is:

$$\text{cin} = \text{cValue} * \sin(2\pi \cdot \text{FREQ} \cdot t) \tag{8.8}$$

The natural period of oscillation estimated from the damped transient (Fig. 8.8) is about 0.08 s, hence the natural frequency is about 12.5 Hz. At a much lower frequency we can expect that output will follow the input closely. To check this, we simulate the response of the system to a unit sinusoid forcing of 1.0 Hz. The simulation interval is taken as 5 s and the output interval as 0.01s. The result (Fig. 8.10) is that the input and the output signals are nearly indistinguishable.

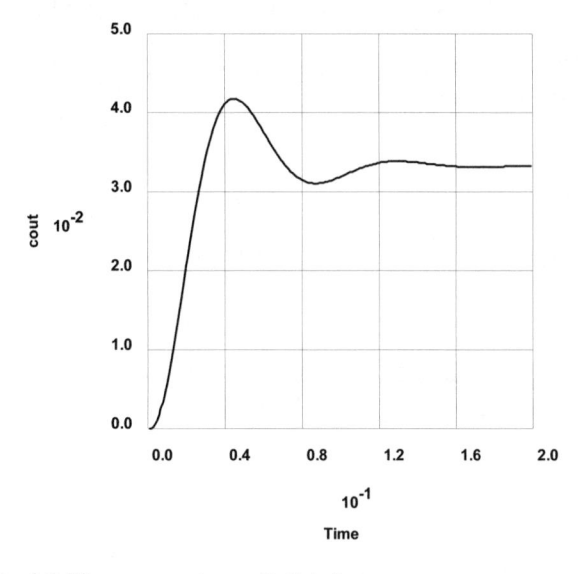

Fig. 8.9. The response to a unit disturbance

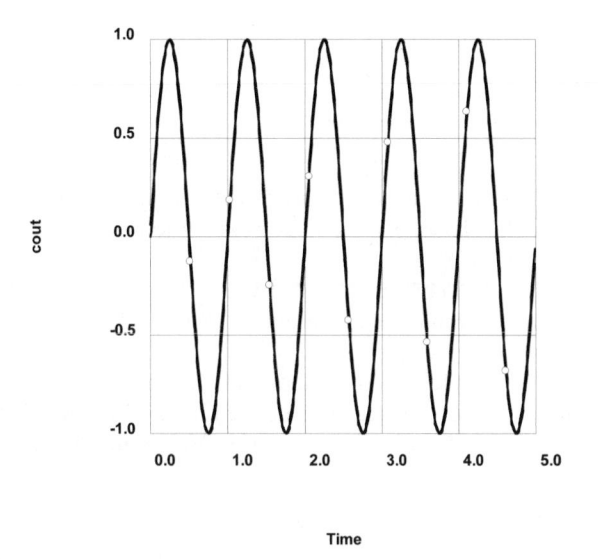

Fig. 8.10. The frequency response at 1.0 Hz

We repeat this simulation with an input frequency equal to the estimated natural frequency of 12.5 Hz. Because we are interested mostly in the phase shift between the output and input, we select a simulation interval of 0.5 s and a rather short output interval of 0.0001 s.

Simulating this situation (Fig. 8.11) results in a time difference where the input and output sinusoids cross the time axis is about 0.0198 s. Hence, the phase lag is $2 \cdot \pi \cdot 12.5 \cdot 0.0198 = 0.4952\pi$, which is close to the phase lag at a resonance of $\pi/2$. Estimation of the resonance frequency thus is quite good. This demonstrates that estimation of the natural frequency based on the damped period of oscillation is quite close to the exact value (for a linear system).

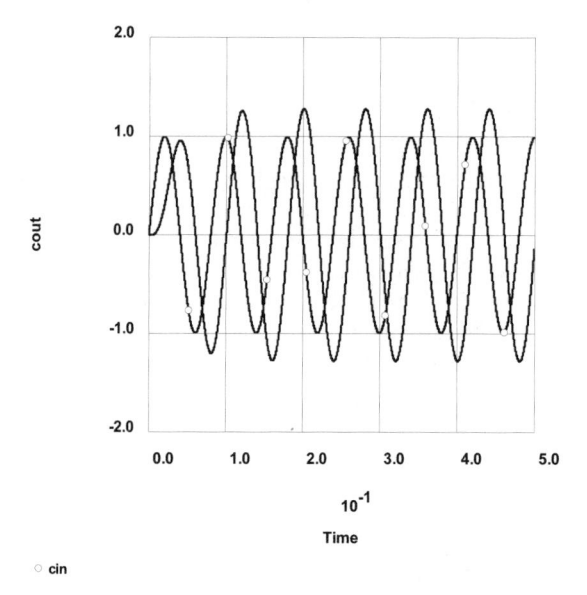

Fig. 8.11. The frequency response at 12.5 Hz (near resonance)

8.3 Control Systems Modelling

Control systems described by block diagrams can readily be modelled using the techniques of block diagram components. As shown in the previous section, it first is necessary to translate them to the time domain. Using a technique similar to analog-computer modelling, time domain relations are described by block diagram components, such as inputs, summators, function generators, integrators, and nodes. The output variables are fed to components used to display simulation results as x *vs.* t plots or x *vs.* y plots. The same technique can be used to process signals extracted from bond graph components. (We used this in most all of the bond graphs discussed in the previous two chapters.)

There is an elementary component that has not been used thus far: the *differentiator* (Sec. 2.6.5). The main reason for introducing this component is to represent often-used controllers based on laws such as the *proportional-integration-derivative* (*PID*) law and their variants. The *PID* control law is commonly defined by:

$$\text{cout} = K \cdot \text{cin} + K_i \int_0^t \text{cin} \cdot dt + K_d \frac{d\text{cin}}{dt} \tag{8.9}$$

Here, K, K_i, and K_d are referred to as *proportional, integrator,* and *derivative* constants (gains), respectively. There are other forms of PID controller laws [1]. The integrator and derivative constants are often expressed as:

$$\left.\begin{aligned} K_i &= K / T_i \\ K_d &= K \cdot T_d \end{aligned}\right\} \tag{8.10}$$

The constants T_i and T_d are, then, referred to as *integrator* and *derivative time,* constants respectively.

The PID model (Fig. 8.12) consists of several components: the input node for branching signals; a summator for the summation of the proportional, integrator, and differentiator actions in Eq. (8.19); function components denoted as K, K_i, and K_d used for multiplication by the corresponding constants; an integrator; and, finally, a differentiator component denoted as D/Dt.

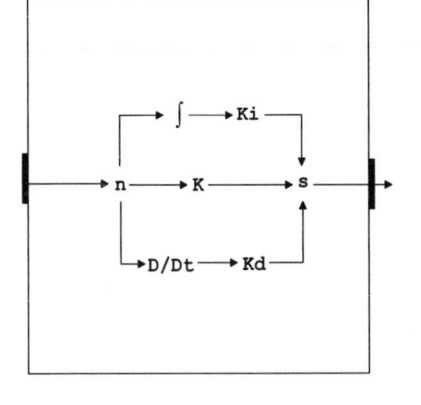

Fig. 8.12. The model of the PID controller

We do not use numerical methods to approximate differentiation with respect to time, as is often the case with simulation tools; rather, we describe differentiation as given in Eq. (2.42). The equation is then appended to the rest of model and solved together as a system of DAEs (Chap. 5).

One of the central problems in the application of PID control is *tuning* the proportional, integrator, and derivative constants. The goal is to obtain a good quality system response. The classical result is the well known and often cited *Ziegler-Nichols tuning* [2,3]. According to that method, the constants are given by:

$$\left. \begin{aligned} K &= 0.6 \cdot K_u \\ T_i &= 2.0 / T_u \\ T_d &= 0.125 \cdot T_u \end{aligned} \right\} \tag{8.11}$$

K_u is the so-called *ultimate gain* of the pure proportional controller when the system starts oscillating, and T_u is the corresponding oscillation period.

Instead of forcing the system to oscillate, the proportional constant usually is increased until the "amplitudes" of the resulting damped oscillations decrease at a ratio of ¼, i.e. any amplitude is ¼ of the immediately preceding amplitude. Denoting the corresponding proportional gain as $K_{1/4}$, the ultimate gain is approximated as $K_u = 2 \cdot K_{1/4}$. The period of steady oscillations is approximated by the period of damped oscillations.

Ziegler-Nichols tuning is referred to as sub-optimal and is used only as the starting point for finer controller tuning. The usual approach is to change the proportional and derivative constants, while holding the integrator constant equal to zero, until a satisfactory transient response to the step input is achieved. This is characterised by a good response time with little or no overshoot. The integrator constant then is varied until satisfactory sensitivity of the output to external disturbances is achieved. Several iterations typically are required to get satisfactory performance. It should be stressed that *PID* control tuning is not always easy to achieve; this is particularly true when the system includes non-linearities, such as the saturation behaviour often found in servo-drives.

Application of *PID* control and parameter tuning is illustrated with the control system example of the last section.

We begin by copying the Simple control system project to a new project file named PID Control system. We then disconnect and delete the proportional controller K and insert the PID controller model described in Fig. 8.12. The resulting system is shown in Fig. 8.13.

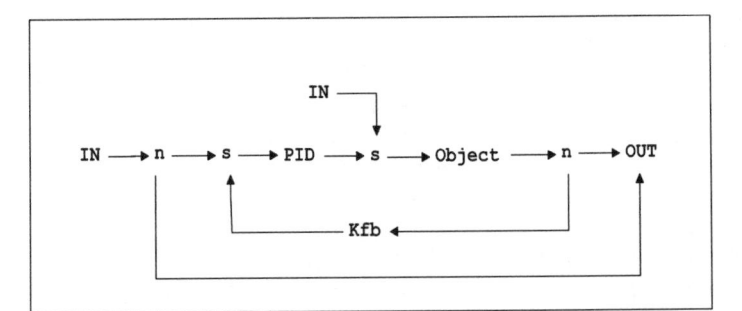

Fig. 8.13. The control system of Sect. 8.2 with PID controller

To tune the controller gains, the integrating and differentiating constants are set to zero and the proportional constant is increased until the appearance of damped oscillations with the ratio of the amplitudes of adjacent waves equal to 0.25. In the present example, this occurs at a proportional constant value of K = 60. The ultimate gain, Ku, thus is 120.

According to Eq. (8.11), the proportionality constant K = 0.6·120 = 72. The estimated period of steady oscillation Tu is 0.058 s; and thus the derivative constant Kd = 0.52.

Using these values, the response exhibits an overshoot that is slightly too large; we therefore increase Kd to 0.8. The new transient (Fig. 8.14) has a settling time of about 0.025 s and the overshoot is only 5%.

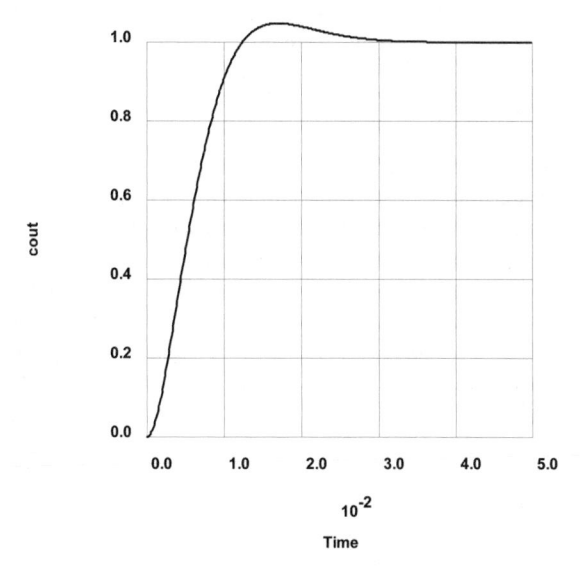

Fig. 8.14. Step response of the system for K = 72, Ki = 0, and Kd = 0.8

The integrator constant, according to Eq. (8.11), is perhaps too high (2480). Thus, we check the system response to disturbance for various values of this constant. At Ki = 0 there is a steady-state error of 0.0137. Increasing Ki causes the output to settle to zero. The settling time increases with the value of this constant. This, however, influences the response to the reference input as well. Fig. 8.15 and 8.16 show response plots for K = 72, Ki = 1400, and Kd = 0.8.

The settling time now is much longer—about 0.1 s—with an overshoot of 11%. The response to disturbance also is near zero after about 0.1 s.

Tuning a *PID* controller is known to be a time-consuming task. In [1] a type of *PID* controller was proposed that enables separate adjustment of the system response and the disturbance rejection capability. In most cases the selection of controller gains is a matter of compromise. The situation becomes much more difficult when non-linearities are involved.

Simulated system response data obtained from the BondSim program are displayed in Table 8.1. The simulation interval was 0.2 s and the output interval was 0.001 s.

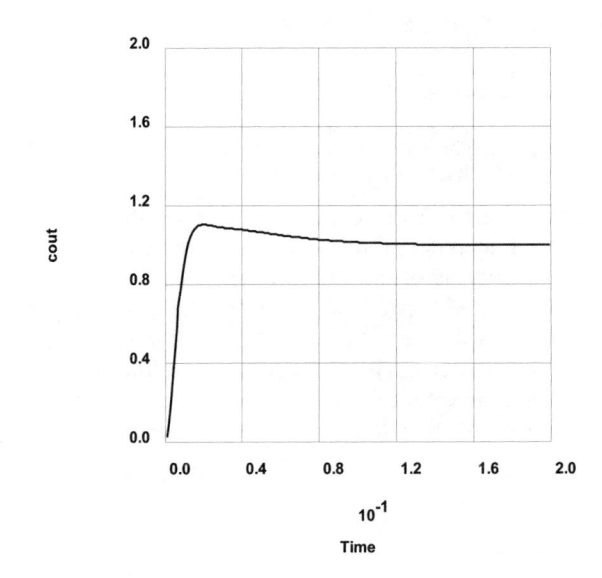

Fig. 8.15. Step response of the system for K = 72, Ki = 1400, and Kd = 0.8

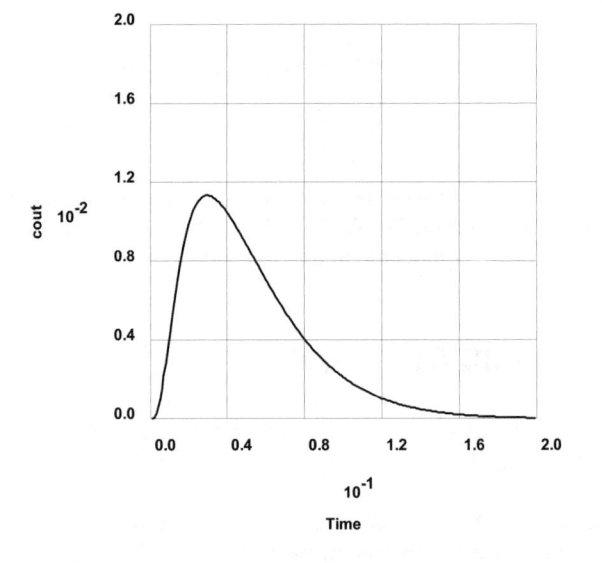

Fig. 8.16. The response to a disturbance with K = 72, Ki = 1400, and Kd = 0.8

These values are compared to those found by applying the inverse Laplace transform to the output. The Laplace transform of the step response is found first. According to Eqs. (8.9), the transfer function of the *PID* controller is given by

$$G_c(s) = \frac{1}{s}(K_d s^2 + Ks + K_i) \tag{8.12}$$

Hence, the system output, to a unit step 1/s (Fig. 8.13), is:

$$Cout(s) = \frac{G_c G_{0b}}{1 + k_{fb} G_c G_{ob}} \cdot \frac{1}{s} \tag{8.13}$$

The transfer function of the object is given in Eq. (8.3). It was difficult to find the inverse Laplace transform in analytical form because the denominator of the system's close-loop transfer function is of the fourth order. Hence, a numerical technique was used. A short FORTRAN program was written that calls a function for the evaluation of the inverse Laplace transform of the function given by Eq. (8.13), and then prints the result to a file. The DINLAP routine from the IMSL Mathematical Library [4] was used to evaluate the inverse transforms. Calculation was done with double-precision arithmetic with a relative error tolerance of $1 \cdot 10^{-6}$.

Table 8.1. System response data

Time (s)	Output values
0.002	0.114764
0.004	0.345686
0.006	0.584282
0.008	0.781502
0.010	0.923752
0.015	1.084201
0.020	1.106306
0.050	1.066662
0.100	1.015413

It is interesting that these are *exactly* the same values, rounded to six decimal places, as those obtained using BondSim (Table 8.1). Thus, the DAEs approach is at least as useful, and is also readily applicable to a much wider class of problem.

8.4 Permanent Magnet DC Servo System

The last section of this chapter deals with position servo systems. This example shows the power of the combined bond graph-block diagram approach to modelling mechatronics systems.

The scheme of the system, for the control of the angular position of a body is given in Fig. 8.17. This is a simplified example of the control of robotic arms. The servo loop consists of a servo driver and a permanent magnet DC motor that drives the body (arm) through a gearbox. The angular position of the body is measured by a sensor, the output of which is fed back to the servo-driver input. The servo driver consists of a *PID* controller input stage and an output stage that converts the low-power controller output to the higher power needed to drive the load.

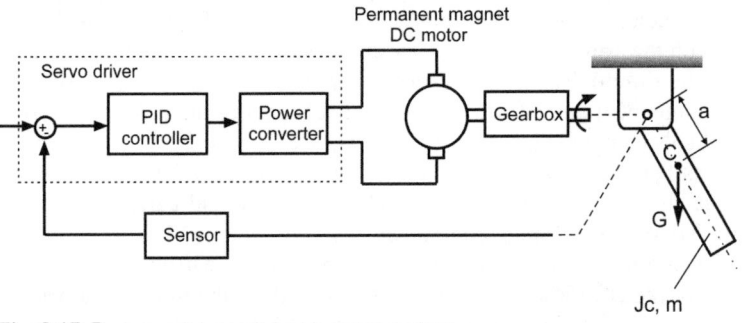

Fig. 8.17. Permanent magnet motor servo system

Model development proceeds by identifying the system components and connecting them as they are in the real system. The example project is named DC Servo System. The system level model is given in Fig. 8.18.

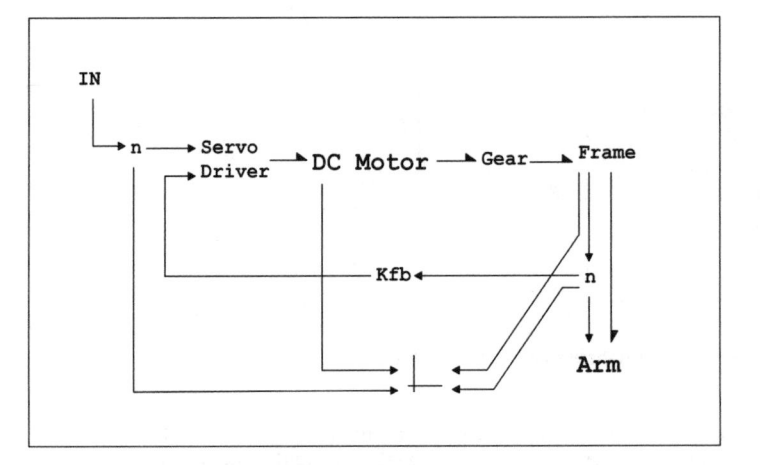

Fig. 8.18. Model of the permanent magnet motor servo system

The Servo Driver component has two control-input ports, one for the input of the reference and one for the feedback signal. It also has a power port in which the transfer of electrical power to the motor takes place.

The permanent magnet motor is represented by the DC Motor component. It has two power ports, one electrical and one mechanical. The gearbox is represented by the transformer (TF) component (renamed Gear). The arm shaft rotates in a bearing fixed in the frame. The motor and gearbox body typically are fixed in the frame, and the gearbox shaft is connected to the arm shaft by a clutch.

Following the approach to modelling mechanical systems presented in Section 2.7.3, we use the Frame object to model the connection of the arm shaft to the frame, and the Arm object to model the dynamics of the body. Thus, the power generated by the motor is transferred through the gear and the Frame to the Arm.

Information on the arm's angular velocity and position are measured at its shaft and fed out of the Frame. The position sensor is represented by a simple constant gain function component named Kfb. The reference signal of the position servo is generated by the IN component.

The variables that are of interest for observing system behaviour, such as the reference input, the arm angular velocity, and the position and current drawn by the motor, are fed to an x-y display component.

We next develop models of the main servo components, starting with the motor. The model of the DC Motor (Fig. 8.19) corresponds to the model of a permanent magnet DC motor usually found in the literature [5, 6]. To show this we have added variables to the bond graph (normally stored in the ports).

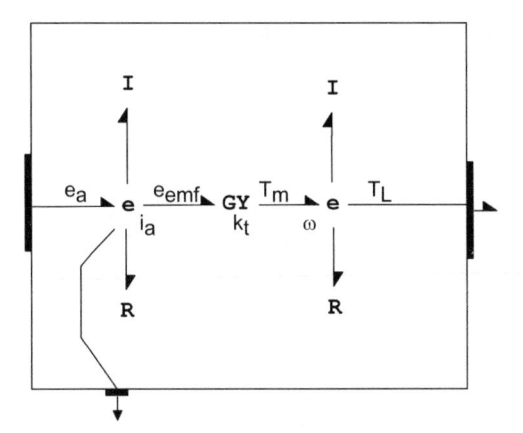

Fig. 8.19. Model of the permanent magnet DC motor

Gyrator GY describes the basic electromechanical conversion in the motor relating the *back emf* e_{emf} and the *armature current* i_a at the electrical side to the *torque* acting on the rotor T_m and its *angular velocity* ω

$$\left. \begin{matrix} T_m = k_t \cdot i_a \\ e_{emf} = k_t \cdot \omega \end{matrix} \right\}$$

(8.14)

The coupling coefficient k_t is known as the *torque constant*. The coupling coefficient in the second equation, usually denoted k_a and called the *back emf constant*, is, in fact, the same coefficient. This is a consequence of the cross-coupling between variables in the electromechanical conversion and the conservation of power in the conversion.

The electrical process in the armature winding is commonly described in terms of the armature resistance R_a and the self-inductance L_a. In the bond graph model of Fig. 8.19 it is represented on the electrical side by resistive and inertial elements, respectively, joined at a common effort (current) junction. Thus, the relation between the *armature voltage* e_a across the electrical motor terminals and the armature current i_a through it reads:

$$e_a = L_a \frac{di_a}{dt} + R_a i_a + e_{emf} \tag{8.15}$$

Similarly, the process at the mechanical side is described by a resistive element that represents linear friction with coefficient B_m, and an inertial element that describes the rotation of the rotor of mass moment of inertia J_m. They are joined at the effort (angular velocity) junction. The corresponding relation reads

$$T_m = J_m \frac{d\omega}{dt} + B_m \omega + T_L \tag{8.16}$$

Eqs. (8.14) – (8.16) are the familiar equations used to describe motor dynamics [2, 5]. Taking Laplace transformations and eliminating the torque and back emf using Eq. (8.14) yields

$$\left. \begin{array}{l} E_a = (L_a s + R_a) I_a + k_t \Omega \\ k_t I_a = (J_m s + B_m) \Omega + T_L \end{array} \right\} \tag{8.17}$$

If we then eliminate the armature current from these equations, we have

$$E_a = \left[\left((L_a s + R_a)(J_m s + B_m) + k_t^2 \right) \Omega + (L_a s + R_a) T_L \right] / k_t \tag{8.18}$$

Neglecting the external (load) torque, we get the familiar motor transfer function

$$\frac{\Omega(s)}{E_a(s)} = \frac{k_t}{(L_a s + R_a)(J_m s + B_m) + k_t^2} \tag{8.19}$$

This is the transfer function used in Sec. 8.2. The time constant appearing in Eq. (8.3), denoted here as T_m, is given by

$$T_m = \sqrt{\frac{L_a J_m}{R_a B_m + k_t^2}} \tag{8.20}$$

Similarly, the motor damping ratio ζ_m is given by

$$\zeta_m = \frac{L_a B_m + R_a J_m}{2\sqrt{L_a J_m (R_a B_m + k_t^2)}} \tag{8.21}$$

and the static gain by

$$K_m = \frac{k_t}{R_a B_m + k_t^2} \tag{8.22}$$

We can also find an equation relating the load torque and the armature current by eliminating the motor angular velocity from Eqs. (8.20). This equation is

$$T_L = \left[\left((L_a s + R_a)(J_m s + B_m) + k_t^2 \right) I_a - (J_m s + B_m) E_a \right] / k_t \qquad (8.23)$$

Eq. (8.18) shows that the load torque reflects at the electrical side; its effect should be taken care of by a servo driver. In the same vein, voltage across the motor terminals is reflected at the mechanical side and affects the torque delivered to the mechanical object that the motor drives. These two effects open various possibilities for motor control. Thus, it is possible to control the angular velocity by controlling the armature voltage and treating the torque affect as a disturbance. Another possibility is to control the torque by regulating the armature current. Both approaches have advantages and disadvantages [6,7].

The servo driver component in Fig. (8.18) is modelled as in Fig. (8.20). It consists of: a summator that outputs the difference of the signals transmitted at the reference and feedback input ports; the *PID* controller already discussed in Sec. 8.3; and a controlled source effort that models the power output driver stage. Thus, control of the DC motor is accomplished by controlling the armature voltage.

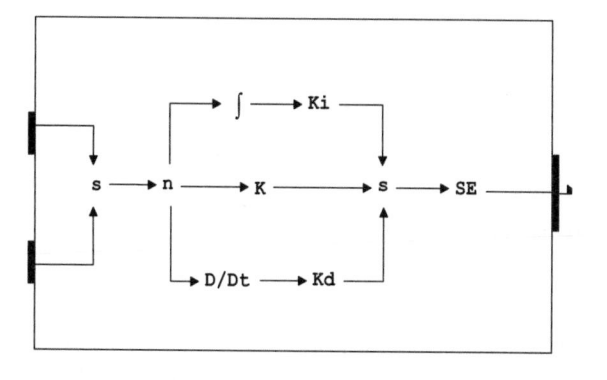

Fig. 8.20. The servo driver

The arm can be modelled as in Sec. 2.7.3 and Fig.2.21 (Platform). The arm has a single port only, the shaft. We thus disconnect and remove the other two ports and all the components connected to them (Fig. 8.21). This model describes the arm as a rigid body with its centre of mass CM translating and rotating about it.

The model of the FRAME (Fig. 8.18) is a simplified variant of the FRAME object in Sec. 2.7.3; in the current model we are not interested in the reaction forces at the arm shaft. In this model (Fig. 8.22) the component Shaft Translation describes the effect of the bearing that fixes the arm shaft axis with respect to the frame. It consists of two source efforts that impose zero velocities of translation in two orthogonal directions. Shaft Rotation is described as in Fig. 8.23. The effort junction e, the angular velocity node, is connected to through the document port to the gearbox port (shaft). On the other side it is connected to the other external port; this restricts the arm motion to rotation. The resistive element describes friction in the bearing.

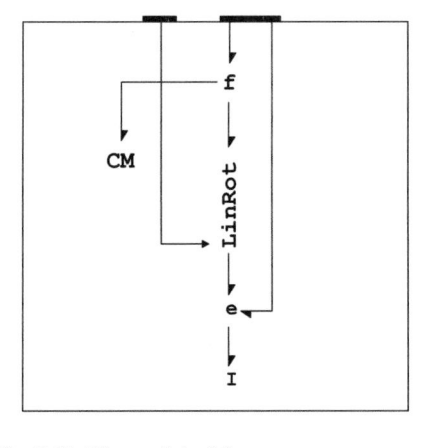

Fig. 8.21. The model of the arm

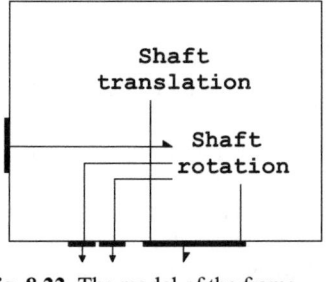

Fig. 8.22. The model of the frame

This completes the model of the servo system. The system parameters used are as follows:

Permanent magnet DC motor:

Armature resistance	R_a = 2 ohm
Armature inductance	L_a = 0.004 H
Torque constant	k_t = 0.360 N·m/A
Rotor mass moment of inertia	J_m = 9.5·10^{-4} kg·m^2
Rotor friction constant	B_m = 1.0·10^{-5} N·m·s
Peak current	i_{amax} = 50 A
Continuous current	i_a = 5 A

Gearbox:

Reduction ratio	1:20

Arm:

Mass	m = 10 kg
Mass moment of inertia about mass centre	$J_c = 0.4$ kg·m^2
Distance of mass centre from the shaft	a = 0.1 m

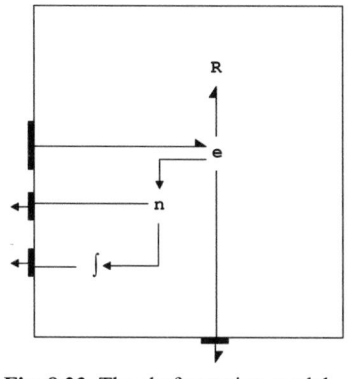

Fig. 8.23. The shaft rotation model

Before beginning the simulation, several important facts that bear on the interpretation of the results must be noted (Fig. 8.17). First, the arm is a non-linear object because the moment of its weight about the axis of rotation is equal to a·m·g·sinθ, where θ is the angle of the arm longitudinal axis to the vertical axis of the co-ordinate system. Thus, changes in the moment of weight are adequately approximated by a linear relation only for small angle changes, e.g. |θ| ≤ 0.1. Because of this, the dynamic behaviour of the servo system can differ for small and large arm angles.

On the other hand, at any position—except the vertical—there is a moment of weight that the motor must supply. Hence, even at a steady-state position of the arm, there is some current flowing through the armature winding of the motor. This is a well-known problem that arises in robotics. It has motivated development of various weight-compensation schemes designed to minimise this effect. We will investigate how much this is a problem in the present example.

We first analyse the system response to a quarter-turn (π/2) step. The *PID* constants are set to K = 2160, K_d = 40, and K_i = 0. The system was simulated for 0.2 s with an output interval of 0.001s. The response (Fig. 8.24) features a settling time of about 0.06 s with an overshoot of 9.8%. During the transient phase, however, the system draws enormous current. After the transient disappears, the output reached 1.56954 rad, about -0.08 % of the input value. The current drawn at steady state was 1.363 A, which can be checked to correspond to the moment of weight that the motor supplies.

We also checked the response to a small input of 0.1 rad. This resulted in no apparent change in behaviour (Fig. 8.25): The settling time and the overshoot were approximately the same, but the system now draws much less current.

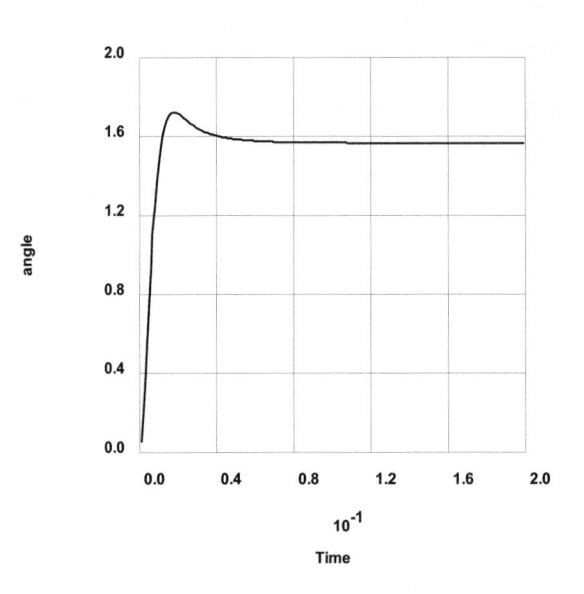

Fig. 8.24. System response to a quarter-turn step (K = 2160, K_d = 40 and K_i = 0)

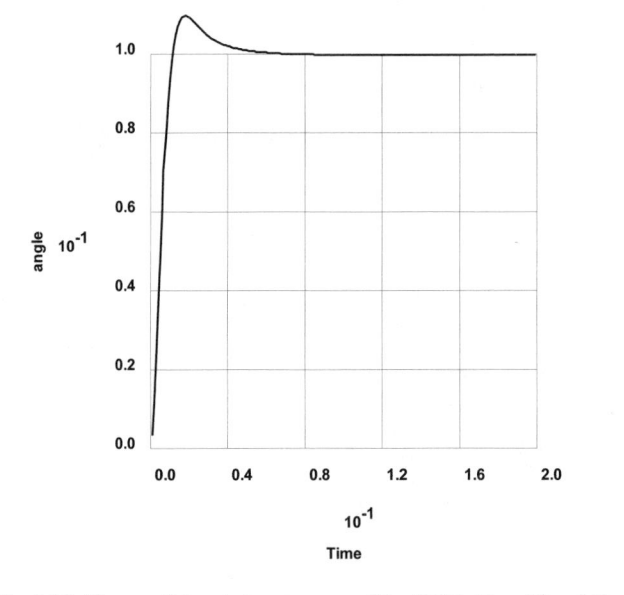

Fig. 8.25. The small input step response (K = 2160, K_d = 40 and K_i = 0)

To limit the amount of current drawn from the source, the supply voltage was limited in the controlled voltage source **SE** (Fig. 8.20). It was necessary to limit it to ± 70 V so that the maximum current was reduced to a safe level for the motor

(50 A). The resulting system output did not overshoot, but the settling time was about 0.25 s (Fig. 8.26). The current change during the transient phase is illustrated in Fig. 8.27. The response to a small input of 0.1 rad was monotonic, but with a settling time less than 0.1 s.

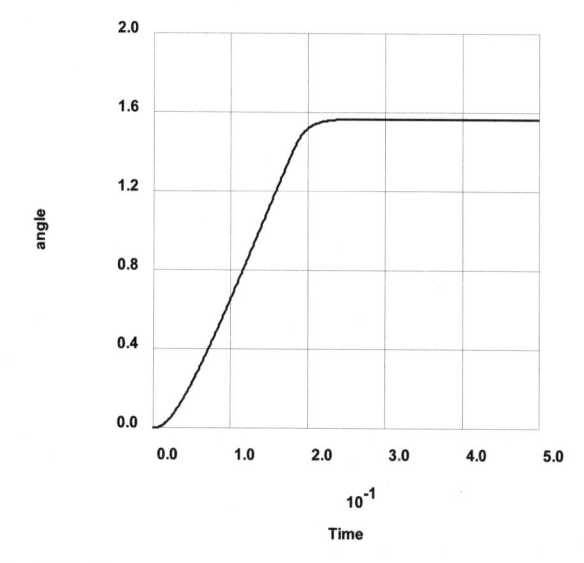

Fig. 8.26. The response with limiting the supply voltage to 70 V

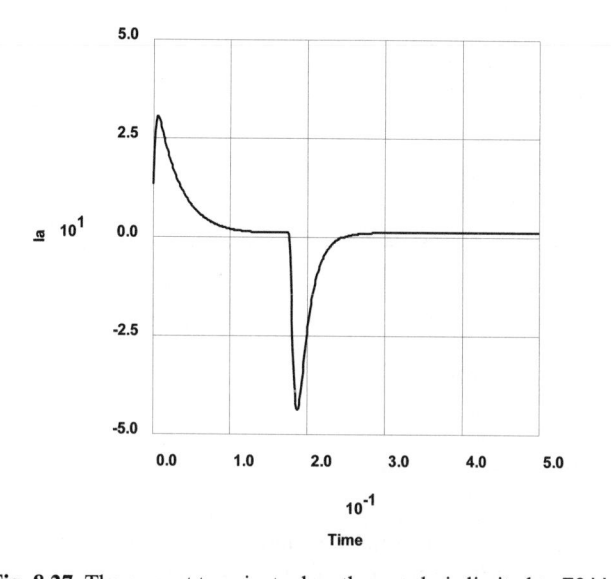

Fig. 8.27. The current transient when the supply is limited to 70 V

The effect of the integrator constant of the PID controller was also examined. Fig. 8.28 shows the response for K_i = 2000 over a simulation time of 5 s. The integrator constant contributes to reducing the steady-state error at the expense of larger reset times. Thus, for K_i = 2000, the integrator time is of the order of 1 s. This is seen clearly in Fig. 8.28: The settling time—to ±2% of the steady-state value—is 2s!

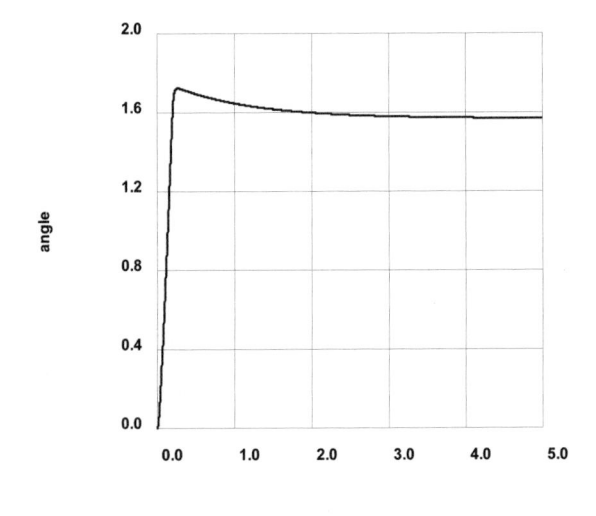

Time

Fig. 8.28. Effect of the integrator constant (K = 2160, K_d = 40 and K_i = 2000)

One conclusion drawn from this simulation is that the simplified model of the controller and, in particular, its power stage can provide only a general impression of how such a system behaves. A much more thorough evaluation requires application of techniques presented in Chapter 7 to develop a physical model of the servo driver based on its detailed electronic design (e.g. as MOS H-bridge).

References

1. RM De Santis (1994) A Novel PID Configuration for Speed and Position Control. Trans. ASME J. of Dynamic Systems, Measurement and Control, 116:542-549
2. HV Vu and RS Esfandiary (1998) Dynamic Systems : Modeling and Analysis. McGraw-Hill, New York
3. W Peesen (1994) A New Look at PID-Controller Tuning. Trans. ASME J. of Dynamic Systems, Measurement and Control, 116:553-557
4. Visual Numerics (1997), IMSL Math/Library Vol. 1 and 2, Houston, USA
5. SE Lyshevski (1999) Electromechanical Systems, Electric Machines, and Applied Mechatronics. CRS Press, Boca Raton

Chapter 9 Multibody Dynamics

9.1 Introduction

There is an extremely large body of literature dealing with the modelling and simulation of multibody systems, e.g. [1–7]. The importance of multi-body systems is also recognized in robotics where different approaches have been developed taking into account the control aspect as well [8,9]. The modelling of multibody systems has attracted attention in bond graph theory, too. The models are based on field multiport elements and multibonds [10–12].

In this section we describe the modelling and simulation of rigid multibody systems using the component model approach. In mechatronics, the problem is not only the mechanical part, but the complete system including the controls and the interaction with the environment as well. The general component model approach developed in this book can be applied readily to such complex systems.

The bond graph approach normally leads to the representation of the multibody system with system constraints described at the velocity level, not positional [13]. This is not specific to bond graphs, but is a characteristic property of the dynamics of systems that are described by the classical Newton-Euler approach In this respect it corresponds more closely to the elegant approach of [7]. We will show that it is a viable approach not only from the modelling point of view, but also from the simulation aspect, as well. The component modelling approach enables the systematic development of the model, starting from the physical components and modelling the structure of the system. In this way the resulting model is more easily understood. Visual representation of the model helps this too.

We start with planar multibody systems first and develop a component model of body dynamics. Then the basic joints—such as revolute and prismatic joints—are analysed and the corresponding models developed. It is shown on an example of a quick return mechanism how the simulation model of mechanisms can be developed systematically. The system behaviour is analysed by simulation.

To show the applicability of the approach to more complicated systems, the well-known Andrews' squeezer mechanism [13] is analysed. The accuracy of the simulation results is compared to the published results [13,14]. It is shown that the simulation times and accuracy achieved are good, at least for engineering needs. As the last example of planar multibody dynamical systems, an engine torsional vibration problem is analysed.

The last two sections deal with modelling of space multibody systems. An approach to modelling of such systems is described and space component models of bodies and basic joints are developed. In the last section of this chapter, the method and components developed are applied to the modelling and simulation of a complete robot system. A three-degree of freedom robotic manipulator is analysed. This has a wrist carrying a tool, which is pressed onto, and moved across, a wall. A hybrid force-position control strategy is applied and the simulation of a typical working cycle is shown.

9.2 The Modelling of a Rigid Multibody System in a Plane

9.2.1 The Component Model of a Rigid Body in Planar Motion

Recall from engineering mechanics that the term *plane motion* denotes motion of a body in which all points move in parallel planes. Referring to Fig. 9.1, motion of the complete body can be represented by the motion of a body section in its plane. The plane that selected for representing the body motion is the one that contains the body mass centre. In mechatronic applications the bodies in question are usually the members of a mechanism assembled by connecting the bodies by suitable joints. The complete mechanism undertakes plane motion only if the joints allow motions in which all the members move in the same plane. The term multibody system in a plane refers to such a case. Otherwise the problem refers to motion in space. Plane motion of a rigid body has already been discussed in some detail in Sec. 2.7.3. For completeness we give here the essential points again.

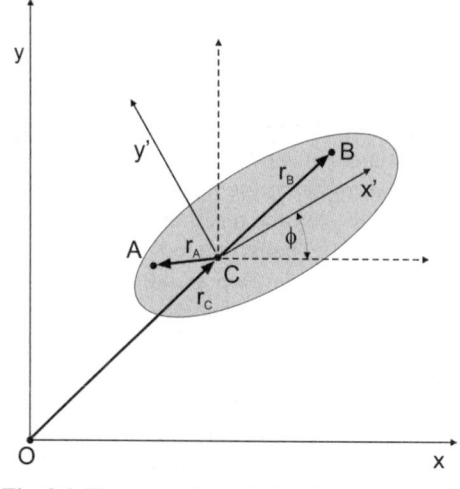

Fig. 9.1. Representation of a body motion in a plane

According to the classical approach of engineering mechanics a plane body motion consists of a translation determined by the motion of the body mass centre and a rotation about an axis through the mass centre that is orthogonal to the plane of motion. To describe the motion of the body, a base (inertial) frame Oxy is defined (Fig. 9.1). The translation part of the motion can be described by the position vector $\mathbf{r}_C$ of its mass centre C in the base frame. Similarly, to describe the rotational part of the motion, a body frame Ox'y' is defined, the origin of which is taken at the mass centre, and which is moving with the body. The rotation of the body can be described by the angle ϕ, which the body frame makes with respect to the base frame.

The bond graph method uses velocities as fundamental quantities for the kinematic description of body motion. This is consistent with Newton's 2nd Law, as well as with other fundamental laws of body dynamics. Positional quantities are found from velocities by integration. Thus we take as the fundamental kinematical variables the vector of mass centre velocity

$$\mathbf{v}_C = \begin{pmatrix} v_{Cx} \\ v_{Cy} \end{pmatrix} \tag{9.1}$$

and the angular velocity ω of the body.

Velocities of any other point P, such as A or B in Fig. 9.1, can be written as

$$\mathbf{v}_P = \mathbf{v}_C + \mathbf{v}_{CP} \tag{9.2}$$

i.e. as the sum of the velocity of the mass centre and the relative velocity component of rotation of the body around the mass centre. We are interested mostly in points where the body is joined to other bodies. These points are normally defined in the body frame by the corresponding coordinates. The rotational part in Eq. (9.2) (from Eqs. (2.87) and (2.88)) is given by

$$\mathbf{v}_{CP} = \mathbf{T}\omega \tag{9.3}$$

Here $\mathbf{T}$ is a matrix describing the transformation of rotational velocities to linear velocities and is defined by

$$\mathbf{T} = \begin{pmatrix} -x'_{CP} \sin\phi - y'_{CP} \cos\phi \\ x'_{CP} \cos\phi - y'_{CP} \sin\phi \end{pmatrix} \tag{9.4}$$

Assuming that there are two points in a body that serve for the connection to other bodies, a body can be represented by a component model (Fig. 9.2) having two power ports.

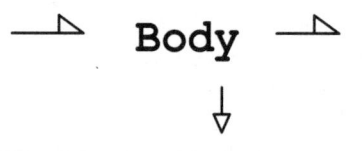

Fig. 9.2. Bond graph representation of a body

The force that one body exerts on another can be represented by a resultant force vector at the connection point and a resultant moment about that point. To represent the forces and velocities at such points, the power ports of the body component model should be compounded. Power variables at the ports can be represented by 3D efforts

$$\mathbf{e}_P = \begin{pmatrix} \mathbf{F}_P \\ \mathbf{M}_P \end{pmatrix} \qquad (9.5)$$

and 3D flows

$$\mathbf{f}_P = \begin{pmatrix} \mathbf{v}_P \\ \omega \end{pmatrix} \qquad (9.6)$$

The first part of such a port serves for the transfer of force $\mathbf{F}_P$ and linear velocity $\mathbf{v}_P$. The other part serves for the transfer of moment $\mathbf{M}_P$ and body angular velocity ω. The flow vector components in Eq. (9.6) are given by Eq. (9.2) and (9.3). Similar relations hold for the effort components of Eq. (9.5). They can be developed by applying the equivalent force and moment laws of engineering mechanics. We develop these by evaluating the power that is transferred to the body as a result of mechanical action as we did in Sec. 2.7.3. This alternative approach is convenient, for it simplifies the bond graph representation.

Using vector notation, the power transferred at the ports is given by

$$\mathbf{f}_P^{\mathsf{T}} \mathbf{e}_P = \mathbf{v}_P^{\mathsf{T}} \mathbf{F}_P + \omega \mathbf{M}_P \qquad (9.7)$$

By substituting from Eqs. (9.2), and (9.3) we get

$$\mathbf{v}_P^{\mathsf{T}} \mathbf{F}_P + \mathbf{M}_P \omega = \mathbf{v}_C^{\mathsf{T}} \mathbf{F}_P + (\mathbf{T}^{\mathsf{T}} \mathbf{F}_P + \mathbf{M}_P) \omega \qquad (9.8)$$

Hence the force at a port not only tends to push the body mass centre, but also affects the rotation of the body. Term $\mathbf{T}^{\mathsf{T}} \mathbf{F}_P$ represents the moment of the force $\mathbf{F}_P$ at the port P about the body mass centre. The matrix $\mathbf{T}^{\mathsf{T}}$ that describes this transformation is the transposed matrix of Eq. (9.4). This term taken together with Eq. (9.3) describe the transformation between linear quantities at the port – the force and the linear velocity components, and the angular quantities referred to the body mass centre – moment about the mass centre and the body angular velocity.

Using the foregoing equations the model of the body moving in a plane can be represented by the bond graph of Fig. 9.3. We can see clearly the compounded structure of the document ports (corresponding to the component ports of Fig. 9.2).

The components denoted by f describe the summation of the velocity vectors in Eq. (9.2). This component consists of two flow junctions that describe the summation of the x- and y-components. These also serve as junctions of the force components at the ports. Similarly the e component represents the body mass centre junctions and describes the summation of the forces at the mass centre. The effort junction e in the middle of the model corresponds to the body angular velocity. It also describes the summation of the moments about the body mass centre. The components LinRot represent the transformations of the linear and angular quantities already discussed. The transformation matrix is given by Eq. (9.4). The matrix depends on the body rotation angle which is, in planar body motion, related to the body angular velocity by

$$\omega = \frac{d\phi}{dt} \qquad (9.9)$$

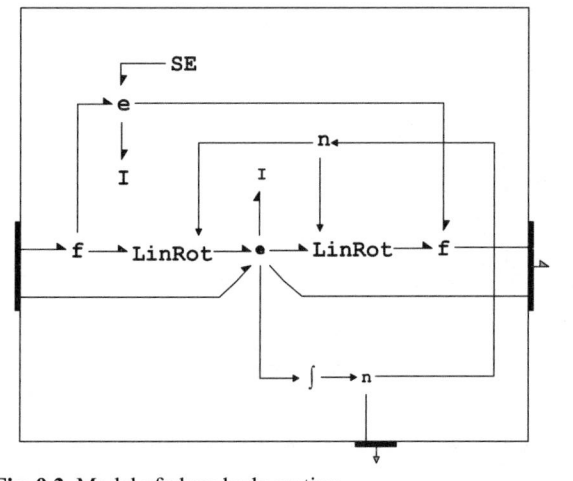

Fig. 9.3. Model of plane body motion

Hence, the rotation angle is given by

$$\phi = \phi_0 + \int_0^t \omega \, dt \tag{9.10}$$

where ϕ_0 is the initial value of the angle. The rotation angle is obtained in the model of Fig. 9.3 by an integrator component that, as input, has a signal taken from the angular velocity junction. The integrator output branches through nodes to the LinRot components, and is also taken out of the body component as a signal if required.

The transformation of the LinRot component is represented as shown in Fig. 9.4. According to Eq. (9.3) and (9.4), and to the $\mathbf{T}^T\mathbf{F}_P$ term in Eq. (9.8), it consists of two transformers and an effort junction. The transformation ratio is defined by the rows of the transformation matrix of Eq. (9.4).

To complete the model, it is necessary to add the dynamics of body motion. This consists of the translational dynamics governed by the body mass centre motion and the dynamics of body rotation about it. The first is given by the equations

$$\left. \begin{array}{l} \dfrac{d\mathbf{p}_C}{dt} = \mathbf{F}_C \\[2mm] \mathbf{p}_C = m\mathbf{I}\mathbf{v}_C \end{array} \right\} \tag{9.11}$$

$\mathbf{F}_c$ is the resultant force at the mass centre, $\mathbf{v}_c$ its velocity, m is the body mass, and $\mathbf{I}$ is the 2×2 identity matrix. The translational dynamics is represented in Fig. 9.3 by an $\mathbf{I}$ component connected to the body mass centre junction $\mathbf{e}$. The component consists of two inertial components that describe the momentum law of Eq. (9.11). There is also a SE component connected to it. This is composed of the source efforts representing the x- and y-components of body weight.

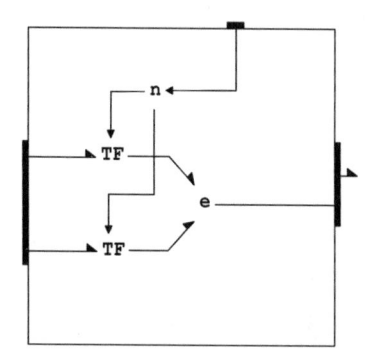

Fig. 9.4. Transformation between the linear and the rotational quantities

Rotational dynamics is simpler and is described by the moment of momentum law

$$\left.\begin{array}{l} \dfrac{dH_C}{dt} = M_C \\[2mm] H_C = J_C\omega \end{array}\right\} \tag{9.12}$$

In these equations M_c is the resultant moment about the mass centre, ω is the body angular velocity and J_c is the centroidal mass moment of inertia of the body. It is represented in Fig. 9.3 by an inertial element connected to the angular velocity junction. This junction is also connected to the lower part of the document ports and, in this way, the moments at the body component ports are transferred directly to the junction.

The model above is described in terms of the absolute angular coordinate, i.e. the rotation angle with respect to the base (inertial) frame. It is possible to develop a model based on the relative angular coordinates, i.e. of a body with respect to a second body. This approach is used in Sect. 9.6 when dealing with robot motion in space.

Another point we wish to stress is the selection of the body frame. We have assumed it is at the origin at the body mass centre. Often it is selected with the origin at one of the connection points, e.g. at point A (Fig. 9.5). This selection affects the coordinates of the transformation matrix of Eq. (9.4) and, hence, the transformer ratios in the LinRot components (Fig. 9.4). These coordinates can be found from the vector relations

$$r_{CA} = -r_{AC} \tag{9.13}$$

and

$$r_{CB} = r_{AB} - r_{AC} \tag{9.14}$$

It is not necessary to change the transformer relations. The relations of Eq. (9.13) and (9.14) can be defined by parameter statements at the level of the LinRot component.

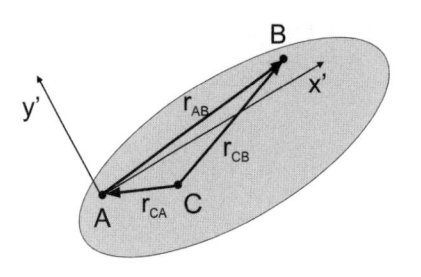

Fig. 9.5. The body frame with the origin not at the mass centre

9.2.2 Joints

Bodies in multibody systems, such as machines and robot manipulators, are joined in ways that restrict their motion. The bodies are connected by *joints*. We represent the joints by separate components that have ports to connect the bodies (Fig. 9.6).

⊸Joint⊸

Fig. 9.6. The joint as a component

Models of joints depend on their type, i.e. which motions are permitted and how the joint is physically designed. We look on a joint as a mass-less component, assuming that its mass has been included in the mass distribution of the bodies that it connects. We develop here models of two basic joints: revolute and prismatic (translational). Others can be developed in a similar way.

The Revolute Joint

This joint connects bodies by a pin or a shaft (Fig. 9.7). We neglect clearances between the pin (shaft) and the bearings. Their axes are represented in the plane of motion by the joint's coincidental central points. In a rigid revolute joint, these points move as single point. The only permitted motion is the relative rotation of the bodies about this point.

A model of the joint is given in Fig. 9.8. The e component consists of effort junctions that represent the x- and y- components of the joint centre velocity. The force is simply transmitted by the joint.

The flow junction describes the relationship between the angular velocities ω_A and ω_B of the bodies

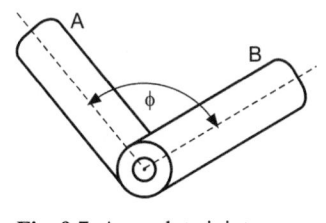

Fig. 9.7. A revolute joint

$$\omega_A - \omega_B - \omega_{AB} = 0 \qquad (9.15)$$

where ω_{AB} is the relative velocity of body B with respect to body A.

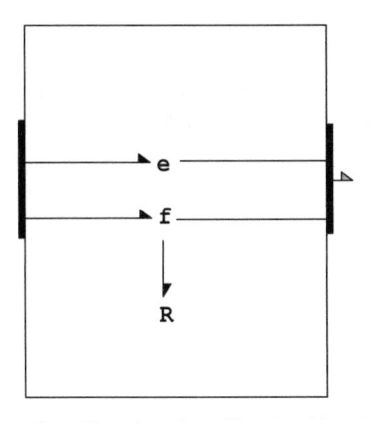

Fig. 9.8. Model of the revolute joint

This relative velocity is important if there is friction at the junction. This is represented in Fig. 9.8 by a resistive element R. It is important also if there is an actuator that drives the bodies about the junction axis, as is often the case in robotics. Otherwise it can be simply removed from the model.[1] In which case the lower bonds in the model of the joined bodies in Fig. 9.3 should be removed as well.

The Prismatic Joint

The prismatic joint connects two bodies - one containing a straight slot and other that has a part that fits precisely into the slot and can slide in it without rotation (Fig. 9.9). The rotation is usually prevented by the form of the slot and the body sliding in it, e.g. both having rectangular cross sections, or by the use of a keyway.

---

[1] This holds for planar motion of the bodies only, for in that case the rotation axis is orthogonal to the plane of the motion.

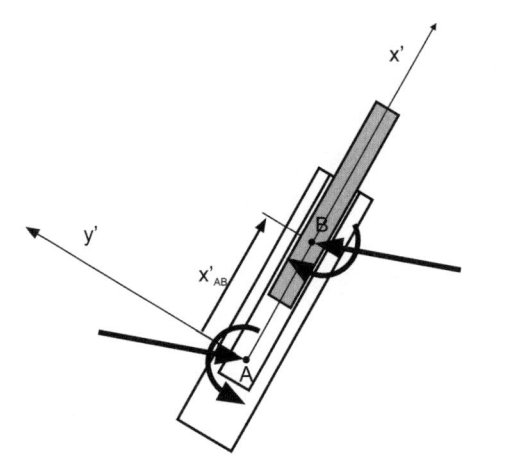

Fig. 9.9. Prismatic joint

The analysis of a prismatic joint is done in a similar way to the body motion in Sect. 9.2.1. We define a joint A$x'y'$ coordinate frame that is fixed in one of the bodies, e.g. the one with the slot (Fig. 9.9). As the origin A of the frame, a convenient point on the centreline of the slot is chosen, for example, it can be the midpoint. The x-axis is directed along the slot axis and the y-axis is orthogonal to it. The other point, B, used for representing the joint also is chosen on the slot centreline, but belongs to the other body.

We assume that at these points the joint is connected to the bodies. They correspond to the ports of the prismatic joint component. Like other body connections, there is a force vector and a moment acting on the joint at one port, and the reactions of other body at the other port. Likewise the ports flow consists of the velocity vectors of the corresponding junction points and the common angular velocity of the joined bodies.

The position vectors of point B and A, with respect to the base frame (not shown in Fig. 9.9), are related by

$$\mathbf{r}_B = \mathbf{r}_A + \mathbf{R}\mathbf{r}'_{AB} \tag{9.16}$$

where $\mathbf{r}'_{AB}$ is the relative position vector of B with respect to A, expressed in the frame of the joint, i.e.

$$\mathbf{r}'_{AB} = \begin{pmatrix} x'_{AB} \\ 0 \end{pmatrix} \tag{9.17}$$

The rotation matrix $\mathbf{R}$ of the joint frame with respect to the base frame (Sect. 2.7.3) is given by

$$\mathbf{R} = \begin{pmatrix} \cos\phi & -\sin\phi \\ \sin\phi & \cos\phi \end{pmatrix} \tag{9.18}$$

Thus Eq. (9.16) reads

$$\mathbf{r}_B = \mathbf{r}_A + \begin{pmatrix} x'_{AB} \cos\phi \\ x'_{AB} \sin\phi \end{pmatrix} \tag{9.19}$$

Here x'_{AB} represents the joint displacement coordinate and ϕ is the angle of the slot axis to the base x-axis. Both of these can change with time. Differentiation of Eq. (9.19) with respect to time gives

$$\mathbf{v}_B = \mathbf{v}_A + \begin{pmatrix} -x'_{AB} \sin\phi \\ x'_{AB} \cos\phi \end{pmatrix}\omega + \begin{pmatrix} \cos\phi \\ \sin\phi \end{pmatrix}v'_{ABx} \tag{9.20}$$

In the last equation

$$\omega = \frac{d\phi}{dt} \tag{9.21}$$

is the angular velocity of the joint and

$$v'_{ABx} = \frac{dx'_{AB}}{dt} \tag{9.22}$$

is the velocity of the relative displacement of one of the junction parts with respect to the other along the joint axis. Note that the other velocity component

$$v'_{ABy} = 0 \tag{9.23}$$

Eq. (9.20) is similar to Eq. (9.2) and (9.3) but has an additional term that comes from the relative translation of the bodies that make up the joint. The matrix of the angular velocity term in Eq. (9.20) can be also denoted as $\mathbf{T}$, but is slightly simpler and is given by

$$\mathbf{T} = \begin{pmatrix} -x'_{AB} \sin\phi \\ x'_{AB} \cos\phi \end{pmatrix} \tag{9.24}$$

Thus Eq. (9.20) can be written as

$$\mathbf{v}_B = \mathbf{v}_A + \mathbf{T}\omega + \begin{pmatrix} \cos\phi \\ \sin\phi \end{pmatrix}v'_{ABx} \tag{9.25}$$

The expression for the power transferred at point (port) B is found as follows. From Eq. (9.25) we have

$$\mathbf{v}_B^T\mathbf{F}_B + M_B\omega = \mathbf{v}_A^T\mathbf{F}_B + (\mathbf{T}^T\mathbf{F}_B + M_B)\omega + v'_{ABx}(\cos\phi \quad \sin\phi)\mathbf{F}_B \tag{9.26}$$

or

$$\mathbf{v}_B^T\mathbf{F}_B + M_B\omega = \mathbf{v}_A^T\mathbf{F}_B + (\mathbf{T}^T\mathbf{F}_B + M_B)\omega + \mathbf{v}_{AB}^T\mathbf{F}_B \tag{9.27}$$

Note that the power transferred at port A is $(\mathbf{v}_A)^T\mathbf{F}_A + M_A\omega$. From this we cannot infer that the power transferred at point B is equal to the power input at point A because that would imply that $(\mathbf{v}_{AB})^T\mathbf{F}_B = 0$. This holds, however, only for frictionless and unpowered prismatic joints. From the equilibrium equations of the mass-less joint we have $\mathbf{F}_A = \mathbf{F}_B$ and

$$M_A = M_B + \mathbf{T}^T \mathbf{F}_B \tag{9.28}$$

All these relations become clearer when we represent them by bond graphs!

We now formulate the bond graph model of the prismatic joint. It is given in Fig. 9.10. Component f consists of flow junctions and describes the joint velocity relation given by Eq. (9.25). The junction variables are the components of the force $\mathbf{F}_A$ at the joint's left port, which is equal to the force transferred at other port, i.e. $\mathbf{F}_B$. The effort junction at the top corresponds to the relative velocity of the joint expressed in the joint frame. The corresponding position coordinate can be evaluated from the integrator that takes as its input the relative velocity of the joint. The component Rot describes the transformation of the x'-components of the relative velocity and the force at the joint to the base frame. It uses the first column of the matrix in Eq. (9.18) only. The source effort defines the force acting along the joint. It is simply zero if the friction is negligible, or can be replaced by a resistive element otherwise. This component can also be used to simulate an actuator driving the junction.

The effort junction e on the right corresponds to the joint angular velocity. It describes the balance of moments as implied by Eq. (9.28). The LinRot component describes the transformation between the linear and angular quantities. The corresponding transformation matrix, as given by Eq. (9.24), needs information on the joint rotation angle and of the position coordinate of the joint. The angle of rotation is found from the integrator, which integrates the angular velocity taken from the corresponding junction.

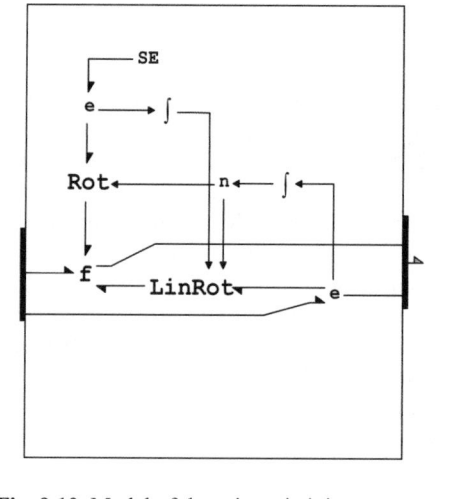

Fig. 9.10. Model of the prismatic joint

9.2.3 Modelling and Simulation of a Planar Mechanism

We apply the modelling approach described to the quick-return mechanism of Fig. 9.11. The mechanism is relatively simple, but it contains all the elements that we have discussed so far – the bodies and the rotational and prismatic joints. More complex problems are analysed in the sections that follow.

The mechanism consists of a crank that rotates about a joint at O_1 with angular velocity ω_0. The end of the crank is connected by a revolute joint at O_2 to a block that can slide along another member, which, in turn, can rotate about the joint at O. This simple mechanism generates an oscillatory motion of the driven member with different forward and return times. We develop a model of the system and then simulate its motion.

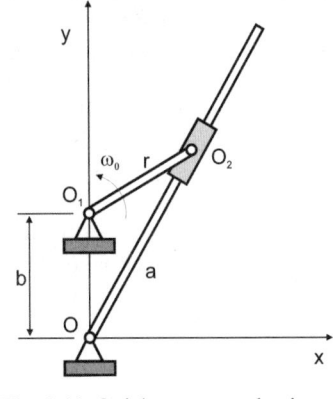

Fig. 9.11. Quick return mechanism

We start, as usual, by creating a project Quick Return Mechanism. We then create component models corresponding to every physical component of the mechanism – Crank, another member Body, revolute joints O, O1 and O2 and prismatic JointT. The direction of power flow is taken from the driver at O1 through the Crank and then across JoinT and the Body to the revolute joint O.

The system level model of the mechanism is shown in Fig. 9.12. Components Crank and Body are created using the body model of Figs. 9.2 and 9.3. To that end the components are copied from the library and inserted into the document. Some minor adjustments are necessary. Thus, the default name Body for crank component is changed to Crank, but is retained in the other member component. [2] Weights of the bodies are not included in the model. Thus the SE components of Fig. 9.3 are removed. The members are connected by joints at O and O1 in Fig. 9.11 to the ground represented here as Base. It is defined later.

There are three revolute joints. Hence we create corresponding component models corresponding to Figs. 9.6 and 9.8 by copying from the library. The names

---

[2] In order to change the name the component is disconnected from the bond lines both outside and inside. After the name has been changed, the bonds are redrawn.

are changed however to correspond to the symbols used in the scheme of the mechanism in Fig. 9.11, i.e. O, O1 and O2. Similarly the prismatic joint component JointT is taken from the library and inserted into the system level document (Fig. 9.12).

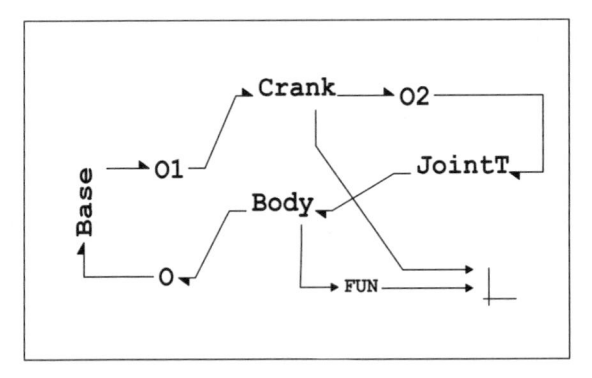

Fig. 9.12. System model of the mechanism of Fig. 9.11

All of the joints are frictionless. Thus, in joints O and O2 the resistive element R of Fig. 9.8 is disconnected and deleted. Also, the flow junctions are disconnected and deleted too. Because there is no moment transferred to nearby bodies, the corresponding bonds to the body angular velocity junctions are removed (including the corresponding ports at the junctions), as in Fig. 9.3. This means that the lower bond of the right document port of the Crank, as well as of the JointT component (Fig.9.10), are removed. In the Body component we must remove the *left* document port lower bond.

Joint O1 is driven at constant angular velocity. Thus the resistive element in the model is replaced by a constant source flow element (Fig. 9.13). The flow source constant is omega, defined by the parameter expression omega = 2*PI*RPS. The PI is π, which is defined at program level, and the crank speed (cycles/sec) is defined as RPS = 1.

Now that all joints are defined, we develop the Base component (Fig. 9.14). It is very simple. The Base fixes the centre of joint O1 and also the angular velocity at the left port (the bearing of the crankshaft). Thus, at the port used for connecting O1 there are zero flow sources. The upper one corresponds to the velocity of the joint centre and is represented as a component consisting of two flow sources. The other source flow fixes the crankshaft bearing. The component at the port that is used for connecting O fixes the position of the centre of the joint. The power flow is opposite to the upper because of the power sense chosen for the system model.

We are interested in the rotation angles of the mechanism. Hence we need a display component. We create such a component as an *X-Y Plotter* using the mechanical components palette (Fig. 9.12).

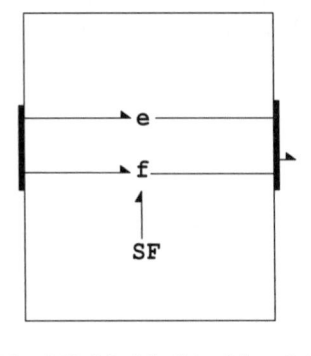

Fig. 9.13. Model of the driven joint O1

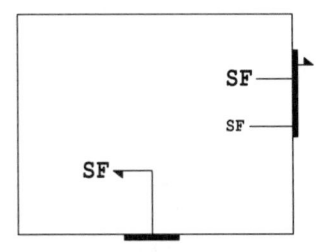

Fig. 9.14. The model of the Base

The display component is created with one input port. Thus we add another one. The upper one is connected to the Crank component output port and is labelled as Phi. The lower one, however, is not simply the body rotation angle, but angle alpha, which the other member (Body) makes with the vertical direction (Fig. 9.15). For this a function component is created that converts the body angle of rotation to the angle required. The angle is calculated as $\alpha = \pi/2 - \phi$, where ϕ is the rotation angle.

To simulate the motion of the mechanism the following geometrical parameters are used: a = 0.8 m, b = 0.4 m and r = 0.2 m. The mass parameters are given in table 9.1.

Table 9.1. Masses and mass moments of inertia with respect to the mass centre

	Crank	Driven member
Mass [kg]	1.5	4
Moment of inertia [kg·m²]	0.005	0.25

Note: The mass centres are located at the mid-points of the member's length

The inital configuration of the mechanism is shown in Fig 9.15. The values of the angles can be calculated from the geometry and are given in Table 9.2. The initial position of the slider is taken at the origin of the joint's frame. These values are used as the initial values of the corresponding integrators. It is assumed that the crank rotates at 1 revolution per second.

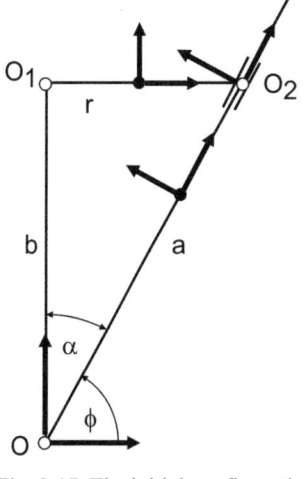

Fig. 9.15. The initial configuration of the mechanism

Table 9.2. Initial values of the angles

Member	Rotation angle [rad]
Crank	0
Body	1.107148718
JointT	1.107148718

Now we can build the mathematical model and start the simulation. The simulation time is set to **2 s**, corresponding to two revolutions of the crank. The output interval chosen is relatively short (**0.001s**) in order to obtain good resolution of the plot. The results are given in Fig. 9.16.

The diagram shows the quick return behaviour of the mechanism. Some interesting points are given in Table 9.3. The data obtained by simulation agree fairly well with the exact data obtained from the geometry of the problem

Table 9.3. Some charactristic data of the quick return mechanism motion

Name	Simulation	Exact value
Forward time [s]	0.666	0.666667
Return time [s]	0.334	0.333333
Ratio of the times	1.994	2.0
Amlitude [rad]	0.523598	0.523599

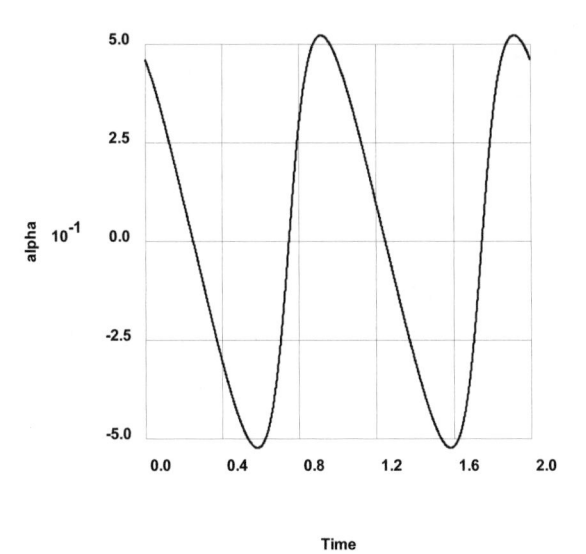

Time

Fig. 9.16. Time history of the angle of the mechanism

9.3 Andrews' Squeezer Mechanism

We now apply the method developed in Sec. 9.2 to the well-known *Andrews' squeezer mechanism* problem. This problem has been promoted as a test of numerical codes [3,13,14]. We take the formulation of the problem as given in [3, 14] and compare the simulation results obtained by the BondSim program to the solution given in [14].

The mechanism (Fig. 9.17) consists of seven bodies that can move in a plane. The bodies are interconnected by revolute joints and also to the base. The arm K1 rotates about the fixed joint at O under the action of the driving torque Md and this pushes, via body K2, the central revolute joint where three bodies—K3, K4 and K6—are connected. Bodies K4 and K6 are further connected via bodies K5 and K7 to another revolute joint A that is fixed to the base. The third body, K3, can rotate about the fixed revolute joint at B. The end D of body K3 is connected to a spring that simulates the squeezer effect. The geometrical parameters of the mechanism are given in Table 9.4. The mass and mass moment of inertia with respect to the mass centre of each body are given in Table 9.5. The spring stiffness and the driving torque are also given. The data were taken from [13].

To develop a simulation model using BondSim, we create a project called Andrews Squeezer Mechanism. All of the bodies—K1 to K7—will be represented by the standard plane body motion component model of Sec. 9.2.1. To create component models of the bodies, the component Body from the library is copied into the document seven times. The components are then moved to positions that approximately correspond to their positions in the mechanism (Fig. 9.18). The component names Body are then changed to the names used in Fig. 9.17, K1 to K7. The weights of the bodies are not included in the component models.

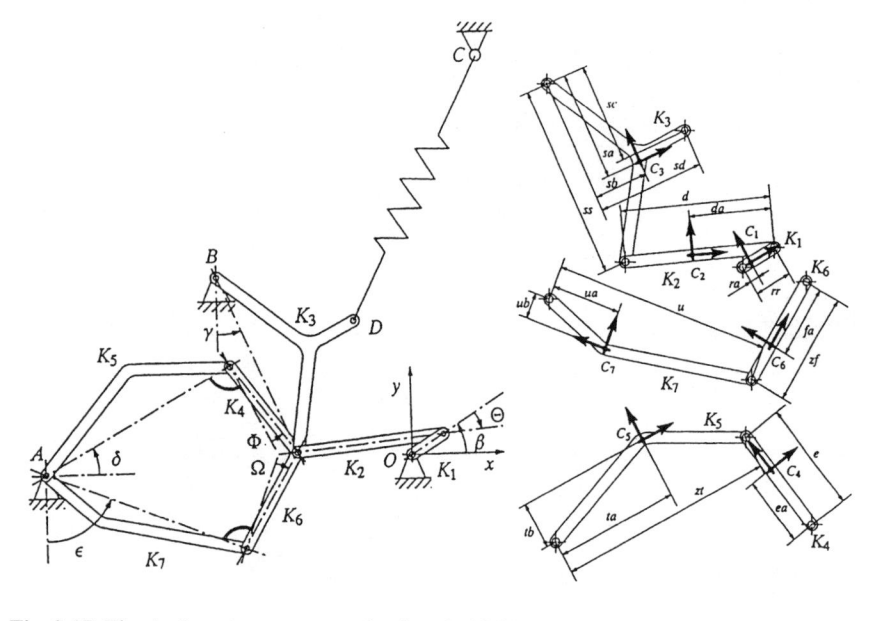

Fig. 9.17. The Andrews' squeezer mechanism (Schiehlen [3], used with permission)

Table 9.4. Geometrical parameters (Fig. 9.17)

Parameter	Value [m]	Parameter	Value [m]
ra	0.00092	ta	0.02308
rr	0.007	tb	0.00916
d	0.028	ss	0.035
da	0.0115	sa	0.01874
zf	0.02	sb	0.01043
fa	0.01421	sc	0.018
el	0.02	sd	0.02
ea	0.01421	xb	-0.03635
u	0.04	yb	0.03273
ua	0.01228	xc	0.014
ub	0.00449	yc	0.072
zt	0.04		

Parameter el corresponds to the length e of body K4 in Fig. 9.17.

The revolute joints are created by copying the standard revolute component model Joint from the library. The names of the joints that are fixed to the mechanism base are changed to the names used in Fig. 9.17, i.e. A, B and O. For the others the default names Joint are retained. To simplify the model, there is no separate base component, as in the last example (Fig. 9.12). Its effect is included directly in the A, B and O components.

Table 9.5. Mechanical parameters

	Mass [kg]	Inertia [kg·m2]	Other
K1	0.04325	2.194e-6	
K2	0.00365	4.410e-7	
K3	0.02373	5.255e-6	
K4	0.00706	5.667e-7	
K5	0.0705	1.169e-5	
K6	0.00706	5.667e-7	
K7	0.05498	1.912e-5	
Spring stiffness k			4530 N/m
Driver torque Md			0.033 N·m

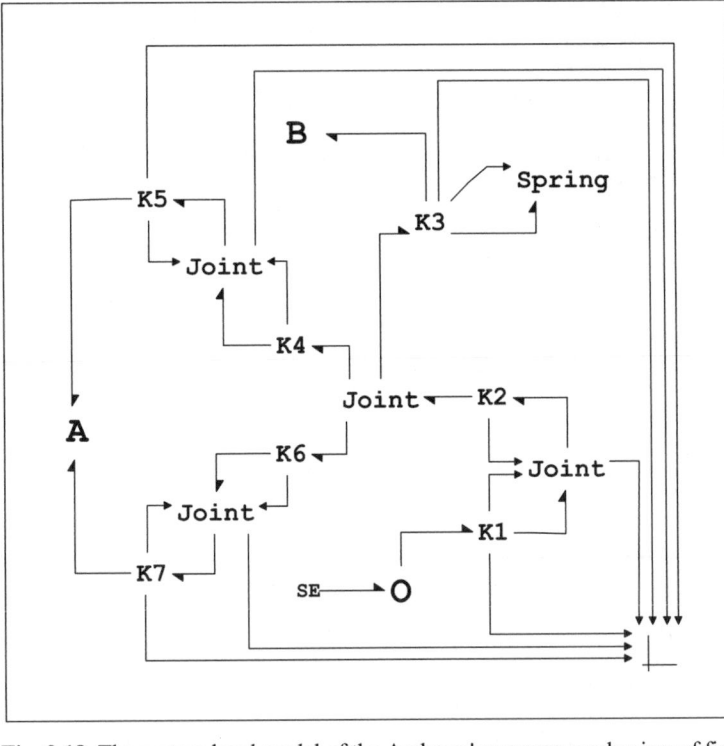

Fig. 9.18. The system level model of the Andrews' squeezer mechanism of fig. 9.17

It was assumed that there is no friction in the joints. Thus, there are no reaction moments between the bodies. Hence, the joints restrict only the movement of the points where the bodies are connected and their models consist of common velocity junctions only. In the same vein, in the body components (Fig. 9.3) there are no

bond lines connecting external ports to body angular velocity junctions. The
model of a typical body component is shown in Fig. 9.19. The only exception to
this rule is the joint O and component K1, because there is a driving torque that is
applied through the joint to the body. The driver is represented by a source effort
component connected to the revolute joint O.

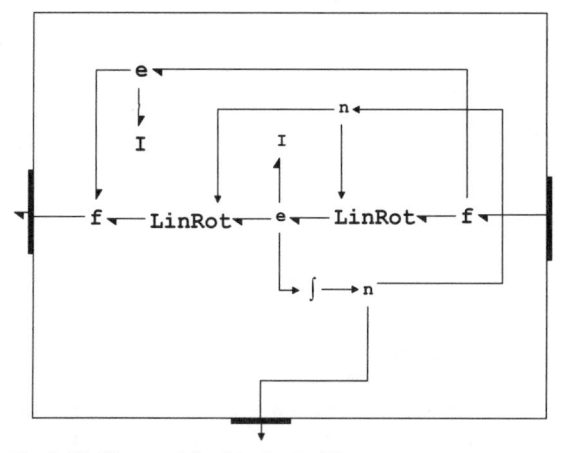

Fig. 9.19. The model of the body K2

All of the bodies except K3 have two ports that are used for connection to the
corresponding revolute joints. Note that bodies K2, K3, K4 and K6 are connected
to the common joint (Joint). Component K3 has an additional port for the attachment of the spring. The spring is represented by a word component model and interacts with the body through a power port. Its model is described later.

Interconnecting the power ports of the bodies to the corresponding joint ports
we get the model of the mechanism as given in Fig. 9.18. The power flow direction is taken from the driver SE through bodies K1 and K2 to the common joint.
The power then branches out on one side to two paths—one through bodies K4
and K5, and one through K6 and K7—to the joint A, and then to the base. On the
other side, it flows through the body K3 and then branches through joint B to the
base and to the spring. This clearly shows that the power generated by the source
is used to move the bodies and to squeeze the spring that is, after all, the purpose
of the mechanism. There are many signals between the components and we explain them next.

The body coordinate frames in the original Schiehlen scheme of Fig. 9.17 exactly correspond to the coordinate frames used in the formulation of the body
bond graph model in Sec. 9.2.1. We use as a base frame, the co-ordinate frame
Oxy of Fig. 9.17. We can look at this as the frame of the base to which the mechanism is jointed by the revolute joints A, O and B. The angles specified in the
scheme, however, are not the absolute, but are relative to the body frames of the
connected bodies. Thus angle β of body K1 is relative to the base frame, but the

angle Θ of the next body K2 is given with respect to the previous body K1 frame, etc. These angles are used as generalized coordinates of the mechanism in [13, 14].

To compare the results we introduce a vector of generalized coordinates defined as in [13]

$$\mathbf{q} = (\beta \quad \Theta \quad \gamma \quad \Phi \quad \delta \quad \Omega \quad \varepsilon)^{\mathsf{T}} \qquad (9.29)$$

By inspection of Fig. 9.17, it is easy to find the relationships between these coordinates and the body rotation angles. The generalized coordinates of bodies K1, K3, K5 and K7 correspond to body rotation angles. Hence

$$q_1 = \phi_1, q_3 = \phi_3, q_5 = \phi_5, q_7 = \phi_7 \qquad (9.30)$$

For the others, these are relative rotation angles

$$\left. \begin{array}{l} q_2 = \phi_2 - \phi_1 \\ q_4 = \phi_4 - \phi_5 \\ q_6 = \phi_6 - \phi_7 \end{array} \right\} \qquad (9.31)$$

These last relationships are represented by summators inside the joint components. This is the reason why the signals from some bodies in Fig. 9.18 are fed back to their common revolute joints. A model typical of such a joint is shown in Fig. 9.20. The output from the summator is taken out of the joint and connected to the display component (the bottom right component in Fig. 9.18). For the others, the outputs from the bodies are fed directly to the display component.

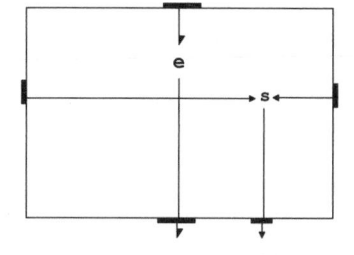

Fig. 9.20. The model of the joint between K6 and K7

We now return to the problem of modelling the spring. The spring is attached between point D of the body and point C of the base. We assume that the attachments are such that the spring extends and contracts without bending. The coordinates of the attachment point D in the base frame are given by (Fig. 9.17)

$$\left. \begin{array}{l} xd = xb + sc \cdot \sin\phi + sd \cdot \cos\phi \\ yd = yb - sc \cdot \cos\phi + sd \cdot \sin\phi \end{array} \right\} \qquad (9.32)$$

Angle ϕ in these equations is the rotation angle of body K3. In this way the coordinates of point D can be evaluated inside component K3 as shown in Fig. 9.21. This is achieved by component D, which consists of two functions that implement

Eq. (9.32). Information on the coordinates is available at the top-right control-output port.

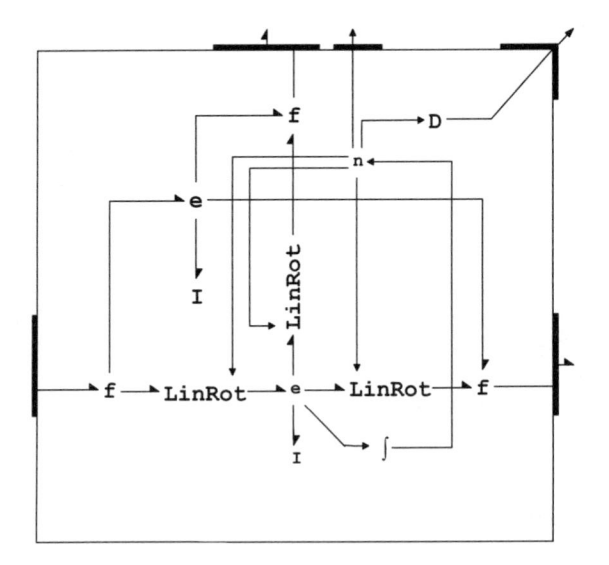

Fig. 9.21. The model of body K3 generating information on the position of point D

At point D there is a force acting on the spring, which is represented by two components in the base coordinate frame (Fig. 9.22). The point moves with some velocity represented by the respective components. The movement of the point is opposed by the spring. To relate these quantities we introduce a coordinate frame attached to the spring. The rotation matrix of the spring frame is given by

$$\mathbf{R} = \begin{pmatrix} \cos\phi_s & -\sin\phi_s \\ \sin\phi_s & \cos\phi_s \end{pmatrix} \tag{9.33}$$

The rotation angle can be found from

$$\left. \begin{aligned} \cos\phi_s &= \frac{xc - xd}{\sqrt{(xc - xd)^2 + (yc - yd)^2}} \\ \sin\phi_s &= \frac{yc - yd}{\sqrt{(xc - xd)^2 + (yc - yd)^2}} \end{aligned} \right\} \tag{9.34}$$

The coordinates of point D in these relations are available at the ports of the component K3, and those of point C are constants given in Table 9.4.

Thus the force and velocity components at point D in the base and spring frames are related by the rotation transformation

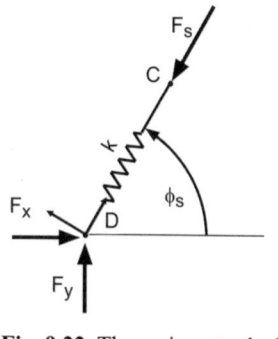

Fig. 9.22. The spring attached to body K3

$$\left.\begin{aligned} \mathbf{F_D} &= \mathbf{R}\,\mathbf{F_D'} \\ \mathbf{v_D'} &= \mathbf{R^T v_D} \end{aligned}\right\} \tag{9.35}$$

The force at D has components in the spring axis direction only, and it is in equilibrium with the spring force. Similarly its velocity has a component because of the spring extension. Hence from Eq. (9.33) the first Eq. (9.35) reads

$$\left.\begin{aligned} F_{Dx} &= F_s\cos\phi_s \\ F_{Dy} &= F_s\sin\phi_s \end{aligned}\right\} \tag{9.36}$$

Similarly from the second equation we get

$$v_s = v_{dx}\cos\phi_s + v_{dy}\sin\phi_s \tag{9.37}$$

Now we develop the model of the Spring component (Fig. 9.23). Effort and flow component transformations are represented by the Rot component and the spring itself by the capacitive element C of stiffness k (Table 9.5).

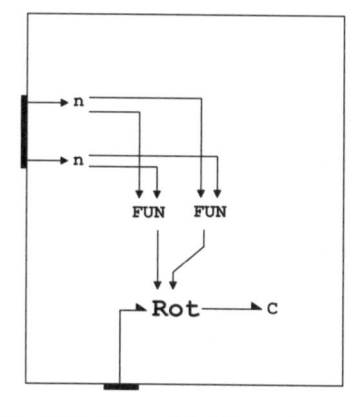

Fig. 9.23. The model of the spring

Inputs to the Rot component are the outputs from two function components giving the cosine and sine of the spring angle as given by Eq. (9.34). The transformations of the Rot component are defined by Eqs. (9.36) and (9.37) and can be represented by two transformers and a flow junction as shown in Fig. 9.24. The transformer ratios are the cosines and sines of the spring angle that is available at the control input port.

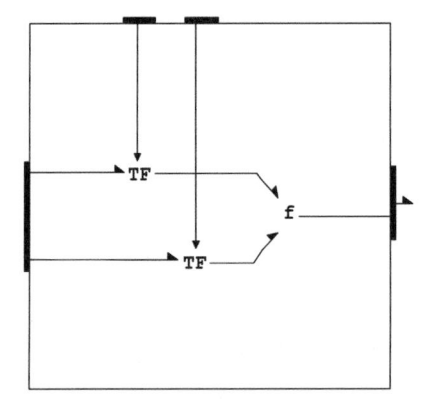

Fig. 9.24. The transformations as a result of the spring rotation

This concludes model development of the Andrews' squeezer mechanism. The reader is advised to explore the project in the BondSim library for more details. The model developed is based on the physical modelling philosophy used in this book. The resulting model equations differ from those developed using other approaches. What the model shows is that, even for a relatively complicated multibody system, a simulation model can be developed systematically using the standard component models from the BondSim program library.

Using the bond graph method the constraints on a body's motion are described at the velocity level. This generally leads to DAE models of index 2 (Chapt. 5). In this case the model consists of 197 implicit equations. The equations are relatively simple, leading to a very sparse matrix of partial derivatives (Jacobian). It has 533 nonzero elements only, i.e. there is on average 2.7 variables per equation. In comparison the mathematical model of the mechanism based on the Lagrange multiplier form of constrained multibody mechanics [13, 14] consists of 21 differential and 6 rather complex algebraic equations. In spite of the great difference in the number of equations, it will be see that the performance of the BondSim package is good.

To complete the model of the mechanism, it is necessary to define its initial configuration. This is defined by knowing the rotation angles of all the bodies and the initial deformation of the spring. For the bodies we use the initial values of the generalized coordinates, as given in [13], and evaluate the corresponding initial rotational angles from Eq. (9.30) and (9.31). These are listed in Table 9.6 and correspond to an initial angle of the crank K1 of zero rad. The initial spring compression is calculated as

$$\Delta_s = l_0 - \sqrt{(xc - xd)^2 + (yc - yd)^2} \qquad (9.38)$$

where l_0 is the unstretched length of the spring. Coordinates xd and yd are given by Eq. (9.32). This is given in the last row of Table 9.6.

Table 9.6. The initial configuration of the mechanism

Parameter	Value
K1 angle [rad]	-6.17138900142764e-002
K2 angle [rad]	-6.17138900142764e-002
K3 angle [rad]	4.5527981916307e-001
K4 angle [rad]	7.10033369709727e-001
K5 angle [rad]	0.487364979543842
K6 angle [rad]	1.00787905438393
K7 angle [rad]	1.23054744454982
Spring displacement [m]	2.51774838892633e-002

The simulations were run under similar conditions as in [14]:

– Simulation interval	0.031 s
– Output interval	0.0001 s
– Maximum stepsize	0.0001 s
– Absolute error tolerance	1e-7
– Relative error tolerance	1e-7

The reference solution in [14] was done on a CRAY computer using the PSIDE parallel software for DAEs. Experiments were also conducted using some other well-known codes including RADAUS [13]. These tests were run on a Silicon Graphics Indy workstation with 100 MHz R4000SC processor. The simulations with BondSim software on the other hand, were done on a laptop with a Pentium III 650 MHz processor and 128 Mb RAM.

Figs. 9.25 – 9.27 give time histories of the angles of the Andrews' squeezer mechanism as defined by Eq. (9.29). The diagrams closely correspond to those of [13 and 14]. Table 9.7 gives the values at t = 0.03 s. The second column gives the values obtained by BondSim using the default error control method, i.e. by controlling the differentiated variables only. The third column lists the reference solution from [14]. To compare the values the number of significant correct digits are calculated as in [14] by determining $-\log_{10}|$relative error$|$. The simulation values agree with the reference solution to about 4 correct digits. Thus the simple error control used in BondSim by default gives relatively good accuracy.

The simulations were also repeated using the local error control of Sect. 5.3.3 (using the default differentiation weight of 1). The results are given in Table 9.8. The values obtained have at least one additional correct digit, that is, they are correct to a minimum of five digits.

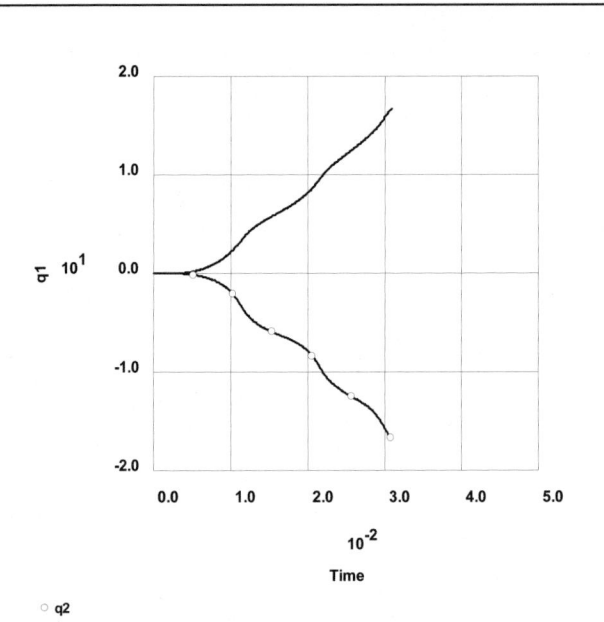

Fig. 9.25. Behaviour of q1 and q2 with time

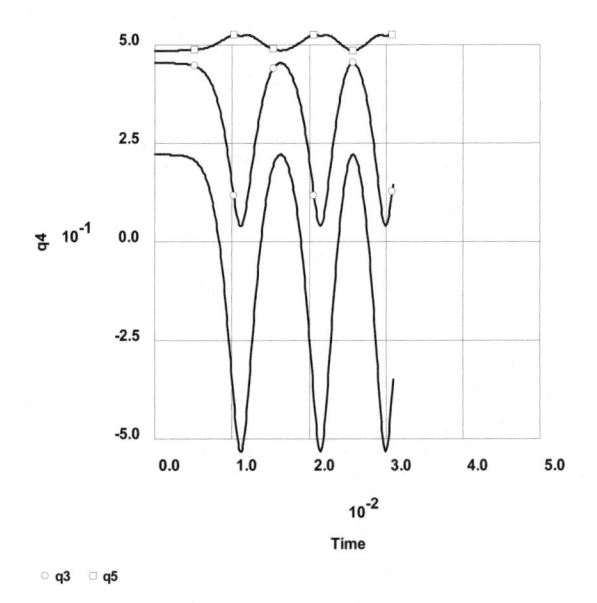

Fig. 9.26. The behaviour of q3, q4 and q5 over time

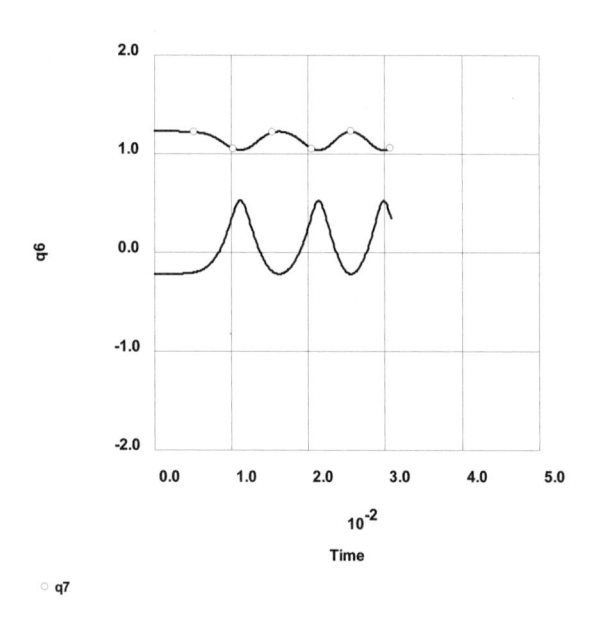

○ q7

Fig. 9.27. The behaviour of q6 and q7 over time

Table 9.7. Values at t = 0.03 s (default error control)

	BondSim	Referent solution [14]	-log(relerr)
q1	15.81163467	15.810771	4.3
q2	-15.7574481	-15.756371	4.2
q3	0.04082927438	0.040822240	3.8
q4	-0.534717734	-0.53473012	4.6
q5	0.5244112271	0.52440997	5.6
q6	0.5347175805	0.53473012	4.6
q7	1.048077964	1.0480807	5.6

Table 9.8. The values at t = 0.03 s (local error control)

	BondSim	Referent solution [14]	-log(relerr)
q1	15.81080905	15.810771	5.6
q2	-15.75641766	-15.756371	5.5
q3	0.0408225647	0.040822240	5.1
q4	-0.5347296137	-0.53473012	6.0
q5	0.5244102524	0.52440997	6.3
q6	0.5347295636	0.53473012	6.0
q7	1.048079848	1.0480807	6.1

Table 9.9 gives some of the simulation statistics. The second and third column gives the statistics when simulations are run using the default and local error control, respectively. The last column lists the values from [14] when solved numerically with the RADAU5 code.

Table 9.9. The simulation statistics

Parameter	Default error control	Local error control	RADAU5 [14]
Scd	3.8	5.1	4.46
Steps	397	484	117
#f	1112	1270	1321
#Jac	720	789	92
CPU [s]	1.24	1.492	0.83

Note : scd is the minimum number of significant correct digits in the solution
steps means the number of integration steps
#f is the number of function evaluations
#Jac is the number of Jacobian matrix evaluations

The local error control is slightly more expensive in terms of CPU time. Generally, the RADAU5 values are better, but the CPU time is not much less. Also its accuracy lies between the BondSim default error control and the local error control. We should also take into account the fact that BondSim evaluates functions and the Jacobian matrix symbolically (Chapt. 5). During the simulation there are also other operations besides the numerical solution. The CPU times are obtained with background plotting, i.e. with the output window closed, because of the very slow graphics of the laptop used. This is generally not such a problem on desktop PCs.

9.4 Engine Torsional Vibrations

The determination of the torsional vibration characteristics of internal combustion engines plays an important role in the design of cars, ships and other vehicles. Traditionally, linear lumped-mass models are used in which the shafts, bearings, couplings and flywheels are modelled as concentrated disks, springs and dampers. The complete engine mechanism assembly of single and multi-cylinder engines is modelled as separate lumped inertias [15]. Such models are often supplied to shipbuilders for the torsional design of ships' propulsion systems. It is well known, however, that the reciprocating mechanisms of the engines are non-linear and, hence, their concentrated mass moments of inertias are not constant, but change during the course of each revolution of the engine crankshaft. How important these variations are is difficult to comprehend a priori and requires detailed analysis of the complete system [16, 17]. There are also some non-linear effects that lumped parameter models cannot predict, e.g. the so-called secondary resonance [17]. Hence, a more detailed model of the engine in the time domain is important.

In this section we develop a model of a complete single-cylinder engine based on multibody models of the reciprocating mechanism. The analysis is based on the plane body motion models of Sec. 9.2 [18, 19]. The scheme of the reciprocating mechanism is given in Fig. 9.28.

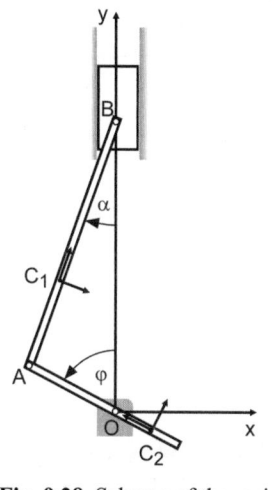

Fig. 9.28. Scheme of the reciprocating mechanism

It consists of a crank that rotates about the bearing at O and which is connected to the engine piston by the connecting rod AB. All connections of reciprocating members are by bearings. We treat bearings as frictionless revolute joints. The piston slides in the cylinder bore and is acted upon by pressure forces developed by the combustion processes in the cylinder chamber. The piston friction is neglected. There is friction in the mechanism however, but this is quite difficult to predict. Thus its overall effect is represented by a linear resistive element (see later). The coordinate frames used are shown in the figure. The detailed geometry is generally not known, but it is reasonable to assume that the crank and the connecting rod mass centres are on their geometrical axes AO and AB, respectively, and the body-fixed axis is taken, in each case, along this axis.

We develop a model to analyse the engine torsional vibrations using the Bond-Sim. A project called EngineTorsional Vibrations is created. A model of the system is developed systematically by creating component models of all the main engine parts and then connecting them by bond lines (Fig. 9.29). The engine parameters are based on [17].

The crank is fixed to the shaft. The shaft is connected on one side to the flywheel and on the other side (the output part) to the load. It is assumed that the engine is unloaded and, hence, the load is represented by a zero effort source SE. The shaft rotates in the bearings fixed in the engine body, which is here represented by the Base component.

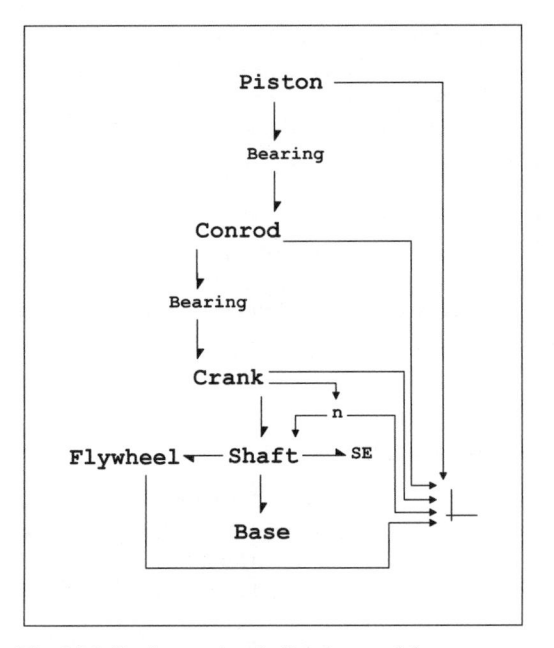

Fig. 9.29. Engine torsional vibration model

The model of the shaft is given in Fig. 9.30. In the centre, component **e** is connected to both power document ports. This component describes the translation of the shaft axis and is represented by an array of effort junctions corresponding to the x- and y-axis motions. It is connected to the **Base** by the lower power port. The shaft cannot translate in the engine body and hence the component **Base** is represented by two zero flow sources. The other effort junction connected to the shaft upper document port is the shaft angular velocity node.

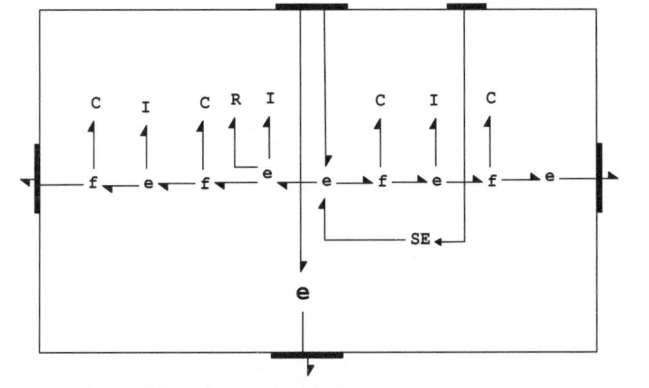

Fig. 9.30. Model of the engine shaft

On each side of the shaft angular velocity junction there are capacitive, inertial and resistive elements that model the shaft torsional dynamics. The model used corresponds to the lumped mass model of [17]. Thus, the first inertial element on the left is the angular inertia of the reduction gear and the camshaft. The resistive element describes the overall friction effect in the engine reduced to the shaft.

A pair of capacitive elements on each side of the centre in Fig 9.30 represents the stiffness of the shaft parts between the crank and the bearings. The bearings are represented by the inertial elements. The last capacitive element on the left of the centreline represents the stiffness of the shaft between the left bearing and the flywheel; that on the right side represents the stiffness of the output portion of the shaft. The controlled source effort models the effect of the combustion forces on the piston, reduced to the shaft. The element is controlled by the angle of shaft rotation. It simulates the load torque by [16, 17]

$$M = M_m (1 + B \cdot \sin(n\varphi)) \tag{9.39}$$

where M_m is the average indicated torque and B and n are appropriate constants.

The Flywheel (Fig. 9.29) is modelled simply by an inertial element connected to an effort junction, from which information on the flywheel angular velocity is obtained. This is fed to the display component (at the right bottom in the document).

The Crank and Conrod components are modelled as rigid bodies in plane motion using the general model for plane motion bodies of Sec. 9.2.1. The components are created simply by copying the Body component from the library. The main difference is the change of name. The ports are moved to the upper and lower parts of the components in order to represent the connections as depicted in Fig. 9.28. Thus, the model of the Crank is as depicted in Fig. 9.31.

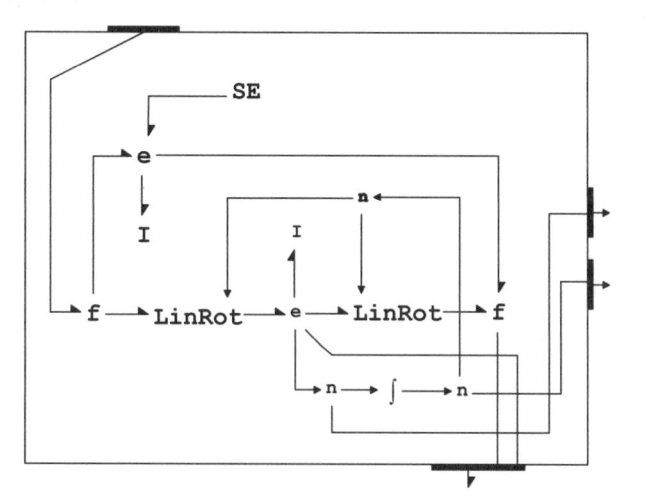

Fig. 9.31. The model of the Crank

Note that, because of the connection through the frictionless bearings, there is no transfer of external moments to the crank. The SE connected to the mass centre joint describes the weight of the component. It can be seen in the lower part of the model that signals of the crank rotation angle and the angular velocity are obtained. These are used to display the crank motion, as well as for simulating the shaft load torque (Fig. 9.29). The model of the Conrod is almost the same, except that only the component rotation angle is picked up. The bearings are modelled by the revolute joint component of Sec. 9.2.2. The Bearing components imply the common velocity condition of the bodies at the connection points (the bearing axes).

The engine piston slides inside the cylinder. This is described by a translation joint. The model is simplified by assuming that the engine body is fixed in the base frame (Fig. 9.32). The left effort junction represents the x-axis piston velocity, which is zero. This condition is implied by a zero flow source. The piston can move in the y-direction and is affected by its inertia and gravity. Note also that the position of the piston is evaluated by integration and the signal is fed out for display.

The effect of the combustion pressure is not represented here but is taken into account by the equivalent load torque reduced to the Shaft as has already been described (see Eq. (9.39)). This analysis of engine torsional vibrations is a common one, for it is well known that the pressure forces can be described by a Fourier series [15]. In Eq. (9.39) the parameter n takes into account different harmonics contained in the piston force and, thus, the effect on the torsional vibration characteristic can be examined by varying it. Alternatively, a model of the combustion process in the cylinder can be developed and included in the model.

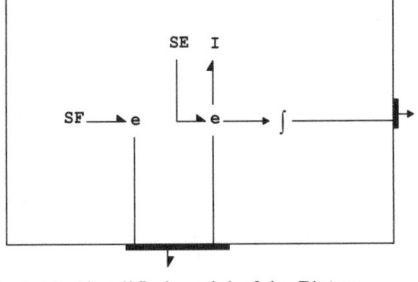

Fig. 9.32. Simplified model of the Piston

The model parameters are summarized in Table 9.10. These are based on the measured data of a small four-stroke single cylinder engine [17]. The last two parameters were chosen in such a way as to achieve, on the one hand, the free running engine velocity of about 1800 rpm, as in the experiments of [17], and on the other hand to achieve relatively good mechanical efficiency (light damping) of the system. The parameter B of Eq. (9.39) was set to 1.

Table 9.10. The engine model parameters

Parameter		Value
Crank	Length OA	0.02491 m
	Ratio OC/OA	0.143
	Mass	0.557 kg
	Inertia	$3.21 \cdot 10^{-4}$ kg·m²
Conrod	Length AB	0.09847 m
	Ratio AC/AB	0.164
	Mass	0.1040 kg
	Inertia	$1.53 \cdot 10^{-4}$ kg·m²
Piston mass		0.1649 kg
Shaft stiffness	The left part	146000 N·m
	The right part	146000 N·m
	The flywheel part	11090 N·m
	The output part	7960 N·m
Shaft inertia	Gear and camshaft	$7.99 \cdot 10^{-5}$ kg·m²
	Bearing left	$5.92 \cdot 10^{-6}$ kg·m²
	Bearing right	$5.92 \cdot 10^{-6}$ kg·m²
Flywheel inertia		$9.35 \cdot 10^{-3}$ kg·m²
Engine friction constant		0.01 N·m·s
Average torque		1.86 N·m

After the model is completed, it is possible to analyse the dynamic behaviour of the engine by simulation. We are interested mainly in the behaviour of the system around the first natural frequency. Thus the time-response of the system to the load torque is found first, then the frequency spectrum of the engine angular velocity is generated. Because the time constant of the system is about 1 s, we need about 5 s to be sure that the velocity settles down to a steady state value. After that, we run the system for an additional 5 s.

Thus the simulation interval is taken to be 10 s. The range of frequency we are interested in is 0 – 1000 Hz. Thus the output interval should be at least 1/2000 s or $5 \cdot 10^{-4}$ s. We choose the default error tolerance of 10^{-6} and the default method. The first resonance appears when parameter n of Eq. (9.39) is about 25. Fig. 9.33 shows transients in the angular velocity for the value n = 25. The output is very noisy because it is on the border of instability. The first signs of instability appears after t = 0.2 s. It is interesting to look at the simulation statistics (Table 9.11). If the simulation is repeated with an error tolerance of 10^{-5} all the numbers are scaled down 10 times, but there are not significant differences in the plots generated.

Table 9.11. The simulation statistics for the Engine Torsional Vibrations problem

Parameter	Error tolerance $1 \cdot 10^{-6}$	Error tolerance $1 \cdot 10^{-5}$
Number of steps	1 863 385	141 735
Function evaluations	3 795 519	348 034
Jacobian matrix evaluations	1 933 949	206 307
CPU time s	1490	168

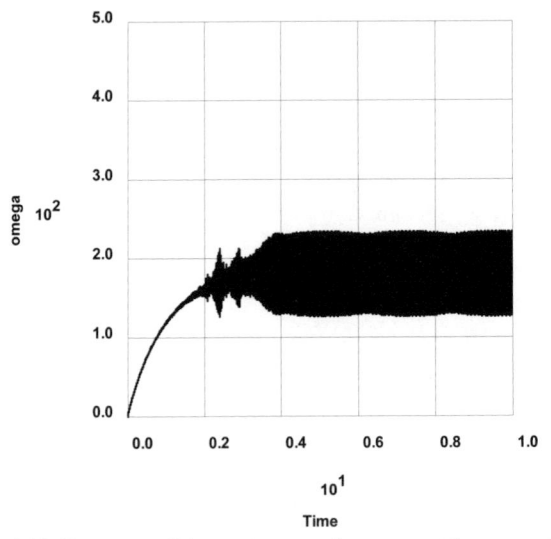

Fig. 9.33. Response of the engine near the resonant frequency (n = 25)

We get the frequency spectrum by clicking with the right mouse on the time plot and select *Frequency spectrum* from a drop-down menu. In the dialogue that appears we select a time window of 5 − 10 s and, for better effect, click on the *Hamming window* button. The resulting spectrum calculated by Fast Fourier Transform is shown in Fig. 9.34. As can be seen, there is a large component at 0 frequency of amplitude 180.348 rad/s (1722 rpm).

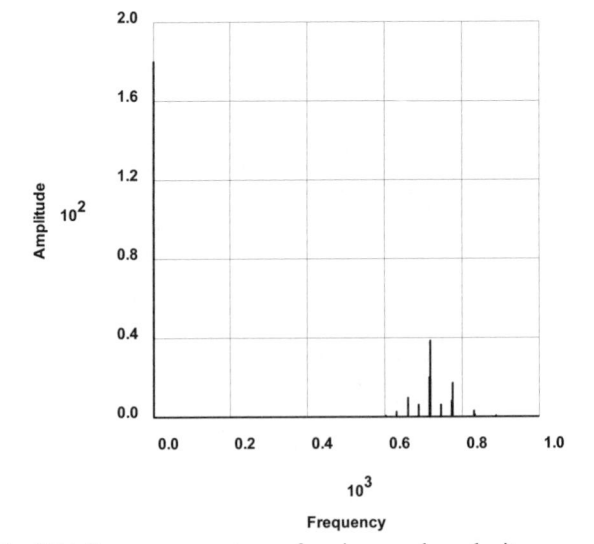

Fig. 9.34. Frequency spectrum of engine angular velocity

We expand the low frequency region, Fig. 9.35, by clicking with the right mouse button on the frequency plot, and then selecting the *Expand* command from a drop-down menu. In a similar way the high frequency region can be expanded. This is shown in Fig. 9.36.

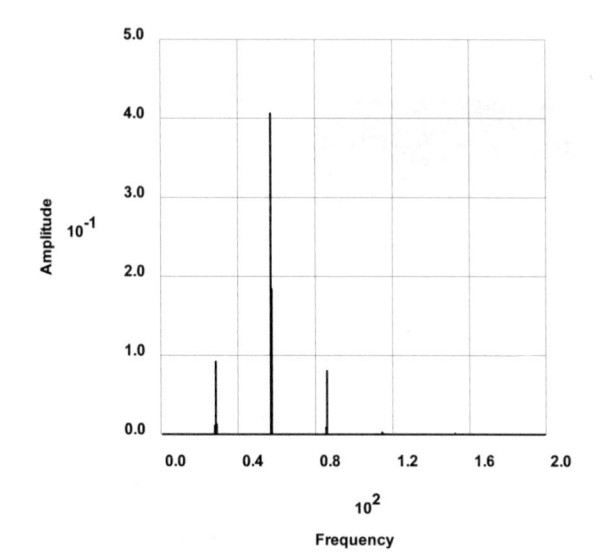
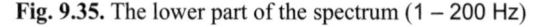

Fig. 9.35. The lower part of the spectrum (1 – 200 Hz)

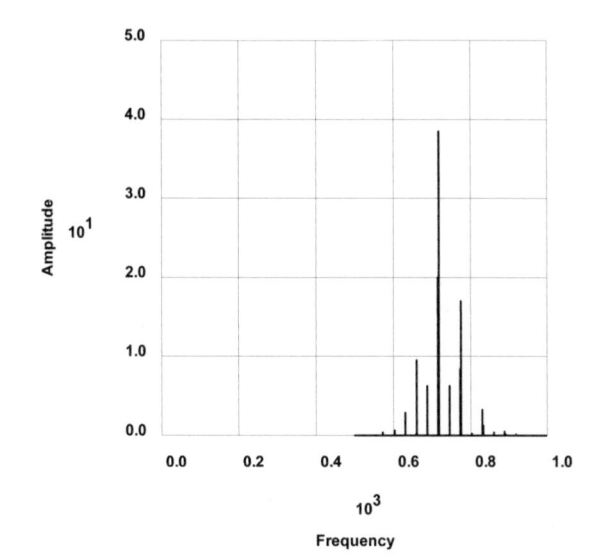
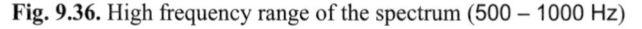

Fig. 9.36. High frequency range of the spectrum (500 – 1000 Hz)

The low frequency region shows characteristic harmonics that appear at multiples of the fundamental frequency of 28.7 Hz (1722 rpm) because of the inertial effects in the engine. It can be seen that the second order harmonic is largest, and then the first, followed by the third order. The higher order harmonics are much lower in amplitude and there is no response until frequencies in the range 600 – 900 Hz. The first natural frequency is estimated at 717.7 Hz with an amplitude of 38.6 rad/s. There are also amplitude peaks on both sides of the resonant frequency, which are displaced by twice the fundamental frequency, i.e. 57.4 Hz. This is termed the *secondary resonance* and is the result of non-linear intercoupling in the engine's reciprocating mechanism [17]. The spectrum diagrams, as well as the characteristic frequencies, agree well with the experimental results reported in [17].

9.5 Motion of Constrained Rigid Bodies in Space

9.5.1 Basic Kinematics

To describe motion of a body in space we use two fundamental coordinate frames (Fig. 9.37) – a base frame Oxyz and a body frame Cx'y'z moving with it, as we did in Sec. 9.2. There can be a number of bodies and, hence, a number of body frames that are used for their description. On the other hand, there is a single base (inertial) frame. In robotics it is often convenient to introduce other frames, as well [9]. All of the frames are 3D Cartesian coordinate frames.

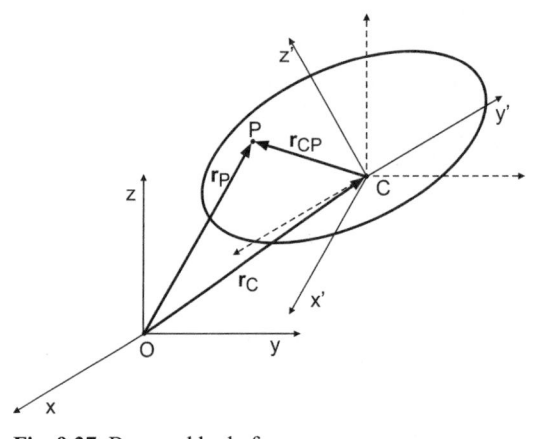

Fig. 9.37. Base and body frames

We assume that the *position* of a body frame with respect to the base is defined by a position vector r_C of the origin C of the body frame. The point C is the reference point for describing the body position – a body centre (pole). As in the planar case, it could be the body mass centre, but it could be also some other point. The body *orientation* is defined by the rotation matrix **R** composed of the direction cosines of the body axes with respect to the base axes (see e.g. [2]). The position of any point P fixed in the body with respect to the base is given by

$$\mathbf{r}_P = \mathbf{r}_C + \mathbf{R}\mathbf{r}_{CP}' \tag{9.40}$$

where $\mathbf{r}_{CP}'$ is the vector of its *material* coordinates, i.e. the coordinates with respect to the body frame.

During motion, the position of the body centre $\mathbf{r}_C$ and its orientation—represented by matrix $\mathbf{R}$—change with time. The material coordinates of a point fixed in the body do not change, but the coordinates change with respect to the base frame. The velocity of the point P, as seen from the base frame, can be found by differentiating Eq. (9.40) with respect to time, i.e.

$$\mathbf{v}_P = \mathbf{v}_C + \frac{d\mathbf{R}}{dt}\mathbf{r}_{CP}' \tag{9.41}$$

The relative velocity of the point P with respect to the origin of the body frame C is

$$\mathbf{v}_{CP} = \mathbf{v}_P - \mathbf{v}_C \tag{9.42}$$

and thus the velocity of point P is

$$\mathbf{v}_P = \mathbf{v}_C + \mathbf{v}_{CP} \tag{9.43}$$

From Eq. (9.41) the relative velocity is owed to the body orientation changes

$$\mathbf{v}_{CP} = \frac{d\mathbf{R}}{dt}\mathbf{r}_{CP}' \tag{9.44}$$

Velocity can be represented as a vector in any frame by its rectangular coordinates. The relative velocity representations in the body frame $\mathbf{v}_{CP}'$ and in the base frame are related by a coordinate transformation

$$\mathbf{v}_{CP} = \mathbf{R}\mathbf{v}_{CP}' \tag{9.45}$$

The rotation matrix is orthogonal and thus we have for its inverse

$$\mathbf{R}^{-1} = \mathbf{R}^T \tag{9.46}$$

where the superscript T is the transposition operator. From Eqs. (9.45) and (9.46),

$$\mathbf{v}_{CP}' = \mathbf{R}^T\mathbf{v}_{CP} \tag{9.47}$$

Substituting from Eq. (9.44) we get

$$\mathbf{v}_{CP}' = \mathbf{R}^T\frac{d\mathbf{R}}{dt}\mathbf{r}_{CP}' \tag{9.48}$$

But

$$\mathbf{r}_{CP} = \mathbf{R}\mathbf{r}_{CP}' \tag{9.49}$$

Thus, we have

$$\mathbf{v}_{CP} = \frac{d\mathbf{R}}{dt}\mathbf{R}^T\mathbf{r}_{CP} \tag{9.50}$$

It is a simple matter to show that the factors of the position vector on the right side of Eqs. (9.48) and (9.50) are skew-symmetric. From Eq. (9.46) it follows that

$$\mathbf{R}^T\mathbf{R} = \mathbf{R}\mathbf{R}^T = \mathbf{I} \tag{9.51}$$

where $\mathbf{I}$ is the identity matrix. Hence, by differentiating with respect to time we get

$$\frac{d\mathbf{R}^T}{dt}\mathbf{R} + \mathbf{R}^T\frac{d\mathbf{R}}{dt} = \mathbf{0} \tag{9.52}$$

or

$$\mathbf{R}^T\frac{d\mathbf{R}}{dt} = -\frac{d\mathbf{R}^T}{dt}\mathbf{R} = -\left(\mathbf{R}^T\frac{d\mathbf{R}}{dt}\right)^T \tag{9.53}$$

And, similarly

$$\frac{d\mathbf{R}}{dt}\mathbf{R}^T = -\mathbf{R}\frac{d\mathbf{R}^T}{dt} = -\left(\frac{d\mathbf{R}}{dt}\mathbf{R}^T\right)^T \tag{9.54}$$

The vector product of two vectors α and β defined by their components in a Cartesian frame

$$\boldsymbol{\alpha} = (\alpha_x \ \alpha_y \ \alpha_z)^T \tag{9.55}$$

and

$$\boldsymbol{\beta} = (\beta_x \ \beta_y \ \beta_z)^T \tag{9.56}$$

reads

$$\boldsymbol{\alpha} \times \boldsymbol{\beta} = \begin{pmatrix} \alpha_y\beta_z - \alpha_z\beta_y \\ \alpha_z\beta_x - \alpha_x\beta_z \\ \alpha_x\beta_y - \alpha_y\beta_x \end{pmatrix} = \begin{pmatrix} 0 & -\alpha_z & \alpha_y \\ \alpha_z & 0 & -\alpha_x \\ -\alpha_y & \alpha_x & 0 \end{pmatrix}\begin{pmatrix} \beta_x \\ \beta_y \\ \beta_z \end{pmatrix} \tag{9.57}$$

From [7] we denote by $\alpha\times$ a skew-symmetric tensor that is, through ordinary vector product operation, associated with vector α. Vector α is the *axial* vector associated with a skew-symmetric tensor. Eq. (9.57) also can be written as

$$\boldsymbol{\alpha} \times \boldsymbol{\beta} = \begin{pmatrix} \alpha_y\beta_z - \alpha_z\beta_y \\ \alpha_z\beta_x - \alpha_x\beta_z \\ \alpha_x\beta_y - \alpha_y\beta_x \end{pmatrix} = -\begin{pmatrix} 0 & -\beta_z & \beta_y \\ \beta_z & 0 & -\beta_x \\ -\beta_y & \beta_x & 0 \end{pmatrix}\begin{pmatrix} \alpha_x \\ \alpha_y \\ \alpha_z \end{pmatrix} \tag{9.58}$$

and, hence,

$$\boldsymbol{\alpha} \times \boldsymbol{\beta} = -\boldsymbol{\beta} \times \boldsymbol{\alpha} \tag{9.59}$$

Thus, we have from Eq. (9.50)

$$\boldsymbol{\omega}\times = \frac{d\mathbf{R}}{dt}\mathbf{R}^T \tag{9.60}$$

and

$$\boldsymbol{\omega} = \text{axial}\left(\frac{d\mathbf{R}}{dt}\mathbf{R}^{\mathrm{T}}\right) \tag{9.61}$$

Similarly, from Eq. (9.48), we have

$$\boldsymbol{\omega}' \times = \mathbf{R}^{\mathrm{T}}\frac{d\mathbf{R}}{dt} \tag{9.62}$$

and

$$\boldsymbol{\omega}' = \text{axial}\left(\mathbf{R}^{\mathrm{T}}\frac{d\mathbf{R}}{dt}\right) \tag{9.63}$$

From Eqs. (9.60) and (9.62),

$$\frac{d\mathbf{R}}{dt} = \boldsymbol{\omega} \times \mathbf{R} \tag{9.64}$$

and

$$\frac{d\mathbf{R}}{dt} = \mathbf{R}\boldsymbol{\omega}' \times \tag{9.65}$$

Vectors $\boldsymbol{\omega}$ and $\boldsymbol{\omega}'$ are the base frame and the body frame representations of the body angular velocity. These representations are related by

$$\boldsymbol{\omega} = \mathbf{R}\boldsymbol{\omega}' \tag{9.66}$$

Thus, the relative velocity expressions of Eq. (9.48) and (9.50) read

$$\mathbf{v}'_{CP} = \boldsymbol{\omega}' \times \mathbf{r}'_{CP} \tag{9.67}$$

and

$$\mathbf{v}_{CP} = \boldsymbol{\omega} \times \mathbf{r}_{CP} \tag{9.68}$$

The velocity equations of Eq. (9.43) now can be written as

$$\mathbf{v}_P = \mathbf{v}_C + \boldsymbol{\omega} \times \mathbf{r}_{CP} \tag{9.69}$$

and

$$\mathbf{v}'_P = \mathbf{v}'_C + \boldsymbol{\omega}' \times \mathbf{r}'_{CP} \tag{9.70}$$

We transform these equations a little more using Eq. (9.59). Thus, Eq. (9.69) can be written as

$$\mathbf{v}_P = \mathbf{v}_C + \left(-\mathbf{r}_{CP} \times \boldsymbol{\omega}\right) \tag{9.71}$$

Similarly, from Eq. (9.70), we have

$$\mathbf{v}'_P = \mathbf{v}'_C + \left(-\mathbf{r}'_{CP} \times \boldsymbol{\omega}'\right) \tag{9.72}$$

W can now represent the body velocity as a *6D* generalized vector composed of linear velocity vectors of the body centre and the body angular velocity. In the base frame the vector is given by

$$\mathbf{f}_C = \begin{pmatrix} \mathbf{v}_C \\ \boldsymbol{\omega} \end{pmatrix} \tag{9.73}$$

and similarly in the body frame as

$$\mathbf{f}_C' = \begin{pmatrix} \mathbf{v}_C' \\ \boldsymbol{\omega}' \end{pmatrix} \tag{9.74}$$

We denote such vectors by **f** because, as will be shown later, these are simply the flows at the bond graph component ports.

In a similar way, the generalized velocity of any other point P fixed in the body can be represented in the base frame by

$$\mathbf{f}_P = \begin{pmatrix} \mathbf{v}_P \\ \boldsymbol{\omega} \end{pmatrix} \tag{9.75}$$

and in the body frame as

$$\mathbf{f}_P' = \begin{pmatrix} \mathbf{v}_P' \\ \boldsymbol{\omega}' \end{pmatrix} \tag{9.76}$$

From Eq. (9.71) we have

$$\mathbf{f}_P = \begin{pmatrix} \mathbf{I} & -\mathbf{r}_{CP} \times \\ \mathbf{0} & \mathbf{I} \end{pmatrix} \mathbf{f}_C \tag{9.77}$$

Also, we have

$$\mathbf{f}_C = \begin{pmatrix} \mathbf{R} & \mathbf{0} \\ \mathbf{0} & \mathbf{R} \end{pmatrix} \mathbf{f}_C' \tag{9.78}$$

From Eqs.(9.77) and (9.78) we have

$$\mathbf{f}_P = \begin{pmatrix} \mathbf{I} & -\mathbf{r}_{CP} \times \\ \mathbf{0} & \mathbf{I} \end{pmatrix} \begin{pmatrix} \mathbf{R} & \mathbf{0} \\ \mathbf{0} & \mathbf{R} \end{pmatrix} \mathbf{f}_{Ci}' \tag{9.79}$$

or

$$\mathbf{f}_P = \begin{pmatrix} \mathbf{R} & -\mathbf{r}_{CP} \times \mathbf{R} \\ \mathbf{0} & \mathbf{R} \end{pmatrix} \mathbf{f}_C' \tag{9.80}$$

Matrix

$$\mathbf{C} = \begin{pmatrix} \mathbf{R} & -\mathbf{r}_{CP} \times \mathbf{R} \\ \mathbf{0} & \mathbf{R} \end{pmatrix} \tag{9.81}$$

can be termed a representation of the body frame *configuration tensor* [7]. Thus, we have

$$\mathbf{f}_P = \mathbf{C}\mathbf{f}_C' \tag{9.82}$$

We can also express Eq. (9.72) as

$$\mathbf{f}_P' = \begin{pmatrix} \mathbf{I} & -\mathbf{r}_{CP}' \times \\ \mathbf{0} & \mathbf{I} \end{pmatrix} \mathbf{f}_C' \tag{9.83}$$

From Eq. (9.78) we find

$$\mathbf{f}_C' = \begin{pmatrix} \mathbf{R}^T & \mathbf{0} \\ \mathbf{0} & \mathbf{R}^T \end{pmatrix} \mathbf{f}_C \tag{9.84}$$

After substituting in Eq. (9.83) we find the body frame configuration representation in the same frame

$$\mathbf{C}' = \begin{pmatrix} \mathbf{R}^T & -\mathbf{r}_{CP}' \times \mathbf{R}^T \\ \mathbf{0} & \mathbf{R}^T \end{pmatrix} \tag{9.85}$$

Thus we have the inverse transformation

$$\mathbf{f}_P' = \mathbf{C}' \mathbf{f}_C \tag{9.86}$$

In this section some of the basic kinematic quantities have been developed. They are generalizations of the corresponding expressions for planar body motion developed in Sec. 9.2.1.

9.5.2 Bond Graph Representation of a Body Moving in Space

Now we can proceed with the development of a bond graph model of a body moving in space. It is similar to the one used for bodies in plane motion in Sec. 9.2.1. The main difference lies in the more complex relationship that governs motion of bodies in space.

To develop a fundamental bond graph representation we consider a single body interacting with its environment. The environment consists of bodies to which the analysed body is connected. We model the body and its environment by two component objects – Body and Base (Fig. 9.38).

We assume that the body is connected to other bodies at two points. We also assume that one body in the environment acts on *the body* through a force and a moment. Likewise *the body* acts on a third body in the environment. This way there is transfer of power from the environment to the body and from the body back to the environment. In order to represent these interactions we assume that the Body component has *three* ports. Two of these, the bottom and the upper one in Fig. 9.38, correspond to the power interactions just described. We have added the third one (the side port) to correspond to the body mass centre. It is possible to develop a model without the explicit use of mass centre ports, but its use simplifies the description of the body dynamics.

As explained previously, velocities seen at a port can be represented by the 6D *flow* vectors of Eq. (9.75) and (9.76). Similarly we can use 6D *efforts* to represent the resultant force and moment at a port. Such efforts can be represented in the base frame by

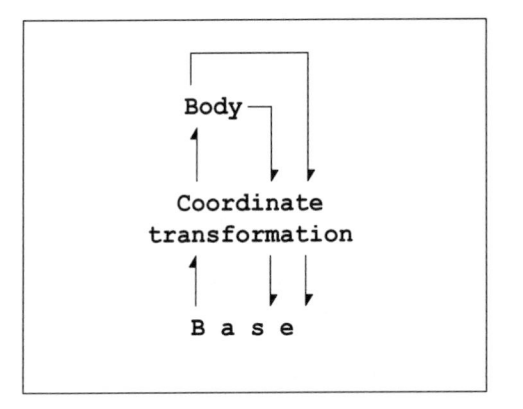

Fig. 9.38. Representation of a body moving in space

$$e_P = \begin{pmatrix} \mathbf{F}_P \\ \mathbf{M}_P \end{pmatrix} \tag{9.87}$$

And, similarly, in the body frame as

$$e_P' = \begin{pmatrix} \mathbf{F}_P' \\ \mathbf{M}_P' \end{pmatrix} \tag{9.88}$$

Here $\mathbf{F}_P$ and $\mathbf{F}_P'$ are 3D representations of the resultant force at a connection point P and $\mathbf{M}_P$ and $\mathbf{M}_P'$ the corresponding representation of the resultant moment.

To represent these 6D vector quantities, the ports of the Base and Body components are assumed to be compounded (Sec. 2.4). The first sub-port corresponds to the linear velocity and resultant force pair, and the other to the body angular velocity and resultant moment. This way the ports can be used to access efforts and flows at the port by two ordered 6D efforts and flows. At the mass centre ports however, as will be seen later, a 3D representation is sufficient. The flow is the mass centre velocity and the effort is the resultant of the forces acting on the body at other ports (interconnection points) reduced to the mass centre. The gravity force (the body weight) acting there is not accounted for and is taken into account when dealing with the body dynamics. The reason for this is that body weight can be easily described in the base frame. This is not so in the body frame, because it is generally rotated with respect to the first.

We represent effort and flow vectors at body ports as seen in the body, i.e. with respect to a body fixed frame. Similarly at the base ports the corresponding quantities are expressed with respect to the base frame. Transformations between these two representations are given in Fig. 9.38 by the Co-ordinate transformation component. This component transforms effort and flow at a body port to the corresponding base representation and vice versa. The transformation between efforts is given by

$$e_P = \begin{pmatrix} \mathbf{R} & \mathbf{0} \\ \mathbf{0} & \mathbf{R} \end{pmatrix} e_P' \tag{9.89}$$

and similarly for the flows

$$f_P = \begin{pmatrix} R & 0 \\ 0 & R \end{pmatrix} f_P' \qquad (9.90)$$

where R is the rotation matrix of the body frame with respect to the base.

We use Cartesian 3D frames only. Thus the transformation matrix is orthogonal as shown in the first part of this section. Thus from Eq. (9.90) we get the inverse transformation

$$f_P' = \begin{pmatrix} R^T & 0 \\ 0 & R^T \end{pmatrix} f_P \qquad (9.91)$$

It follows, then, that the transformation does not change the power transfer between the Body and the Base

$$f_P'^T e_P' = f_P^T e_P \qquad (9.92)$$

The transformation of port quantities can be represented by the 6D transformation component of Fig. 6.39a. This component transforms both 3D parts of the 6D effort-flow pairs of corresponding ports, according to Eq. (9.89) and (9.91), as shown in Fig. 6.39b. The transformations are represented by the Rot components that describe the transformation of 3D effort-flow vectors using the rotation matrix R. The transformation can be represented using transformer elements TF (see Sec. 9.6).

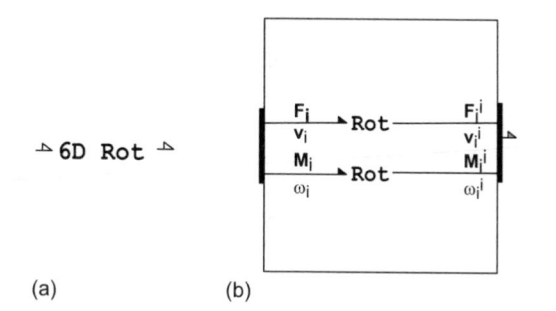

(a)   (b)

Fig. 9.39. 6D transformation. (a) The component representation, (b) The structure

Velocities at a port are related to the velocity of the body mass centre by Eq. (9.82). To find the relation between forces and moments, we evaluate the power transfer between the ports as in Sec.9.2.1. From Eq. (9.83) we get

$$f_P'^T e_P' = f_C'^T \begin{pmatrix} I & 0 \\ r_{CP} \times & I \end{pmatrix} e_P' \qquad (9.93)$$

After substituting from Eqs. (9.74), (9.76) and (9.88), then evaluating, we get

$$v_P'^T F_P' + \omega'^T M_P' = \begin{pmatrix} v_C'^T & \omega'^T \end{pmatrix} \begin{pmatrix} F_P' \\ r_{CP}' \times F_P' + M_P' \end{pmatrix} \qquad (9.94)$$

or

$$v_P'^{\,T} F_P' + \omega'^{\,T} M_P' = v_C'^{\,T} F_P' + \omega'^{\,T} (r_{CP}' \times F_P' + M_P')\qquad(9.95)$$

The last equation is the generalization of the planar case Eq. (9.8). It describes, jointly with Eq. (9.82), the basic velocity and force relationships for rigid bodies.

The basic structure of the Body component, defined by Eqs. (9.83) and (9.95), is given in Fig. 9.40. It is very similar to that of the planar body model in Fig. 9.3, but involves three-dimensional quantities.

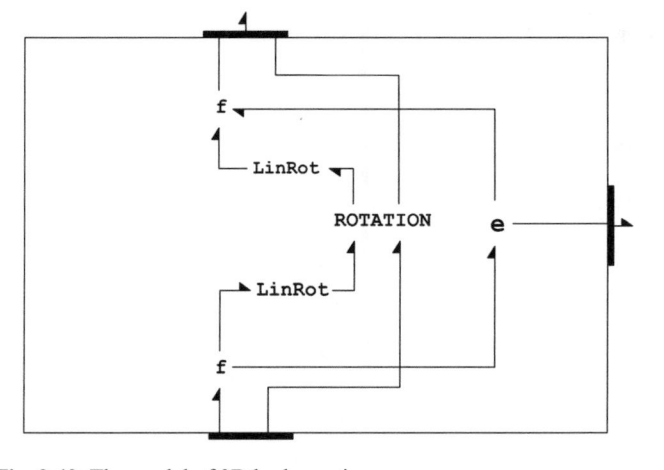

Fig. 9.40. The model of 3D body motion

Here the component e is also an array of effort junctions corresponding to the centre of mass velocity. Instead of a simple effort junction for the body angular velocity, the component Rotation is used. This is because the rotational dynamics in 3D are much more involved then in the planar case. (We discuss this component later.) Component f represents an array of flow junctions that describe the port velocities relationship given by the first 3D row of Eq. (9.83). It gives the linear velocity at a port as the sum of the velocity of the body mass centre and the relative velocity of the port owing to rotation of the body about the mass centre. The joint variable is the force at the port. The body angular velocity, however, is a property of the body; as such—according to the bottom row of Eq. (9.83)—it is directly connected to the corresponding Rotation component port.

The LinRot components represent transformations between linear and angular quantities. That is, between the relative velocity at the port with respect to the centre of mass of the body and the body angular velocity on the one hand, and of the force at the port and its moment about the mass centre, on the other. These transformations are defined by the skew-symmetric tensor operations in Eq. (9.83) and Eq. (9.95) and can be represented by transformers, as in Fig. 9.41.

Every transformer in Fig. 9.41 corresponds to a nonzero entry of tensor $-r_{CP}'\times$ or its transpose. Thus, the flow junctions on the left evaluate the relative velocities at the left port as a result of the multiplication of the angular velocity at the right port by matrix $-\mathbf{r}_{CP}'\mathbf{x}$. Likewise, the effort junctions on the right give the moment at the right port of the force at the left port by multiplying it by the $\mathbf{r}_{CP}'\mathbf{x}$ matrix. The ratios of the transformers are the material coordinates of a body point corresponding to the port with respect to the mass centre. Thus, they are parameters that depend on the geometry of the rigid body and don't change with its motion.

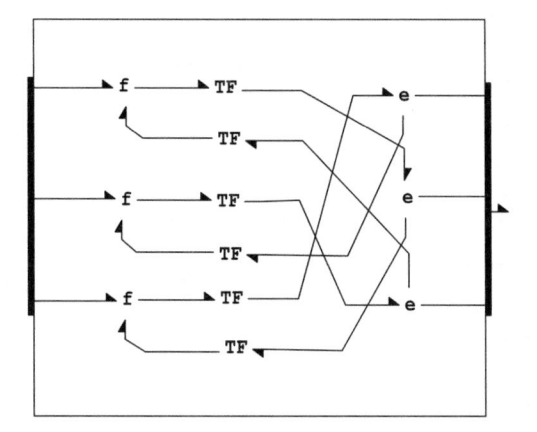

Fig. 9.41. The representation of LinRot transformations

9.5.3 Rigid Body Dynamics

To complete the model, we need a dynamic equation governing rigid body motion. The simplest form of such an equation is given with respect to axes translating with the body mass centre (Fig. 9.42). We assume that the base frame is an inertial frame.

The translational part of the motion can be described by

$$\left.\begin{array}{c} \mathbf{p} = m\mathbf{I}\mathbf{v}_C \\[2mm] \dfrac{d\mathbf{p}}{dt} = \mathbf{F} \end{array}\right\} \tag{9.96}$$

Here, m is the body mass and $\mathbf{F}$ is the resultant of the forces reduced to the mass centre. We represent the dynamics of body translation in the **Base** (Fig. 9.43a) by the CM component that describes the motion of the mass centre of the body. This last component (Fig. 9.43b) consists of effort junctions corresponding to the x, y and z components of the body mass centre velocity with respect to the base frame. These junctions are connected to the **Base** port to which the resultant of the forces acting on the body is transferred.

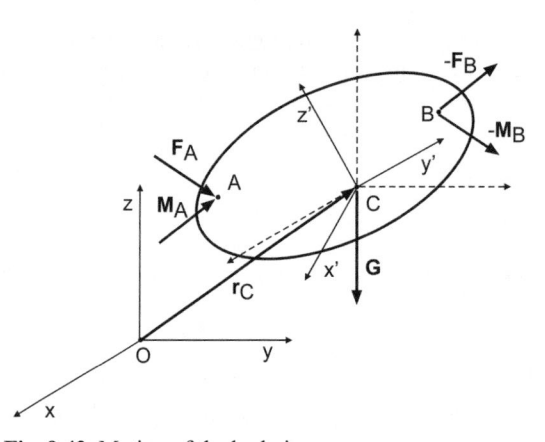

Fig. 9.42. Motion of the body in space

Note that the weight of the body is added here. This force is represented by source efforts defining the weight components with respect to the base frame axes. The momentum law in Eq. (9.95) is represented by the inertial elements I with the body mass as a parameter. Because it is used both in the SE and I components, it can be defined just once, at the level of the CM document (Fig 9.43b).

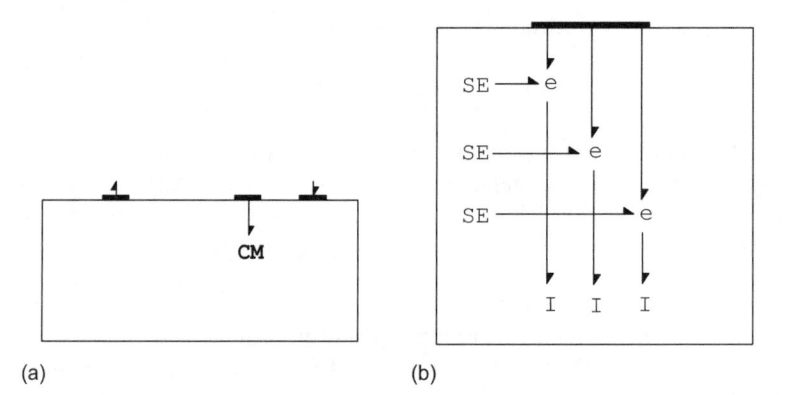

(a)   (b)

Fig. 9.43. Rigid body translation. (a) Representation in the **Base** (b) The **CM** component

The rotational part of the body motion is commonly described in a frame translating with the body mass centre and that is parallel to the base frame. The moment of momentum law with respect to such a frame reads

$$\left.\begin{aligned} \mathbf{H} &= \mathbf{J}\boldsymbol{\omega} \\ \frac{d\mathbf{H}}{dt} &= \mathbf{M} \end{aligned}\right\} \tag{9.97}$$

where $\mathbf{J}$ is the mass inertia matrix and $\mathbf{M}$ is the resultant moment about the body mass centre. Because of the rotation of the body, the inertia matrix changes during the motion. This is because the equations of rotational dynamics are as a rule (at least for rigid bodies) represented with respect to a frame fixed to the body. It is not difficult to transform Eq. (9.97) to this form.

In the body frame the body moment of momentum is given by

$$\mathbf{H} = \mathbf{R}\mathbf{H}' \tag{9.98}$$

Thus from the first Eq. (9.97) we have

$$\mathbf{H}' = \mathbf{J}'\boldsymbol{\omega}' \tag{9.99}$$

where

$$\mathbf{J}' = \mathbf{R}^{\mathsf{T}}\mathbf{J}\mathbf{R} \tag{9.100}$$

is the body mass inertia matrix with respect to the centroidal body axes. Multiplying the second Eq. (9.97) by $\mathbf{R}^{\mathsf{T}}$ and substituting from Eq. (9.98), we get

$$\mathbf{R}^{\mathsf{T}}\frac{d}{dt}(\mathbf{R}\mathbf{H}') = \mathbf{M}' \tag{9.101}$$

or

$$\frac{d\mathbf{H}'}{dt} + \mathbf{R}^{\mathsf{T}}\frac{d\mathbf{R}}{dt}\mathbf{H}' = \mathbf{M}' \tag{9.102}$$

Using Eq. (9.62), the last expression can be written as

$$\frac{d\mathbf{H}'}{dt} + \boldsymbol{\omega}' \times \mathbf{H}' = \mathbf{M}' \tag{9.103}$$

These are the famous Euler equations of body rotation in which the rate of change of the moment of momentum is represented by its local change and a part convected by the body rotation. This other part has a very elegant representation in a bond graph setting. This is the celebrated *Euler Junction Structure* (EJS) [10, 11].

Now we can describe the ROTATION component of Fig. 9.40, which represents the body rotation about the mass centre with respect to the body frame (Fig. 9.44). The Torque balance component in the middle consists simply of an array of three effort junctions that corresponds to the x, y and z components of the body angular velocity with respect to the body frame. The left and right ports transfer the moment of the forces and of the moments acting on the body. In the centre it is connected to I and EJS components. The first of these consists of an array of inertial elements that describes the local rate of change of the moment of momentum. There is a control-out port that serves for the transfer of information on the moment of momentum vector $\mathbf{H}'$ that the EJS component needs. This can be seen in Eq. (9.103). The EJS component is shown in Fig. 9.45.

It consists of three gyrators connected in a ring. Note that the power circulates inside the structure, thus there is no net power generation or dissipation. The gyrators are modulated by the body moment of momentum rectangular components. The EJS can be slightly simplified if the body axes correspond to the inertia tensor principal axes. We do not assume this, as selection of the body axes is usually based on how the body is connected. Thus, we assume that the body inertia matrix is fully populated. We do not recalculate the moment of momentum in the EJS component, but rather take it from the inertial component.

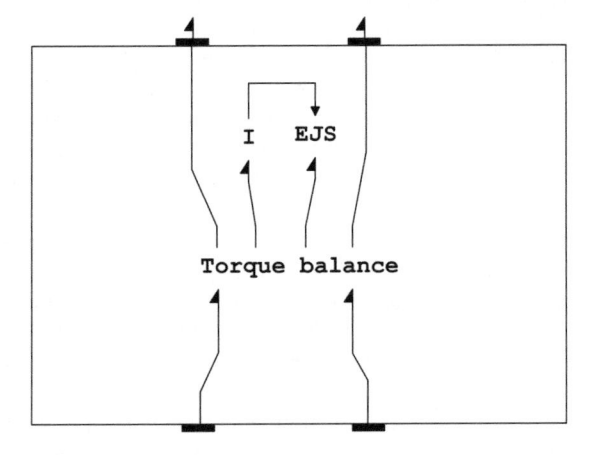

Fig. 9.44. Model structure of the ROTATION component of Fig. 9.40

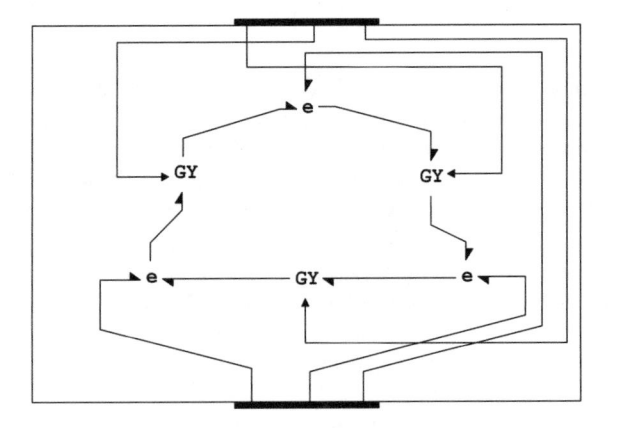

Fig. 9.45. The Euler Junction Structure (EJS)

This completes development of the component model of a body moving in space. Now we turn our attention to modelling the interconnections between bodies in space.

9.5.4 Modelling of Body Interconnections in Space

Typical mechatronics systems, such as robots, consist of manipulators regulated by controllers. The manipulators are multibody systems consisting of several members (links) interconnected by suitable joints. They are powered by servo-actuators. In the previous sections we developed a fairly general component model of bodies that can be used for the representation of manipulator links. Now we develop models of the joints. Two types of joint are considered – revolute and prismatic. Based on these component models, a typical robotic manipulator can be represented in a way similar to which real manipulators are assembled. In the next section we apply the methods of this section to the modelling of a complete robotic system, including its controller.

The approach is applicable to other multibody systems as well. Robot manipulators are chosen for several reasons. These are fascinating systems that have influenced development in many fields such as multibody mechanics, conventional and intelligent control, sensors and actuator technology, and have promoted mechatronics as a design philosophy. The modelling and simulation of such complex systems is not an easy task. The problem of modelling manipulators as multibody systems is only one part of it; there is also the problem of the control of such complex space systems, particularly when there are interactions with the environment. The bond graph method is a good candidate for solving such multidisciplinary problems.

Revolute Joints

Revolute joints have already been discussed in Sec. 9.2.2. The basic difference with those discussed earlier is that, because bodies connected by a revolute joint can move in three-dimensional space, the axis of the joint is not confined to specific motions but can move freely. To describe this effect, we analyse the bodies A and B connected by a revolute joint, as shown in Fig. 9.46. The bodies could, for example, be two links of a robot manipulator joined by a revolute joint. We assume that the z-axis of the coordinate frame $O_A x_A y_A z_A$ of body A is directed along the joint axis. We assume further that there is a body B frame $O_B x_B y_B z_B$. The precise positions of the frames are not prescribed and in a specific multibody system they can be defined as is most convenient, e.g. using the Denavit-Hartenberg convention [9]. The frame $Oxyz$ is the base frame.

Let P be the centre point of the revolute joint used as the reference connection point. The joint is represented by word model components as in Sec. 9.2.2 (Fig. 9.6). The ports are assumed compounded such that the port variables are 6D flows and efforts at the connection of the joint to body A and B These are expressed in their respective frames. At A (power-in) these are

$$\mathbf{f}_P^A = \begin{pmatrix} \mathbf{v}_P^A \\ \boldsymbol{\omega}_A^A \end{pmatrix} \tag{9.104}$$

and

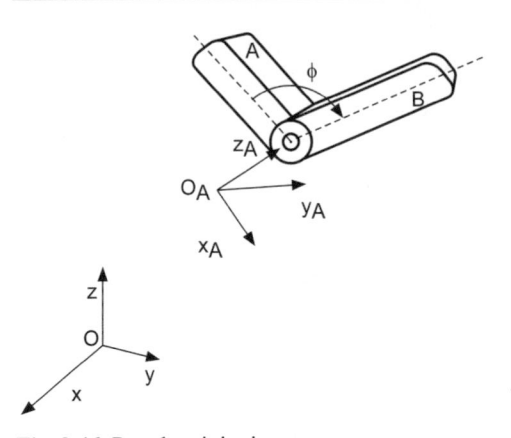

Fig. 9.46. Revolute joint in space

$$\mathbf{e}_P^A = \begin{pmatrix} \mathbf{F}_P^A \\ \mathbf{M}_P^A \end{pmatrix} \tag{9.105}$$

Similarly at B (power-out) we have

$$\mathbf{f}_P^B = \begin{pmatrix} \mathbf{v}_P^B \\ \boldsymbol{\omega}_B^B \end{pmatrix} \tag{9.106}$$

and

$$\mathbf{e}_P^B = \begin{pmatrix} \mathbf{F}_P^B \\ \mathbf{M}_P^B \end{pmatrix} \tag{9.107}$$

The linear velocities of bodies A and B at the connection point are common to both of the bodies. Hence

$$\mathbf{v}_P^A = \mathbf{R}_B^A \mathbf{v}_P^B \tag{9.108}$$

where $\mathbf{R}_B{}^A$ is the rotation matrix of the body B frame with respect to the body A frame.

A similar relation holds for the forces. Because the linear velocities at common point are equal, the same is true for the power transferred across the joint during its translation, i.e.

$$(\mathbf{v}_P^A)^T \mathbf{F}_P^A = (\mathbf{v}_P^B)^T \mathbf{F}_P^B \tag{9.109}$$

By substituting from Eq. (9.108) we have

$$\mathbf{F}_P^B = \mathbf{R}_A^B \mathbf{F}_P^A \tag{9.110}$$

Eqs. (9.108) and (9.110) describe the relationships between the linear part of the flow-effort 3D vectors at the joint ports. We develop next the relationships between the angular parts, i.e. angular velocities and moments.

Rotation matrices of the bodies A and B frames with respect to the base frame are related by the composition of rotations

$$\mathbf{R}_B = \mathbf{R}_A \mathbf{R}_B^A \tag{9.111}$$

By differentiating with respect to time we get

$$\frac{d\mathbf{R}_B}{dt} = \frac{d\mathbf{R}_A}{dt}\mathbf{R}_B^A + \mathbf{R}_A \frac{d\mathbf{R}_B^A}{dt} \tag{9.112}$$

From Eqs. (9.64) and (9.65) we have

$$\boldsymbol{\omega}_B \times \mathbf{R}_B = \mathbf{R}_A \boldsymbol{\omega}_A^A \times \mathbf{R}_B^A + \mathbf{R}_A \boldsymbol{\omega}_{AB}^A \times \mathbf{R}_B^A \tag{9.113}$$

where

$$\boldsymbol{\omega}_{AB}^A = \text{axial}(\frac{d\mathbf{R}_B^A}{dt}\mathbf{R}_A^B) \tag{9.114}$$

is the relative angular velocity of body B with respect to body A expressed in body A frame. Simplifying Eq. (9.113) we get

$$(\mathbf{R}_A)^T \boldsymbol{\omega}_B \times \mathbf{R}_A = \boldsymbol{\omega}_A^A \times + \boldsymbol{\omega}_{AB}^A \times \tag{9.115}$$

Hence (see Eqs. (9.64) and (9.62))

$$\boldsymbol{\omega}_B^A \times = \boldsymbol{\omega}_A^A \times + \boldsymbol{\omega}_{AB}^A \times \tag{9.116}$$

The last equations imply that

$$\boldsymbol{\omega}_B^A = \boldsymbol{\omega}_A^A + \boldsymbol{\omega}_{AB}^A \tag{9.117}$$

In addition we have

$$\boldsymbol{\omega}_B^A = \mathbf{R}_B^A \boldsymbol{\omega}_B^B \tag{9.118}$$

Note also that

$$\boldsymbol{\omega}_{AB}^A = \begin{pmatrix} 0 & 0 & \dot{\phi} \end{pmatrix}^T \tag{9.119}$$

where is ϕ is the joint's angle of rotation.

We use the component in Fig. 9.47 to represent a revolute joint. The component has an additional port that corresponds to the relative rotation of the joint. This port can be used for actuation of the joint. There are also two signal ports for the joint's relative velocity and the rotation angle output.

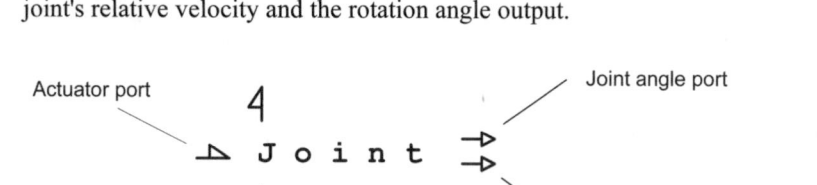

Fig. 9.47. Revolute joint component model

We can now represent the model of a revolute joint. It consists of two main parts (Fig. 9.48). One, a component termed **Revolute**, represents the basic relations between the port variables in the frame of a body connected to the lower port (body A). The other component, **Joint Rotation**, transforms the port variables from the frame of the other body (body B) to the frame of the lower (body A).

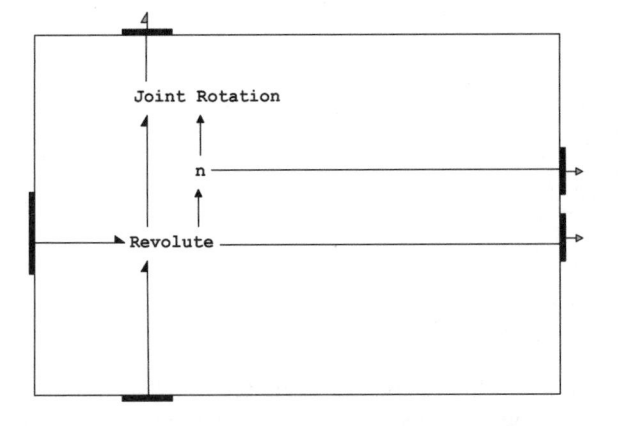

Fig. 9.48. Structure of the revolute joint component model

The **Revolute** component itself consists of two components (Fig. 9.49). The **Tr** component represents the translation part of the joint model. It consists solely of separate effort junctions for the translation in the direction of the three axes. This ensures that the joined ends of the bodies move with a common velocity and that the corresponding forces are the same.

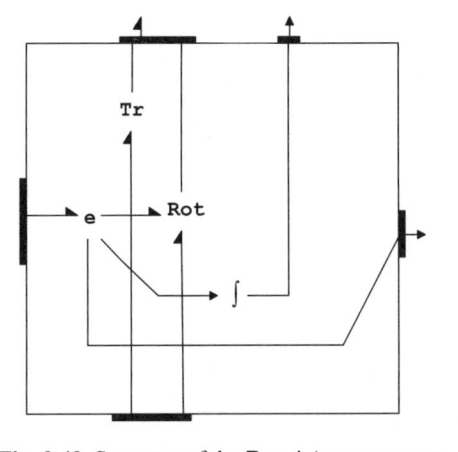

Fig. 9.49. Structure of the Revolute component

The other component, Rot, represents the relationship between the angular velocities as given by Eqs. (9.117) and (9.119). The model of the component (Fig. 9.50) is very simple and consists of two effort junctions in the x- and y- directions in which the angular components are the same for both of the joined bodies. There is also a flow junction that corresponds to the relative rotation about the z-axis.

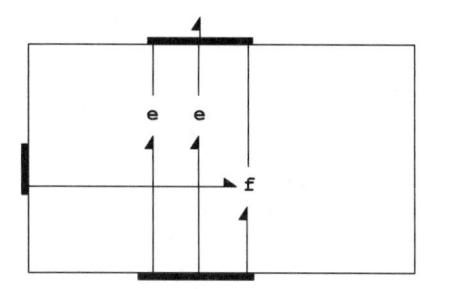

Fig. 9.50. The representation of the joint rotation

The effort junction on the left of the component in Fig 9.49 corresponds to the relative angular velocity of the joint. It is used to extract information about joint's angular velocity and to evaluate joint's angle by integration. The joint's angle is used for the rotation transformations as shown in Fig. 9.48 and is often for control.

An important function of the joint is the rotation transformation between the two link frames (Fig. 9.48). This is depicted in Fig. 9.51. The transformations are applied to the linear effort-flow parts, and separately to the angular. These are represented by components RotAB. The components represent transformations, as given by Eqs. (9.108) and (9.110) for linear variables. The same transformations hold for angular variables.

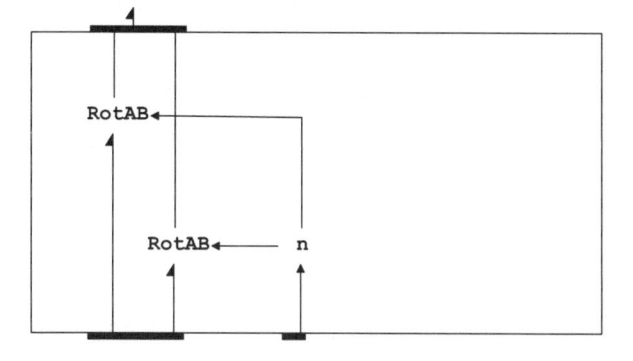

Fig. 9.51. The rotational transformations by the joint

The transformations are usually expressed by transformers TF, which transform flow and effort components. These are then summed as given in Eq. (9.108) and

(9.110). We do not give here the general structure of the component. Later in Sec. 9.6, we meet some specific examples of such transformations.

Revolute joints play an important role in the design of robotic manipulators. They offer the simplest way to change the orientation of robot links. The component model introduced here gives the main functionality of such joints. They are used later for the building of manipulator models. This is illustrated in Sec. 9.6.

Prismatic Joints

Prismatic joints have already been described in Sec. 9.2.2. The basic difference here is that the axis of the joint can be anywhere in space (Fig. 9.52). To describe the effects of prismatic joints on bodies connected by such a joint, we define a body frame $Ax_Ay_Az_A$ attached to one body at point A. The z-axis is directed along the relative displacement of the joint (joint axis). There is also a second body B and a frame attached to it.

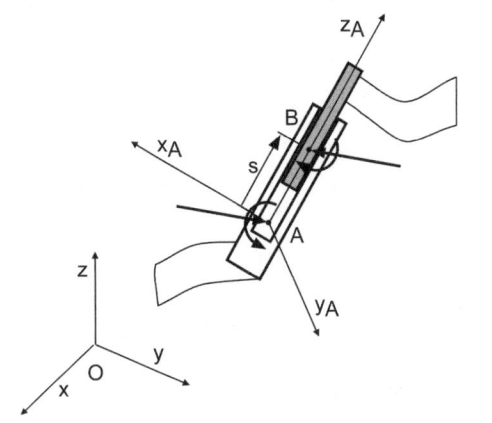

Fig. 9.52. Prismatic joint in space

The precise positions of the frames are not specified and can be defined as is the most convenient, e.g. by using Denavit-Hartenberg convention [9]. The frame Oxyz is the base frame.

Let B be a point on the second body. This body can move along the joint slot. The joint can be represented by a word model component as in Fig. 9.47, but of course its model will be different. The ports are assumed compounded such that the port variables are 6D flows and efforts at the connection of the joint to the bodies (the upper and lower in Fig. 9.47). The side port is used for actuation of the joint. We develop the governing equations first. They are generalizations of the corresponding equations for the planar case in Sec. 3.2.2, but are expressed with respect to the body frame, not the base frame, as we did for plane revolute joints.

The position vector of point B (Fig. 9.52), the reference point of the body that can slide along the joint, is given by

$$\mathbf{r}_B = \mathbf{r}_A + \mathbf{R}_A \mathbf{r}_{AB}^A \qquad (9.120)$$

where $\mathbf{R}_A$ is the rotation matrix of frame A with respect to the base and $\mathbf{r}_{AB}{}^A$ is its relative position with respect to body frame A. By differentiation with respect to time, we get for its velocity

$$\mathbf{v}_B = \mathbf{v}_A + \mathbf{R}_A \mathbf{v}_{AB}^A + \frac{d\mathbf{R}_A}{dt} \mathbf{r}_{AB}^A \qquad (9.121)$$

where

$$\mathbf{v}_{AB}^A = \frac{d\mathbf{r}_{AB}^A}{dt} \qquad (9.122)$$

is the relative velocity of the junction. Substituting from Eq. (9.65) in Eq. (9.121) we get

$$\mathbf{v}_B = \mathbf{v}_A + \mathbf{R}_A \mathbf{v}_{AB}^A + \mathbf{R}_A \boldsymbol{\omega}_A^A \times \mathbf{r}_{AB}^A \qquad (9.123)$$

Multiplying from the left by the transposed rotation matrix $\mathbf{R}_A{}^T$, we find

$$\mathbf{v}_B^A = \mathbf{v}_A^A + \mathbf{v}_{AB}^A + \boldsymbol{\omega}_A^A \times \mathbf{r}_{AB}^A \qquad (9.124)$$

or

$$\mathbf{v}_B^A = \mathbf{v}_A^A + \mathbf{v}_{AB}^A + (-\mathbf{r}_{AB}^A \times \boldsymbol{\omega}_A^A) \qquad (9.125)$$

In addition, we have

$$\mathbf{v}_B^A = \mathbf{R}_B^A \mathbf{v}_B^B \qquad (9.126)$$

Here, $\mathbf{R}_B{}^A$ is the rotation matrix of the frame of the second body with respect to the body frame of A. Note that in prismatic joints this is a constant matrix.

In the body frame A, the relative position vector is very simple

$$\mathbf{r}_{AB}^A = \begin{pmatrix} 0 & 0 & s \end{pmatrix}^T \qquad (9.127)$$

and hence the relative velocity of the joint is

$$\mathbf{v}_{AB}^A = \begin{pmatrix} 0 & 0 & \dot{s} \end{pmatrix}^T \qquad (9.128)$$

We also introduce a matrix $\mathbf{T}$ as in Sec. 9.2.2 defined by

$$\mathbf{T}_A^A = -\mathbf{r}_{AB}^A \times \qquad (9.129)$$

From Eqs. (9.57) and (9.127) we have

$$\mathbf{T}_A^A = \begin{pmatrix} 0 & s & 0 \\ -s & 0 & 0 \\ 0 & 0 & 0 \end{pmatrix} \qquad (9.130)$$

Eq. (9.125) now reads

$$\mathbf{v}_B^A = \mathbf{v}_A^A + \mathbf{v}_{AB}^A + \mathbf{T}_A^A \boldsymbol{\omega}_A^A \qquad (9.131)$$

This is similar to Eq. (9.25) for the planar case. We can expect that the model of the space prismatic joint is just a generalization of the model of Fig. 9.10.

The representation of the model of the prismatic joint is similar to the revolute joint of Fig. 9.48 and is shown in Fig. 9.53. The Joint Rotation component transforms efforts and flows between the two body frames. This is similar to the revolute joint (Fig. 9.51), but here the matrix is constant.

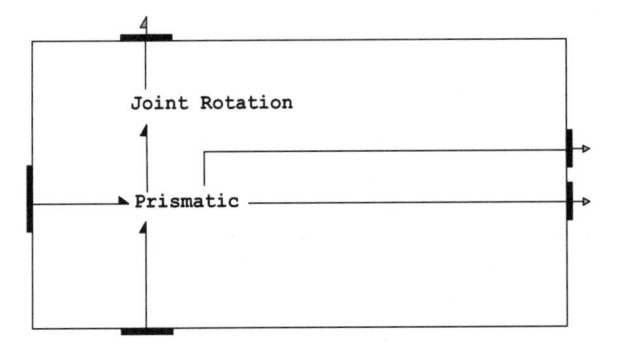

Fig. 9.53. The structure of the prismatic joint component model

The model of the Prismatic component is given in Fig. 9.54. Component f describes the joint the velocity relationship given by Eq. (9.131). It consists of flow junctions that describe the velocity relationship in the direction of the body axes. Component e on the right corresponds to the joint's angular velocity vector. The LinRot component describes the transformation between linear and angular quantities. The corresponding transformation matrix is given by Eq. (9.130). It depends on the joint's relative position, found from the integrator that uses the joint relative velocity from the corresponding junction.

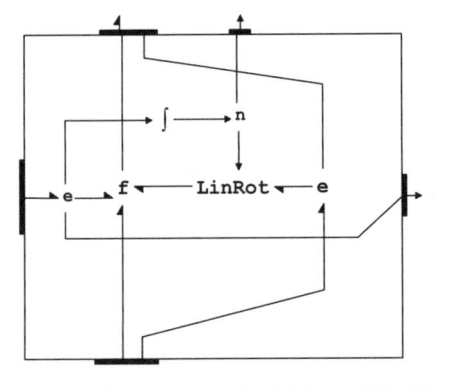

Fig. 9.54. The representation of the model of the space prismatic joint

Note that because of the very simple form of the matrix in Eq. (9.130), there are no z-components generated by the linear-to-rotational transformation. Summation of the velocity components represented by the flow junction that corresponds to Eq. (131) is simplified, too. Thus, we see that the space prismatic junction is not much more complicated than the planar one.

9.6 Motion of an Anthropomorphic Robot Arm Under Hybrid Control

9.6.1 Problem Formulation

In this section we apply the component-modelling technique to a problem of the partially constrained motion of a robot system. The system consists of a robot that presses a tool onto a vertical wall and moves it over the surface of the wall (Fig. 9.55). This type of problem is encountered in the robotic polishing of surfaces. In such applications it is important not only to control the motion of the tool over the surface, but also to maintain a certain value of the contact force.

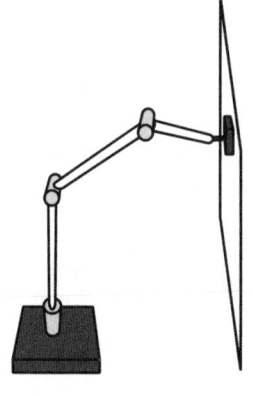

Fig. 9.55. An industrial robot pressing a tool to a wall

The problem is not an easy one. The robot manipulator carrying the tool changes its configuration when the tip of the tool approaches the wall from an open to a closed kinematic chain. When contact is established, the tool tip has to move over the surface. Hence control of the robot must drive the joints in such a way as to force the tool tip *over* the surface and not *into* the surface, thereby damaging it. Using only positional control is not possible, if the manipulator and/or the wall are rigid. In this case, because of deviations from the ideal program trajectory, there will be a component of the tool tip motion in the direction of the surface normal. Because of the high stiffness of the complete structure, very high reaction forces are developed. These forces are much higher than the capabilities of robot drives, and the robot will be locked with the tool penetrating the wall surface. The problem is well known in robotics and different strategies have been used to attack it [9, 20-22].

To help to solve this problem we assume that the robot wrist is equipped with a force sensor and that the overall mechanical system is somewhat compliant. We assume that most of the stiffness comes from the force sensor, but part of it comes from the wall, too. The robot is controlled by a hybrid law: the *PD* position control law in the unconstrained sub-space, and by the *I* force control law in the constrained subspace. For simplicity, we assume that the control is applied to the first three joints only. Thus, we model a three-link robot arm as a multibody system. The wrist and tool together are modelled as a point body. Compliance is modelled in the direction orthogonal to the wall by a Contact component of Sec 6.4.1.

A schematic diagram of the robot arm is given in Fig. 9.56. This is an anthropomorphic arm with three revolute joints powered by servo-actuators. The coordinate frames used for description of the motion are also given in the figure.

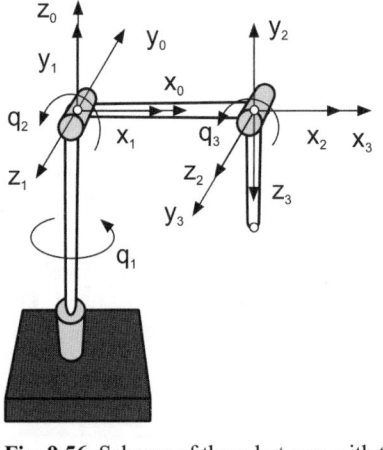

Fig. 9.56. Scheme of the robot arm with the coordinate frames

The geometrical and and inertial parameters are given in Fig. 9. 57 and Table 9.12 and are based on [23, 24]. The distance of the wrist centre to the wall is w = 0.1 m.

Table 9.12. Parameters of the robot arm

	Link 1	Link 2	Link 3	Wrist and tool
L [m]	0.5	0.4	0.3	
Lc [m]	0.25	0.2	0.10	
m [kg]	13	10	5	2
I_{Cxx} [kg·m^2]	0.7	0.15	0.15	
I_{Cyy} [kg·m$^{2]}$	0.35	0.3	0.15	
I_{Czz} [kg·m^2]	0.7	0.3	0.15	

Note: Masses and moments of inertia are of the links, icluding actuators

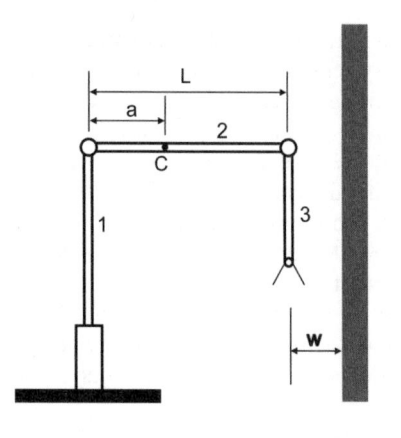

Fig. 9.57. The robot geometry

9.6.2 Model of the Robot System

We next give a brief description of the model. The complete model is held in the BondSim program library under the project name Robot Hybrid Control. The system level representation is shown in Fig. 9.58. It consists of three main components: Controller, Manipulator and Work Space.

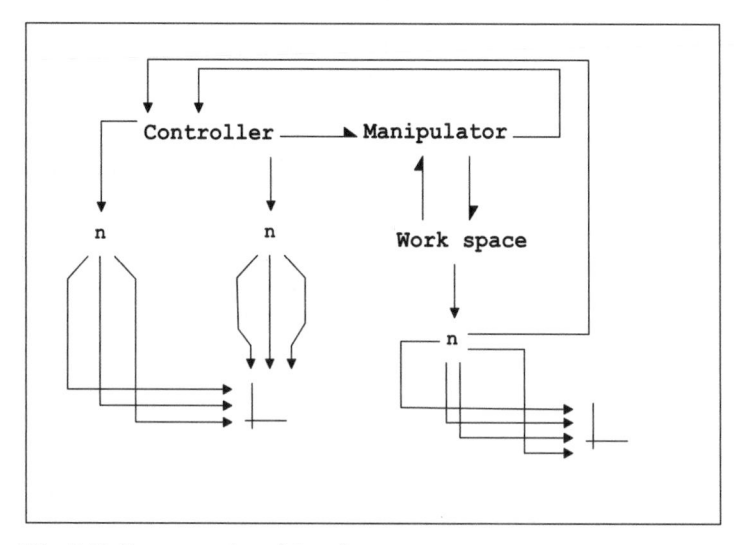

Fig. 9.58. Representation of the robot system

The Controller and Manipulator are connected by a power bond line, which describes the action of the first on the last. This is a multidimensional line that represents the interaction between the manipulator joints and the actuators. The inner signal line connecting the right manipulator port to the Controller is the joint position and velocity feedback. The outer serves for the feedback of the wall force that is taken from the Work space component at the contact of the tool tip and the wall.

There are many signals that are picked up and fed to the display. Because the signals fed out of the controller are packed as 3D bonds, they are first unpacked by n components that simply contain three branching nodes. The left display component is used for displaying the joint variables: angular position (the left ports) and torque (the upper ones). The right component serves for displaying the operational space variables: the coordinates of the position of the tip of the tool and the wall contact force. This way we get a picture of the behaviour of the robot system during simulation. Other variables could be added if necessary.

The model of the Manipulator component is shown in Fig. 9.59. The model reflects the structure of the real manipulator. The joints and links are created using components developed in Sec. 9.5. Thus, all of the links are created as copies of the 3D body component of Fig. 9.40. The names are changed appropriately. Thus they differ only by the names and by values of the parameters. The only exception is the Link 3 component, because there is no torque at the tool tip, only a force. Hence the upper moment line is removed from the component, and all contained components; e.g. the ROTATION component of Fig. 9.44.

The mass moments of inertia of the link components are set according to the values in Table 9.12. The coordinates of the end points of the links with respect to the mass centre of the link in the LinRot components (Fig. 9.40) are set in accordance with the data of Fig. 9.57 and Table 9.12.

The joint components are created as copies of the 3D joint component of Sect. 9.5.4 (Figs. 9.48 – 9.51). But there are some differences from joint to joint. The general structure is the same as given in Fig. 9.48, but the Joint rotation components differ because the rotation matrices of the body (link) frames change from joint to joint. For every joint a specific transformer component is used. We show this for Joint 2, which rotates the frame of Link 2 with respect to the frame of Link1. The rotation matrix of these two frames is given by (Fig. 9.56)

$$\mathbf{R}_{21} = \begin{pmatrix} \cos q_2 & -\sin q_2 & 0 \\ \sin q_2 & \cos q_2 & 0 \\ 0 & 0 & 1 \end{pmatrix} \qquad (9.132)$$

The corresponding transformations are given by (Fig. 9.51)

$$\left. \begin{aligned} \mathbf{f}^1 &= \mathbf{R}_{21}\mathbf{f}^2 \\ \mathbf{e}^2 &= \mathbf{R}_{21}^{T}\mathbf{e}^1 \end{aligned} \right\} \qquad (9.133)$$

These transformations are represented by the components shown in Fig. 9.60. There are four transformers corresponding to the four nonzero elements in the first two rows of the rotation matrix in Eq. (9.132).

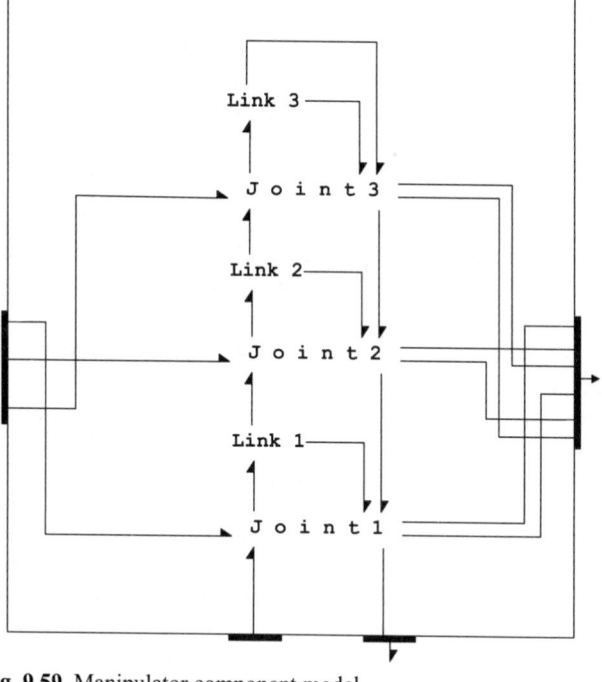

Fig. 9.59. Manipulator component model

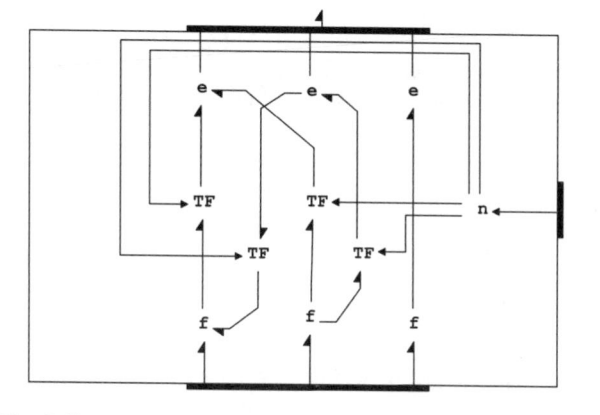

Fig. 9.60. Representation of a transformation between link frames

Every effort and flow vector component in Eq. (9.133) is represented by a separate effort or flow junction, and the transformers are connected between them. Connections to the flow junctions correspond to the matrix operation in the first part of Eq. (9.133), and that of the effort junction to the second equation of Eq.

(9.133). The transformation described by the other rotation matrices can be created in a similar way.

Joints transform not only the 6D flows and efforts of connected ports of nearby links, but other variables, as well. The links describe the rotational dynamics parts only. The translational dynamics parts are represented in the base (the Work space). Thus, the resultant of the link forces reduced to the link mass centre and the velocity of the mass centre have all to be rotated by the joints to the base (or vice versa). In a similar way the forces (and the moments) acting on the tool tip and its velocity need to be transformed from the base (operational space) to the last link frame. Thus, there are usually more transformers of the same type inside a joint rotation component as we go from robot tip to the base, not just two as in Fig. 9.51. The quantities transferred across the joints correspond in Fig. 9.51 to the vertical bonds near the right end of the joint and the link components. The figure clearly shows that these bonds are packed together into a higher dimensional vector structure. Thus the bond connected to the bottom-right port in Fig. 9.51 is really 12-dimensional. It should be kept in mind that elements of this multidimensional connection can be accessed in the reverse order to the order they were created. For example, access could begin with the resultant of the 1st link forces and the corresponding mass centre velocity first; then proceed to that of the 2nd link; and so on, until the tool tip is reached. It is not necessary to pack the bonds this way, but then there would be a forest of bond lines that would be difficult to draw and even more difficult to understand. The component model and the compounded port concepts developed in this book enable the visual representation of complicated tensor quantities and operations in a clear way.

Returning to the manipulator model of Fig. 9.59, we see that the power bonds at the left port are branching out to the joints. In a similar way the joint position and velocity signals are collected at the right port. Thus on the outside (Fig. 9.58) they appear as two bond lines only, a 3 D power line and a 6D feedback line.

The Work Space component of Fig. 9.58 describes the manipulator body and the robot working environment as seen in the base (inertial) frame. As already discussed in Sec. 9.6 the Link component represents the link rotational dynamics. The translation part of the link dynamics is represented in the base, as are all of the interactions of the manipulator with its environment. The model of the Work Space component is shown in Fig. 9.61.

The manipulator body motion is represented by the Body Motion component. Because the manipulator body is fixed with respect to the base coordinate frame, the component consists solely of zero source flow components. In mobile robots these can be used to describe the moving of the body of the robot. The other two components describe the translational dynamics of the manipulator as well as the interaction of the manipulator arm with the environment. The structure of these components reflects how the bonds connected to the right port in Fig. 9.61 are "packed" (see Fig. 9.59). Thus, to unpack a bond we have to proceed in the reverse order to how it was composed. Thus the left bond comes from the first link mass centre. It is connected to the component CM1, which represents the translational dynamics of the first link. It is represented as shown in Sect. 9.5.3 and Fig. 9.43. The only difference is that its mass is defined as given in Table 9.12. We proceed in a similar way. Thus the model of the component denoted as Rest in Fig. 9.61 has a similar structure (Fig. 9.62). Here the translational dynamics of the second link is represented by CM2. The mass parameter of the second link as given in Table 9.12 is specified in the CM2 component.

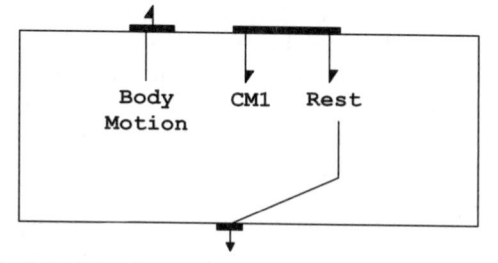

Fig. 9.61. Model of the robot work space

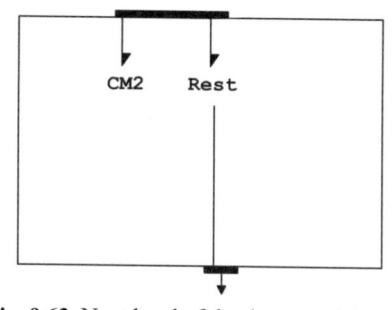

Fig. 9.62. Next level of the decomposition of the component Rest

The last Rest component consists of the CM3 component that represents the last link translational dynamics and a component denoted as Tool (Fig. 9.63). The mass parameter of the CM3 component is defined in Table 9.12.

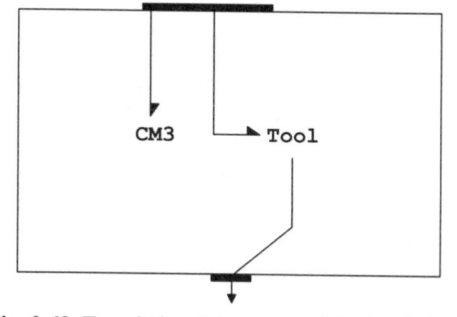

Fig. 9.63. Translational dynamics of the last link

The Tool component, as described at the beginning of this section, incorporates a wrist, a gripper equipped with a force sensor, a tool and a model of the interaction with the wall. Its simplified model is shown in Fig. 9.64. The wrist and all it carries is represented by a particle at the end of the third link (the wrist centre point). Its inertia and weight are represented by an inertial component I and source effort component SE, one per axis of the movement, respectively.

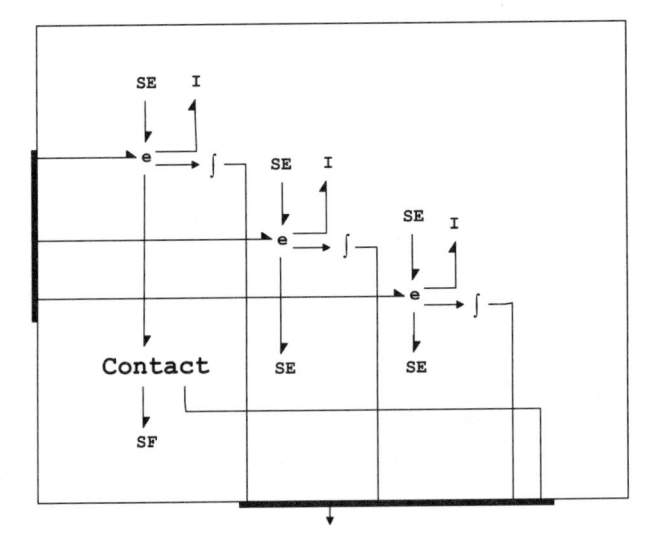

Fig. 9.64. Model of the Tool component

The wall restricts movement of the tool only in the base x –direction. The interaction with the wall is represented by a Contact component (see Sec. 6.4.1 and Figs. 6.53 and 6.54). The SF component defines the wall as being fixed, i.e. its velocity is zero. There is no force on the robot arm when the tool is off the wall. The wall is modelled as a spring-damper system of relatively high stiffness constant and low damping as described in Sec. 6.4.1. The initial tool tip and wall positions are defined in Fig. 9.57 and Table 9.12. The tool tip can move freely over the wall y-z plane. The friction in that plane is represented by two SE components and is assumed to be zero

The integrators shown in the figure are used for evaluating the position of the tool tip and are used only for monitoring its motion. The robot control is based on joints' variables, in addition to the signal of the interaction force taken from the Contact component. Initial values of the integrators are set according to the initial configuration of the manipulator arm in Fig. 9.57, i.e. x = 0.4 m, y = 0 and z = –0.3 m. All four signals are packed into a 4D signal bond (see Figs. 9.63, 9.62, 9.61 and 9.58).

9.6.3 Hybrid Position/Force Control

Now we come to the controller that has to ensure proper regulation of the robot. The model of the controller that we use is shown in Fig. 9.65. It uses hybrid force/position control with velocity feedback of the robot in the base (operational space) frame. The control law is based on the operational space control scheme of [9]. This uses transposed Jacobian control. We do not use gravity compensation.

Starting from the port where the feedback is connected we need to separate first the joint position signals from the velocity. This is achieved by two components n composed of nodes only that are used to extract the signals and then pack them again as 3D vectors of, respectively, joint angles

$$\mathbf{q} = (q_1 \ q_2 \ q_3)^\mathsf{T} \tag{9.134}$$

and velocities

$$\dot{\mathbf{q}} = (\dot{q}_1 \quad \dot{q}_2 \quad \dot{q}_3)^\mathsf{T} \tag{9.135}$$

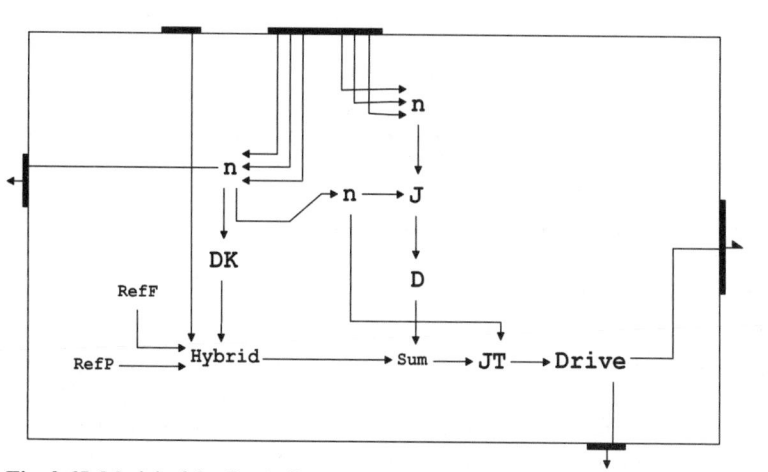

Fig. 9.65. Model of the Controller

The positional part is converted to the vector of tool tip coordinates with respect to the base frame using the relations of direct kinematics of the manipulator. For the anthropomorphic configuration of Fig. 9.56, these are given by

$$\mathbf{r}_{\text{tool}} = \begin{pmatrix} L_2 \cos q_1 \cos q_2 + L_3 \cos q_1 \sin(q_2 + q_3) \\ L_2 \sin q_1 \cos q_2 + L_3 \sin q_1 \sin(q_2 + q_3) \\ L_2 \sin q_2 - L_3 \cos(q_2 + q_3) \end{pmatrix} \tag{9.136}$$

These relationships are represented by the component DK that is composed of function elements and summators that express the coordinates of the tool tip in the base frame as a function of the joint coordinates.

We also need the transformation of the joint velocities to the velocity of the tool tip with respect to the base frame. The relationship follows directly from (6.136) by differentiation

$$\mathbf{v}_{tool} = \mathbf{J}\dot{\mathbf{q}} \tag{9.137}$$

where

$$\mathbf{J} = \partial \mathbf{r}_{tool} / \partial \mathbf{q} \tag{9.138}$$

is the manipulator Jacobian matrix. The Jacobian matrix can be evaluated directly from Eq. (6.136) as

$$\mathbf{J} = (\mathbf{J}_1 \ \mathbf{J}_2 \ \mathbf{J}_3) \tag{9.139}$$

where

$$\mathbf{J}_1 = \begin{pmatrix} -L_2 \sin q_1 \cos q_2 - L_3 \sin q_1 \sin(q_2 + q_3) \\ L_2 \cos q_1 \cos q_2 + L_3 \cos q_1 \sin(q_2 + q_3) \\ 0 \end{pmatrix} \tag{9.140}$$

$$\mathbf{J}_2 = \begin{pmatrix} -L_2 \cos q_1 \sin q_2 + L_3 \cos q_1 \cos(q_2 + q_3) \\ -L_2 \sin q_1 \sin q_2 + L_3 \sin q_1 \cos(q_2 + q_3) \\ L_2 \cos q_2 + L_3 \sin(q_2 + q_3) \end{pmatrix} \tag{9.141}$$

and

$$\mathbf{J}_3 = \begin{pmatrix} L_3 \cos q_1 \cos(q_2 + q_3) \\ L_3 \sin q_1 \cos(q_2 + q_3) \\ L_3 \sin(q_2 + q_3) \end{pmatrix} \tag{9.142}$$

Component J of Fig. 9.65 represents the matrix operation of Eq. (9.137) as shown in Fig. 9.66. It is composed of components J1, J2 and J3, which represent the operations in Eqs. (9.140)–(9.142). These components need information on the joint angles that are fed to the left port of the component J. Components denoted by J*q multiply components of the Jacobian column matrices by the joint velocity components. Finally, the outputs are summed and connected to the output. Component D represents the proportional gain component of the velocity feedback formed of function elements for every input component.

The central point of the controller is the Hybrid component, which represents the hybrid/position control of the robot. The component gets the input from the reference-input components – RefF for the force of the tool tip on the wall in the x-direction, and RefP for its position along the wall (y- and z-coordinates). It also accepts the force feedback and the tool position output from the direct kinematics component. We return to the hybrid control component later. The difference between the hybrid controller output and the velocity feedback is amplified by the component JT and is used to drive the joint servo-actuators. The component JT represents multiplication by the transposed Jacobian and is represented in a similar way to the multiplication by Jacobian operators in Fig. 9.66. The servo-actuators used for the joints are torque-controlled [9] and are represented by the Drive component consisting simply of controlled source efforts.

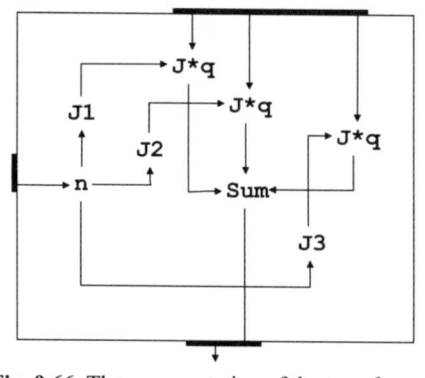

Fig. 9.66. The representation of the transformation of joint velocities

The hybrid controller structure is shown in Fig. 9.67. The top path corresponds to the control of the force and uses *I* control action. The bottom three paths constitute the position loops with *P* controller action. The last function element selects position control for approaching the wall and moving away from it, and the force control during motion of the tool tip over the wall surface. Switching from one control law to the other is based on a time sequence. Thus it is assumed that the tool approaches the wall first and after a sufficient length of time—such that the tool is at the wall surface—the x-axis position control is switched off. The force control is then applied during some period of time during which the tool slides over the wall, e.g. polishing the wall. Finally, at a given time the force control is switched off, the position control is switched on, and the tool is removed from the wall.

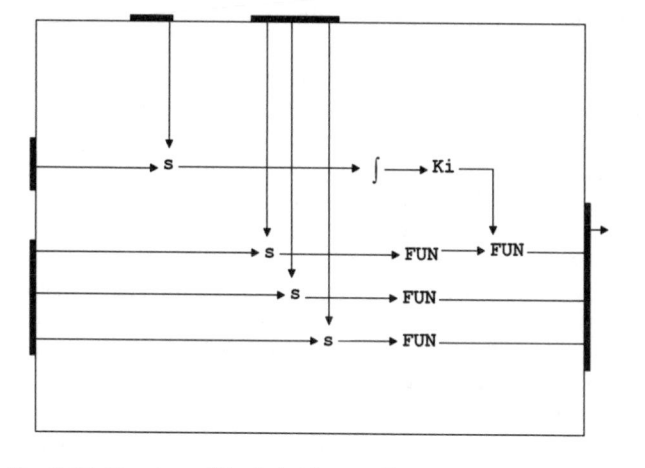

Fig. 9.67. Structure of the hybrid controller

9.6.4 The Simulation of the Robot Motion

We now simulate some typical robot operations consisting of approaching the wall, operation on the wall and returning back from the wall. However, it is necessary to determine the workspace range on the wall, within which it is possible to place the tool tip. In the case of the anthropomorphic robot arm of Figs. 9.56 and 9.57, the distance from the origin of the base frame to a point on the wall is given by

$$L_w^2 = (L_2 + w)^2 + y^2 + z^2 \tag{9.143}$$

This distance is not greater then L_2+L_3 and hence the reachable set on the wall is given by

$$y^2 + z^2 \le (L_3 - w)(2L_2 + L_3 + w) \tag{9.144}$$

This is a circle on the wall if $L_3 > w$ (Fig. 9.68a). At a point in this range the robot arm could be moved to one of two possible postures: elbow down or elbow up (Fig. 9.68b).

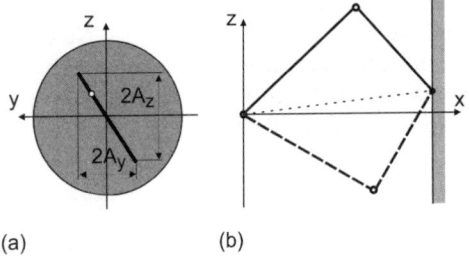

(a)   (b)

Fig. 9.68. The robot workspace on the wall. (a) The reachable region (b) Possible postures

To illustrate the behaviour of the complete system we move the tool tip over the wall along a straight segment as shown in Fig. 9.68a. This can be achieved by

$$\left. \begin{aligned} x &= L_2 + w \\ y &= A_y \sin(2\pi t / PER) \\ z &= A_z \sin(2\pi t / PER) \end{aligned} \right\} \tag{9.145}$$

Using the data in Table 9.12 with $w = 0.1$ m, the radius of the bounding circle is 0.48990 m. Thus, the amplitudes in Eq. (9.145) are taken as $A_y = 0.2$ m and $A_z = 0.3$ m. The period is PER = 5 s.

The program trajectory is defined in the components RefP and RefF of Fig. 9.65 by Eq. (9.146). The parameters are defined as

T0 = 2 s, T1 = 3 s, T2 = 15 s, T3 = 16 s, RET = 0.5 s, and Fw = 50 N.

The yend and zend are the values of the respective coordinates at t = T3.

$$
\left.\begin{aligned}
x &= t < T0?0.4 + 0.05 * t : \\
&\quad (t < T3?0.5 : 0.4 + 0.1 * \exp(-(t - T3)/RET)) \\
y &= t < T1?0 : (t < T3?Ay * \sin(2 * PI * (t - T1)/PER) : \\
&\qquad\qquad yend * \exp(-(t - T3)/RET)) \\
z &= t < T0? - 0.3 + 0.15 * t : (t < T1?0 : (t < T3? \\
&\quad Az * \sin(2 * PI * (t - T1)/PER) : zend * \exp(-(t - T3)/RET))) \\
F &= t < T1?0 : (t < T2?Fw : 0)
\end{aligned}\right\} \qquad (9.146)
$$

Hence, starting from the initial configuration in Fig. 9.57, the manipulator joints are rotated in such way that the tip moves with a linear velocity of 0.05 m/s in the direction of the wall and with a velocity of 0.15 m/s in the z-direction. After 2 s, the tool tip reaches the centre of the wall; after 3 s the controller is switched to the force control. The tool tip is then moved over the surface of the wall, pressing into it with a force of 50 N. After 15 s, the force reference value is lowered to zero and, after 16 s the arm is removed from the wall by applying a simple exponential time decaying trajectory with a decaying time constant of 0.5 s.

Gains of the controller are set to relatively high values, for it is also necessary to compensate for the effects of the arm, the wrist and the tool weights. No direct weight compensation is applied at the manipulator, e.g. by counter weights or springs, or at the controller level by some kind of feed-forward action, such as in [9]. Values of the controller settings are given in Table 9.13.

Table 9.13. Setting of controller gains

Gain	Value
P	100 000 (all axes)
Ki	500
D	10 000 (all axes)

Simulations were run for 20 s at the output interval of 0.01 s. Processing time of a simulation run is 17.8 s, a large part of which occurs during the approach to the wall. The results are shown in Figs. 9.69 – 9.72.

Fig. 9.69 shows time histories of the tool tip motion, and Fig. 9.70 the change of the force at the wall. The tip of the tool follows the program trajectory well. Regarding the force there are sharp peaks at the start, but it quickly drops to the program value of 50 N, around which it oscillates with an amplitude of about 0.047 N.

Fig. 9.71 shows the changes of the joint angles, and Fig. 9.72 shows the torques developed at the joints. Owing to discontinuities in the velocities, there are sharp peaks in the torque curve at the switching points of the programmed trajectory.

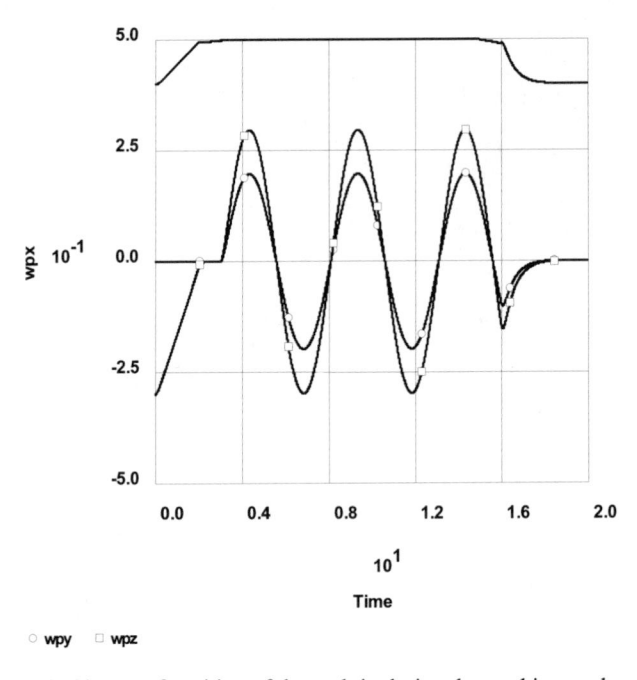

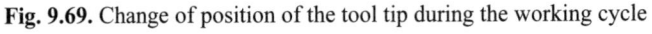

Fig. 9.69. Change of position of the tool tip during the working cycle

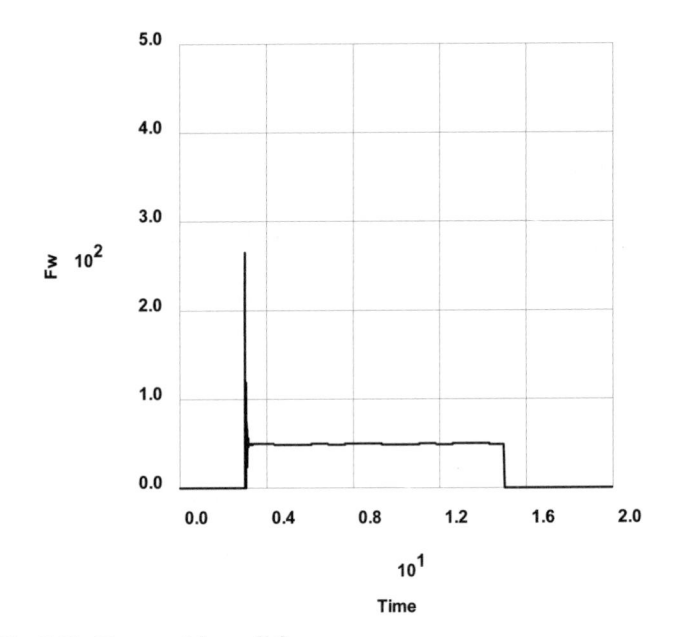

Fig. 9.70. Change of the wall force

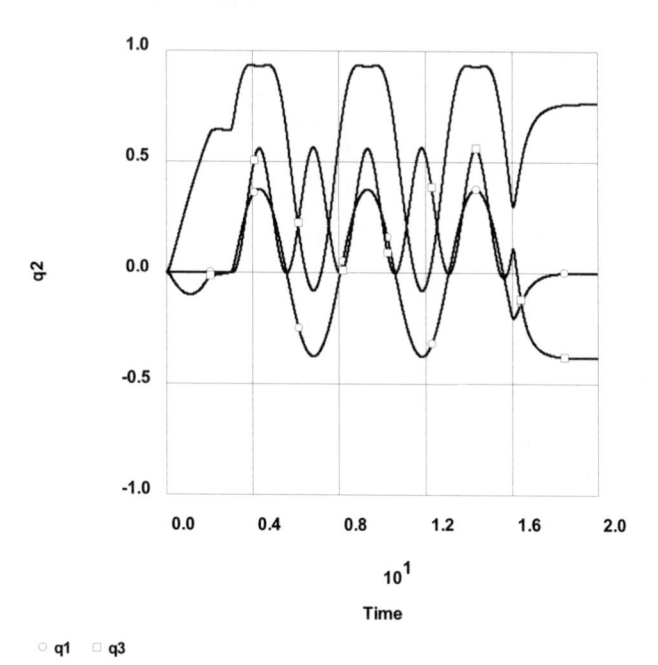

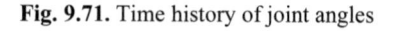

Fig. 9.71. Time history of joint angles

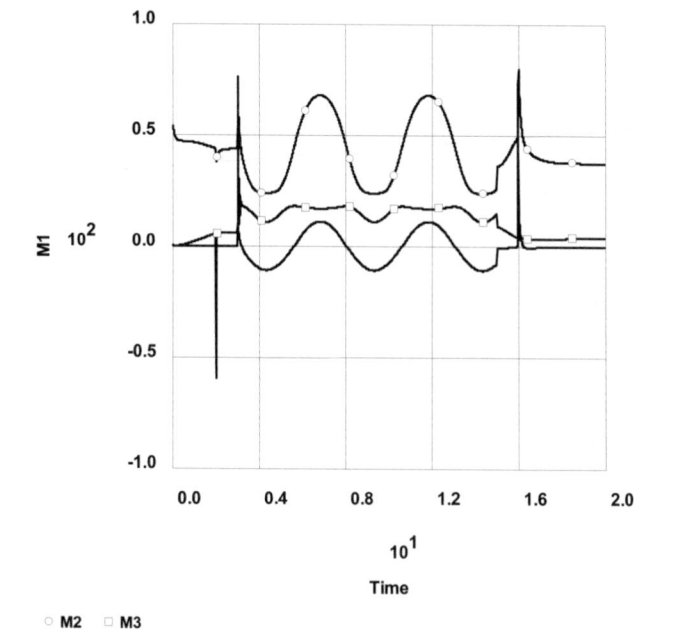

Fig. 9.72. Time histories of the joint torques

References

1. J Wittenburg (1977) Dynamics of Systems of Rigid Bodies. BG Teubner, Stuttgart
2. EJ Haug (1989) Computer-Aided Kinematics and Dynamics of Mechanical Systems. Allyn and Bacon, Boston
3. W Schiehlen (1990) Multibody Systems Handbook. Springer-Verlag, Berlin
4. R von Schwerin (1991) Multibody System Simulation: Numerical methods, Algorithms, and Software. Springer-Verlag, Berlin
5. AA Shabana (1998) Dynamics of Multibody Systems, 2nd edn. Cambridge University Press, Cambridge
6. PJ Rabier and WC Rheinboldt (2000) Nonholonomic Motion of Rigid mechanical Systems From a DAE Viewpoint. SIAM, Philadelphia
7. M Borri, L Trainelli and CL Bottasso (2000) On representation and Parameterizations of Motions. Multibody Systems Dynamics 4: 129-193
8. JJ Craig (1986) Introduction to Robotics: Mechanics and Control.
9. L Sciavicco and B Siciliano (1996) Modeling and Control of Robot Manipulators. McGraw-Hill, New York
10. Pieter C Breedveld (1984) Physical systems Theory in Terms of Bond Graphs, PhD thesis, Technische Hochschool Twente, Entschede
11. Dean C Karnopp, Donal L Margolis and Ronald C Rosenberg (2000) System Dynamics: Modeling and Simulation of Mechatronic Systems, 3rd edn. John Wiley, New York
12. Fahrenthold EP and Wargo JD (1994) Lagrangian Bond Graphs for Solid Continuum Dynamics Modeling. ASME J. of Dynamic Systems, Measurement and Control 116: 178 – 192
13. E Hairer and G Wanner (1996) Solving ordinary Differential Equations II, Stiff and Differential-Algebraic Problems, 2nd Revisited edn. Springer-Verlag, Berlin
14. WM Lioen and JJB Swart Test Set for Initial Value Problem Solvers, Amsterdam. Available at http://www.cwi.nl/cwi/projects/IVPteseset/.
15. JP Den Hartog (1956) Mechanical Vibrations, 4th edn. McGraw-Hill, New York
16. CH Pan and JJ Moskwa (1996) An Analysis of the Effects of Torque, Engine Geometry, and Speed on Choosing an Engine Inertia Model to Minimize Prediction Errors, ASME J. of Dynamic Syst., Meas., and Control, 118:181-187
17. DC Hesterman and BJ Stone (1994) A Systematic Approach to the Torsional Vibration of Multi-cylinder Reciprocating Engines and Pumps, Proc. Instn. Mech. Engers. 208:395-408
18. V Damic and P Kesic (1997) A System Approach to Modelling and Simulation of Multibody Systems Using Acausal Bond Graphs. In P Marovic, J Soric and nN Vrankovic (eds.) Proc. of the 2nd Congress of Croatian Society of Mechanics, Supetar, Croatia, pp 415-422
19. V Damic, J Montgomery and N Koboevic (1998) Application of Automated Modelling in Design. In P Marjanovic (ed.) Proc. 5th International Design Conf., Dubrovnik, Croatia, pp. 111-116
20. G Ferretti, G Magnani and P Rocco (1995) On Stability of Integral Force Control in Case of Contact with Stiff Surface, ASME J. of Dynamic Syst., Meas., and Control, 117:547-553

21. KP Junkowski, HA ElMaraghy (1996) Constraint Formulation for Invariant Hybrid Position/Force Control of Robots, ASME J. of Dynamic Syst., Meas., and Control, 118:290-299
22. J De Schutter, D Torfs, H Bruyninckx and S Dutre (1997) Invariant Hybrid Force/Position Control of a Velocity Controlled Robot with Compliant End Effector Using Modal decoupling, The Int. J. of Robotic Research, 16:340-356
23. V Damic (1987) An Approach to Computer Aided Design of Industrial Robots Using Simulation Package Simulex. In M. Vukobratovic (ed) Proc of 5th Yugoslav Symposium on Applied Robotics and Flexible Automatization, Bled, pp 28-39
24. F Harashima, J Hu, H Hashimoto and T Ichiyama (1986) Tracking Control of Robot Manipulator Using Slide Mode. In Proceedings of 15th International Symposium On Industrial Robotics, Tokyo, pp. 661-662

Chapter 10 Continuous Systems

10.1 Introduction

Continuous systems are important in many engineering disciplines, such as structural mechanics, fluid mechanics, thermal systems, electrical field etc. They are important in mechatronics applications, too, i.e. control of robotic manipulators taking in count flexibility of mechanical structure, sensor design, and micromechanics systems design. Solving such problems requires some form of discretisation. Two general methods are widely used: finite difference and finite element. The first starts with partial differential equations that describe the problem and uses suitable numerical approximation of the equation on a selected grid. Finite element methods, on other hand, are based on disretization of the physical problem domain, dividing a continuous system or a continuous component into finite elements. Motion of elements typically is based on an assumed displacement field used to formulate the governing equations of element motion. The equations are formulated using different methods, such as the Lagrange equations, the D'Alembert-Lagrange principle, the Galerkin method, or others. Finite element methods nowadays dominate the scene; many extremely powerful software packages based on these methods are available to solve various engineering problems. There are also related methods, such as finite-volume and boundary-element methods that are popular in fluid mechanics and the electrical engineering and field theory. We will not go into details of these methods here, but suggest that the interested reader consult suitable references, such as [1–2]. It should be noted that the methods available for mixed problems are not as powerful as for homogenous problems, particularly if interconnection is strong.

The question we would like to ask is where is the place of bond graph methods in this area. The strength of bond graphs lies in its multidisciplinary paradigm and visual expressiveness. In many problems of mechatronics and micro-mechanics, the physical domain is not uniform. There are strong interactions between processes taking place in physically different fields. This is, for example, the case in computer-controlled drives of mechanical links that often are not too rigid, or in sensors where there are strong interactions of solid mechanics, fluid mechanics, electrical and thermal processes. Bond graphs are capable of helping to bridge the gap between these different fields.

In this final chapter we describe an approach for solving the problems dealing with continuous systems based on bond graphs. In the literature there are different approaches for solving continuous systems by bond graphs [3–5]. The approach described here is based on the component model philosophy developed in this book and implemented in BondSim. We start with description of the general approach to modelling continuous systems based on component models. This will be applied first to the problem of modelling electric transmission lines. Next, a bond graph component model of a beam element, based on classic Euler-Lagrange theory, will be developed. It will be used in the solution of two practical problems: package vibration testing and a Coriolis mass-flow meter.

10.2 Spatial Discretisation of Continuous Systems

Continuous systems possess infinitely many degrees of freedom and, as such, are not directly amenable to numerical solution. Their solution requires some sort of discretisation. Various approaches are possible. Here we consider an approach that is compatible with the methods used in this book, i.e. representing the continuous system as an assemblage of bond graph components, then developing the mathematical model as a set of differential-algebraic equations that are solved directly. This kind of approach belongs to a class of methods known as the Method of Lines (MOL) [6]. To represent continuous systems—such as robotic links or transmission lines—by a bond graph component model, it is necessary to discretisize them spatially. One of most powerful ways to do this is by the finite-element method. Other approaches also can be used, notably finite-difference approximations and even *ad hoc* lumping. We are not speaking in favour of any of these, but wish simply to explain how they can be implemented in a bond graph setting.

A continuous system can be discretized by dividing it into a number of finite parts (elements). For example, a beam can be divided into number of small parts (Fig. 10.1). The parts are called finite to distinguish them from the differential elements used in mathematical analysis of the problem. The parts also can be of different forms. Thus, in finite element analysis, triangular or quadrilateral elements often are used in plane problems; tetrahedral and hexahedral (solid) elements for space problems; plate elements for analysis of plates, etc. In finite difference approximations one-, two- or three-dimensional grids are used.

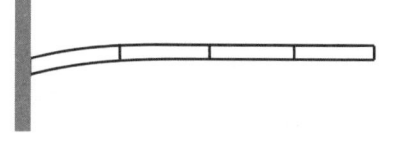

Fig. 10.1. Continuous beam as an assemblage of finite parts

The parts are assumed to be joined only at distinct nodes. Thus, discretisation is based on physical variables defined only at distinct nodes. Values of variables inside the finite parts (elements) are described in terms of nodal variables. In the finite element method one group of variables, such as displacements in deformable body mechanics, fluid velocities in fluid mechanics, or field potential in electric fields, are represented inside the elements by polynomial interpolating functions.

The approximation is smooth over an element, but there can be discontinuities in the derivatives of variables with respect to space coordinates at the transition from one element to the other. The expression for associated variables—e.g., forces, stresses, and currents—are found by applying a suitable method, such as Lagrange equations, the Principle of Virtual Work, and like.

An element of the continuous domain can be represented by a bond graph component (Fig. 10.2a). Depending on the number of external nodes of the element that are used to join neighbouring elements, this component has a suitable number of power ports. These generally are compounded and, hence, can represent efforts and flows at the ports as vector or tensor quantities (Fig 10.2b). Interconnection to other components is made by components consisting of an effort or flow junction, depending on the type of variables at the nodes. In solid continuum domains these typically are displacements. Displacements are not directly supported as port variables in bond graphs; thus, velocities will be used that correspond to the common flows at the effort junction nodes. Bond graph representation also shows assumed power flow thought the domain. The component model of underlying processes can be developed using different approaches, as already noted previously.

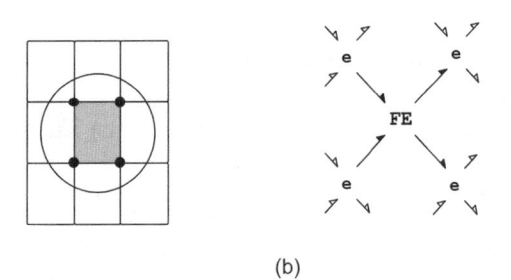

(a)   (b)

Fig. 10.2. Finite element as bond graph component. (a) Four-node finite element, (b) Bond graph component connection diagram

The finite-element method—as well as the finite-volume and boundary-element methods—uses pre-processors and post-processors that are, to a large extent, responsible for their successful application in engineering design and research. They are responsible for element mesh generation and displaying results in a user-friendly way. Such special purpose devices are not used in BondSim; the hierarchical component model approach we have developed can be used to simplify discretisation of continuous components by bond graph finite-element components.

To that end, a component can be defined that is represented by some number of bond graph finite-element components (Fig. 10.3). Such a component plays the role of a "super element". In same way, a lower-level super-element component can be defined that is composed of such higher-level super elements. This way, a complex model consisting of a large number of finite elements can be constructed easily by using the component copy and insertion operations supported by the program. This technique is similar to finite-element sub-structuring [2].

In the next sections we apply this technique to solve some practical continuous system problems.

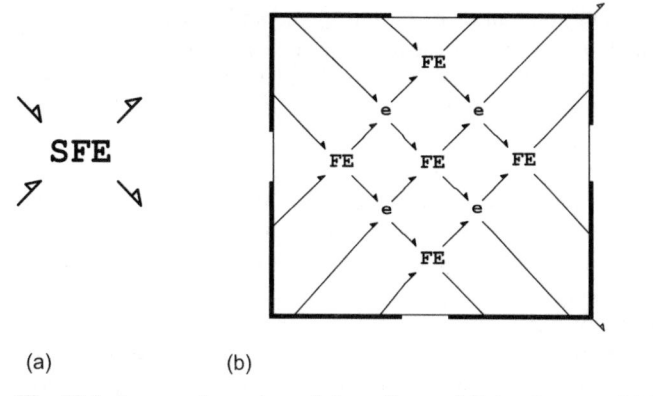

SFE

(a)   (b)

Fig. 10.3. A super element consisting of several finite elements. (a) Component
representation, (b) Structure of the component

10.3 Model of Electric Transmission Line

Electrical transmission lines transmit electromagnetic energy between terminals.
They can be of different forms, such as coaxial cables, parallel wires, strip lines.
They find natural application in telecommunications and electric power engineering, and also as high-speed busses in modern digital computers. There are many
references that treat modelling and analysis of electrical transmission lines [7]. We
will not go into details here, but use some of those results to show how such lines
can be modelled and analysed by bond graphs with BondSim.

A transmission line is characterised by a series resistance R' in ohm/m and an
inductance L' in H/m of both conductors, and by a shunt capacitance C' in C/m and
conductance G' in S/m. A differential element of the transmission line of length dx
is described by the equivalent circuit of Fig 10.4. This is used as a starting point
for deriving the differential equations of the transmission line.

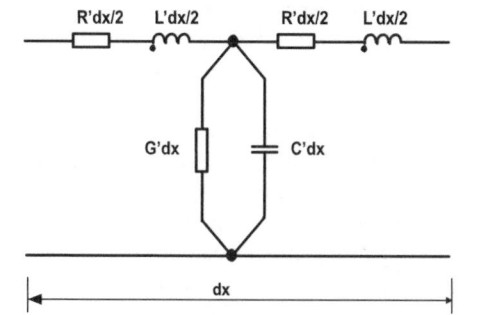

Fig. 10.4. Equivalent model of a transmission line of a length dx

To develop a discrete model of the line, we will approximate a differential line element by a line section of finite length. Such a section is represented by the component in Fig. 10.5a, which is created by the *Transmission Line* icon of the *Electrical Component* palette (accessible from the *Tools* menu). The component name is changed to TSect. Its model is shown in Fig. 10.5b.

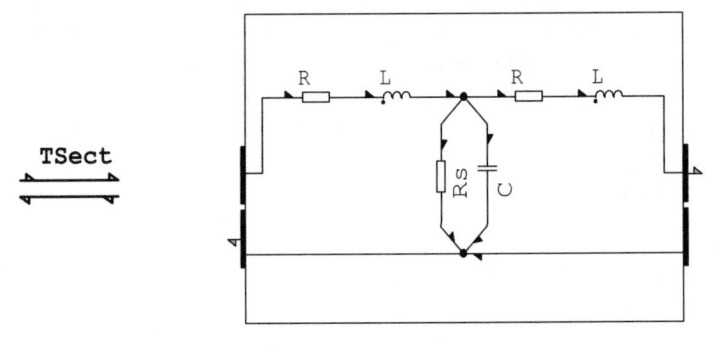

(a)   (b)

Fig. 10.5. Component model of a line section. (a) Representation, (b) Model

The parameters of the model are

1. Section resistances R = R'·Len/2
2. Section inductance L = L'·Len/2
3. Shunt capacitance C = C'·Len
4. Shunt resistance Rs = 1/(G'·Len) or, alternatively, G = G'·Len

where Len is the length of the section.

To simplify generation of a transmission line model consisting of a large number of the sections, we define a component of the same form as in Fig. 10.5a, but denoted as Line, and which consists of five line sections (Fig. 10.6).

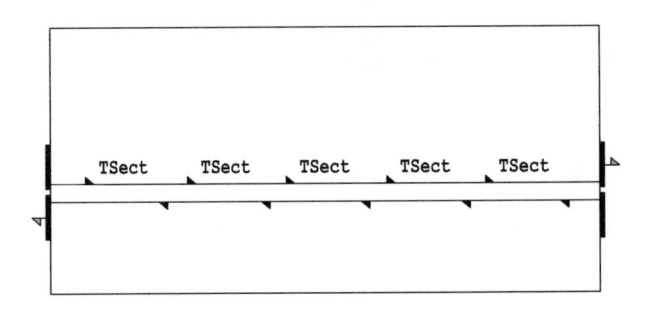

Fig. 10.6. A Line component composed of five line sections

We precede in the same manner by defining a new line component containing, for example, five previously defined Line components. In this way, it is not difficult to create a line component that, in a hierarchical fashion, contains a large number of sections. As the number of sections increases and, hence, their length decreases, the model better approximates the line.

To apply this modelling approach to transmission lines, we analyse the telephone line of [8] and compare the results given therein with those generated by BondSim. The parameters of the line are

1. Length 322 km (200 mi)
2. R' = 0.006304 ohm/m
3. L' = $2.441 \cdot 10^{-6}$ H/m
4. C' = $4.950 \cdot 10^{-12}$ F/m
5. G' = $0.1801 \cdot 10^{-9}$ S/m

The line is divided into ten sections of length 32.2 km (20 mi) each. The parameters of the line segments are

1. R= 101.5 ohm
2. L = 0.0393 H
3. C = $0.159 \cdot 10^{-6}$ F
4. Rs = $0.172 \cdot 10^{6}$ ohm

The line is driven by a sinusoidal voltage source of 1V amplitude, circular frequency of ω = 5000 rad/s, and is loaded by characteristic impedance Z_0.

The model of the Electric Line, including the source and load, is shown in Fig. 10.7. The line is represented by the TL component, consisting of two Line components of Fig. 10.6. Each of the line segments of the 32.2-km length is represented by TSec components of Figs. 10.5 with parameters as given above. We thus have $2 \cdot 5 = 10$ components. We next evaluate the characteristic impedance of the line, which is represented by component Z0.

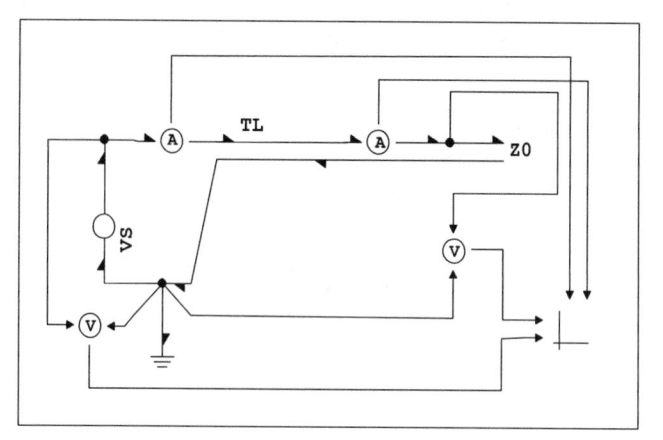

Fig. 10.7. Model of Electric line

The characteristic impedance is given by

$$Z_0 = \sqrt{\frac{R' + j\omega L'}{G' + j\omega C'}}$$ (10.1)

The modulus of the impedance is

$$|Z_0| = \left(\frac{R'^2 + \omega^2 L'^2}{G'^2 + \omega^2 C'^2}\right)^{1/4} = 744.991$$

and its argument is

$$\arg(Z_0) = 0.5 \cdot (\mathrm{a}\tan(\omega L'/R') - \mathrm{a}\tan(\omega C'/G')) = -0.234746 \text{ rad}$$

Hence, the characteristic impedance is

$$Z_0 = 744.991 \cdot e^{-0.2347455} = 724.558 - j173.282$$ (10.2)

This way, the characteristic impedance can be represented by a resistor and a capacitor (Fig. 10.8) with resistance and capacitance, respectively, of

R0 = 724.558 ohm
C0 = $1/(173.282 \cdot 5000) = 1.15419 \cdot 10^{-6}$ F

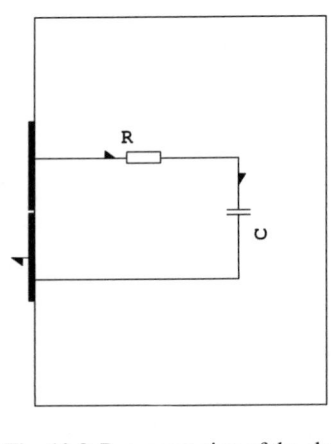

Fig. 10.8. Representation of the characteristic impedance Z0

In the model of Fig. 10.7, an ammeter and a voltmeter are inserted at both the sending and receiving ends of the line to provide output of the current and voltage, respectively. Outputs are fed to an output (display) component at the right. Before proceeding with the simulation, we evaluate another important characteristic of the line, the propagation parameter.

The propagation parameter is given by

$$\gamma = \alpha + j\beta = \sqrt{(R' + j\omega L') \cdot (G' + j\omega C')}$$ (10.3)

The modulus of the propagation factor is

$$|\gamma| = ((R'G' - \omega^2 L'C')^2 + \omega^2(R'C' + G'L')^2)^{1/4} = 1.8439 \cdot 10^{-5} \ 1/m$$

and its argument is

$$\arg(\gamma) = 0.5 \cdot a\tan\left(\frac{\omega(R'C' + G'L')}{R'G' - \omega^2 C'L'}\right) = 1.32954$$

Hence,

$$\gamma = 1.8439 \cdot 10^{-5} e^{j1.32954} = 4.40544 \cdot 10^{-6} + j1.7905 \cdot 10^{-5} \ 1/m$$

This way we get the attenuation factor of the line $\alpha = 4.40544 \cdot 10^{-6}$ 1/m and the phase constant $\beta = 1.7905 \cdot 10^{-5}$ 1/m.

In a transmission line loaded by the characteristic impedance, there are no reflected waves, so voltage amplitudes along the line are given by

$$V(x) = V_s \cdot e^{-\gamma x} \tag{10.4}$$

The amplitude at the receiving end of line x = Len is

$$V_r = V_s \cdot e^{-\alpha Len} = 0.2421 \ V$$

The current at receiving end is

$$I_r = V_r / |Z_0| = 3.249 \cdot 10^{-4} \ A$$

A similar relationship holds for the current amplitudes, i.e.

$$I_r = I_s \cdot e^{-\alpha Len} \tag{10.5}$$

Hence,

$$I_s = I_r \cdot e^{\alpha Len} = 1.342 \cdot 10^{-3} \ A$$

Finally, from Eq. (10.4), we find the time delay for voltage and current waves to reach the other end. This is given by

$$\Delta t = \beta \cdot Len / \omega = 0.001153s$$

Values found above correspond to steady-state sinusoidal wave transmission along the line. The transmission line problem in [8] is based on such steady-state frequency analysis.

The simulation based on the model developed above, on other hand, predicts the complete transient behaviour of the line and thus gives a more complete picture of the processes involved. This is, of course, approximate, as a finite number of lumps are used; but accuracy can be increased by increasing the number of sections.

We simulate processes in the line using a simulation interval of 0.05 s. The period of the supply voltage is $2\pi/\omega = 0.00126$ s. Thus, we use an output interval of 0.00001 s. Results are shown in Figs. 10.9 – 10.11.

There is an initial period of about 0.01 s during which there are transients in the currents and the voltages, before they settle to a steady state. Voltage and current amplitudes at the receiving end are 0.2312 V and 0.3104 mA, respectively. At the sending end, the amplitude of the current is 1.392 mA, as in [8].

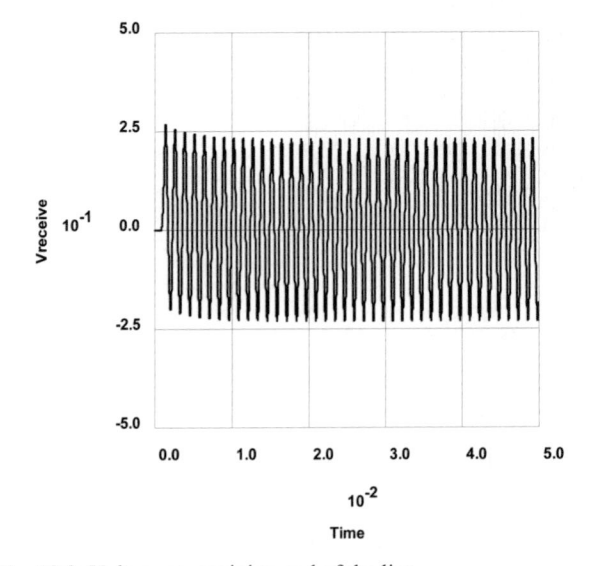

Fig. 10.9. Voltage at receiving end of the line

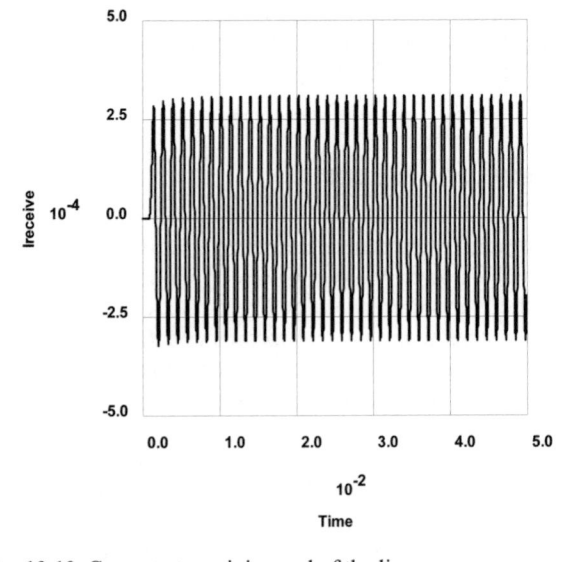

Fig. 10.10. Current at receiving end of the line

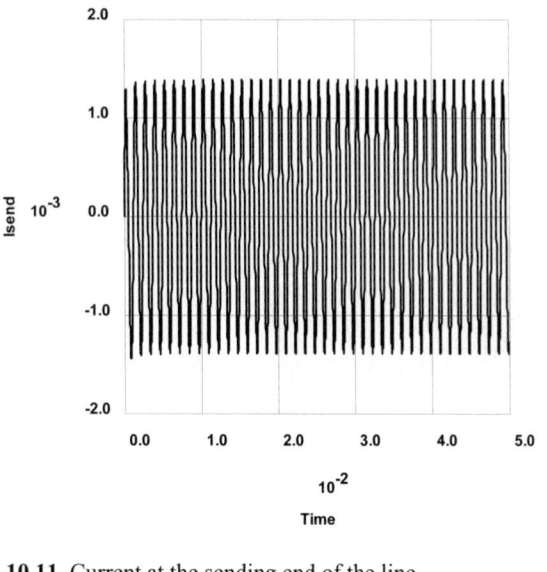

Fig. 10.11. Current at the sending end of the line

Figs. 10.9 and 10.10 show a time delay before the waves reach the other end of the line. It is difficult to determine accurately when current and voltage start to increase from zero. Therefore, the delay is found by the difference in times when the voltage crosses the time axis for the first time. Thus, the delay is 0.001190 s. The values found by simulations are close to those found analytically. Better accuracy can be achieved by increasing the number of sections in which the line is divided. The complete simulation uses 6.11 s of CPU time.

10.4 Bond Graph Model of a Beam

In this section we analyse a simple two-dimensional beam. This is a simple examples of a more general deformable body and often is used to model elastic links in manipulators and other multibody systems. Interested readers are advised to consult any of the numerous references that deal with beams and flexible body dynamics. We point out here [9], which is more or less a standard reference on multibody system dynamics; and [10], because of its elegance and relevance to the component modelling approach used in this book.

We develop a bond graph model of two-dimensional beams using finite-element discretization. The beam will be described using Euler- Bernoulli theory—also known as thin-beam theory—in which shear deformation and rotary inertia are neglected. In spite of its simplicity, it is quite useful for solving various practical problems in flexible body dynamics. We will use it to solve problems of packaging systems vibration testing in Sect. 10.5; and of Coriolis mass-flow meters in Sect. 10.6. Another bond graph approach to Euler-Bernoulli beams is described in [5].

We begin by analysing an element of a beam that moves in a plane with respect to a body frame Oxz, which for this analysis is assumed to be fixed in a base frame (Fig. 10.12a). Fig. 10.12b shows an element of the beam of length L with an attached element frame $O_e x_e z_e$. It will be assumed that the undeformed beam element is straight, having a uniform cross section and parallel to the body frame. Displacement of the element with respect to the body frame is described by two vertical displacements of its ends—w_1 and w_2—and two end slopes, θ_1 and θ_2.

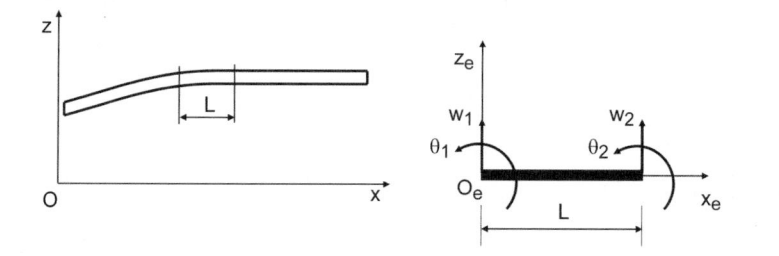

(a)   (b)

Fig. 10.12. A beam in planar motion. (a) An element of the beam, (b) The co-ordinates

A beam element can be represented by a BFE component with two ports, corresponding to the left and the right ends of the element (Fig. 10.13).

$\triangle$ **BFE** $\triangle$

Fig. 10.13. Component model of a beam element

The port flows are represented by vectors composed of linear and angular velocities of element ends

$$\mathbf{f}_1 = \begin{pmatrix} \dot{w}_1 \\ \dot{\theta}_1 \end{pmatrix} \tag{10.6}$$

and

$$\mathbf{f}_2 = \begin{pmatrix} \dot{w}_2 \\ \dot{\theta}_2 \end{pmatrix} \tag{10.7}$$

The effort at the left port consists of force F1 in the transverse direction and moment M1 of action of the left part of the beam on the element

$$\mathbf{e}_1 = \begin{pmatrix} F_1 \\ M_1 \end{pmatrix} \tag{10.8}$$

Similarly, the effort at the right port is composed of a force F2 and a moment M2 of the action of the beam element on the beam part at the right of the element

$$\mathbf{e}_2 = \begin{pmatrix} F_2 \\ M_2 \end{pmatrix} \qquad (10.9)$$

This way, at the right element end there is a reaction force and moment, and power flows through the element from the left to the right port (Fig. 10.13).

In similar way, we introduce generalised displacements of beam element ends

$$\mathbf{q}_1 = \begin{pmatrix} w_1 \\ \theta_1 \end{pmatrix} \qquad (10.10)$$

and

$$\mathbf{q}_2 = \begin{pmatrix} w_2 \\ \theta_2 \end{pmatrix} \qquad (10.11)$$

To simplify model development, we introduce end displacements

$$\mathbf{q} = \begin{pmatrix} \mathbf{q}_1 \\ \mathbf{q}_2 \end{pmatrix} \qquad (10.12)$$

as generalised coordinates of the element. Comparing with Eqs. (10.6) and (10.7), we have

$$\dot{\mathbf{q}} = \begin{pmatrix} \mathbf{f}_1 \\ \mathbf{f}_2 \end{pmatrix} \qquad (10.13)$$

Transverse displacement of a point on the beam axis is described by [2]

$$w = \mathbf{Sq} \qquad (10.14)$$

where $\mathbf{S}$ is the matrix of shape functions

$$\mathbf{S} = \begin{pmatrix} S_1 & S_2 & S_3 & S_4 \end{pmatrix} \qquad (10.15)$$

with

$$\begin{matrix} S_1 = 1 - 3\xi^2 + 2\xi^3 \\ S_2 = L\left(\xi - 2\xi^2 + \xi^3\right) \\ S_3 = 3\xi - 2\xi^3 \\ S_4 = L\left(-\xi^2 + \xi^3\right) \end{matrix} \qquad (10.16)$$

and $\xi = x_e/L$. Differentiating Eq. (10.14) with respect to time, we get the velocity of the beam transverse displacement

$$v = \mathbf{S} \begin{pmatrix} \mathbf{f}_1 \\ \mathbf{f}_2 \end{pmatrix} \qquad (10.17)$$

Next, we develop the dynamic equation of element motion. A convenient way to do this is by using the Lagrange equations

$$\frac{d}{dt}\left(\frac{\partial T}{\partial \dot{\mathbf{q}}}\right) - \left(\frac{\partial T}{\partial \mathbf{q}}\right) = \mathbf{Q} \tag{10.18}$$

where T is kinetic energy of element and $\mathbf{Q}$ is a vector of generalised forces. The generalised forces include effects of element elastic forces, as well as those of forces at the element boundary, i.e. at the ports.

The kinetic energy of a beam element is given by

$$T = \frac{1}{2}\int_0^L \rho A v^2 dx \tag{10.19}$$

where ρ is the mass density of the beam, and A is the element's cross-sectional area. Using Eq. (10.17), we get

$$v^2 = \begin{pmatrix} \mathbf{f}_1^T & \mathbf{f}_2^T \end{pmatrix} \mathbf{S}^T \mathbf{S} \begin{pmatrix} \mathbf{f}_1 \\ \mathbf{f}_2 \end{pmatrix} \tag{10.20}$$

By substituting in Eq. (10.19), the kinetic energy function can be written as

$$T = \frac{1}{2}\begin{pmatrix} \mathbf{f}_1^T & \mathbf{f}_2^T \end{pmatrix} \mathbf{M} \begin{pmatrix} \mathbf{f}_1 \\ \mathbf{f}_2 \end{pmatrix} \tag{10.21}$$

where $\mathbf{M}$ is the consistent mass matrix of the beam element. It is given by

$$\mathbf{M} = m\int_0^1 \mathbf{S}^T \mathbf{S} \, d\xi \tag{10.22}$$

where $m = \rho A L$ is the element mass. Using the shape functions of Eq. (10.16) and evaluating integrals in Eq. (10.22), the mass matrix can be written as [2]

$$\mathbf{M} = \frac{m}{420}\begin{pmatrix} 156 & 22L & 54 & -13L \\ 22L & 4L^2 & 13L & -3L^2 \\ 54 & 13L & 156 & -22L \\ -13L & -3L^2 & -22L & 4L^2 \end{pmatrix} \tag{10.23}$$

We also can represent the mass matrix of Eq. (10.23) in block form

$$\mathbf{M} = \begin{pmatrix} \mathbf{M}_{11} & \mathbf{M}_{12} \\ \mathbf{M}_{21} & \mathbf{M}_{22} \end{pmatrix} \tag{10.24}$$

where 2×2 blocks are found easily by inspection of Eq. (10.23)

If we denote the momentum of the beam element by

$$\mathbf{p} = \partial T / \partial \dot{\mathbf{q}} \tag{10.25}$$

we get, from Eqs. (10.13) and (10.21),

$$\mathbf{p} = \mathbf{M}\begin{pmatrix} \mathbf{f}_1 \\ \mathbf{f}_2 \end{pmatrix} \tag{10.26}$$

Thus,

$$\frac{d}{dt}\left(\frac{\partial T}{\partial \dot{\mathbf{q}}}\right) = \dot{\mathbf{p}} \tag{10.27}$$

In addition, $\partial T/\partial \mathbf{q} = 0$, because the kinetic energy depends only on the velocities.

We now turn to the right side of Eq. (10.18). The part of the generalised forces owed to elastic forces can be found from the strain energy V of the element. The other part is given by efforts at the element ends (ports). Thus, we have

$$\mathbf{Q} = -\frac{\partial V}{\partial \mathbf{q}} + \begin{pmatrix} \mathbf{e}_1 \\ -\mathbf{e}_2 \end{pmatrix} \tag{10.28}$$

Using simple beam theory, the beam element strain energy is given by

$$V = \frac{1}{2}\int_0^L EI\left(\frac{d^2w}{dx^2}\right)^2 dx \tag{10.29}$$

where E is the Young modulus of elasticity and I is the second moment of the beam section. Substituting from Eq. (10.14), we find

$$V = \frac{1}{2}\mathbf{q}^T\mathbf{K}\mathbf{q} \tag{10.30}$$

where $\mathbf{K}$ is the stiffness matrix, given by

$$\mathbf{K} = \frac{EI}{L^3}\int_0^1 \frac{d^2\mathbf{S}^T}{d\xi^2}\frac{d^2\mathbf{S}}{d\xi^2}d\xi \tag{10.31}$$

Using shape function of Eq. (10.16) and evaluating integrals, we get [2]

$$\mathbf{K} = \frac{EI}{L^3}\begin{pmatrix} 12 & 6L & -12 & 6L \\ 6L & 4L^2 & -6L & 2L^2 \\ -12 & -6L & 12 & -6L \\ 6L & 2L^2 & -6L & 4L^2 \end{pmatrix} \tag{10.32}$$

This matrix also can be represented in 2×2 block form

$$\mathbf{K} = \begin{pmatrix} \mathbf{K}_{11} & \mathbf{K}_{12} \\ \mathbf{K}_{21} & \mathbf{K}_{22} \end{pmatrix} \tag{10.33}$$

Now, from Eq. (10.30)

$$\frac{\partial V}{\partial \mathbf{q}} = \mathbf{K}\mathbf{q} \tag{10.34}$$

After substituting in Eq. (10.28), we have

$$\mathbf{Q} = -\mathbf{K}\begin{pmatrix} \mathbf{q}_1 \\ \mathbf{q}_2 \end{pmatrix} + \begin{pmatrix} \mathbf{e}_1 \\ -\mathbf{e}_2 \end{pmatrix} \tag{10.35}$$

Finally, substituting from Eq. (10.25) and (10.35) into Eq.(10.18), we get the governing equation of beam element dynamics

$$\frac{d\mathbf{p}}{dt} + \mathbf{K}\begin{pmatrix} \mathbf{q}_1 \\ \mathbf{q}_2 \end{pmatrix} = \begin{pmatrix} \mathbf{e}_1 \\ -\mathbf{e}_2 \end{pmatrix} \tag{10.36}$$

We further modify this equation by introducing beam damping

$$\frac{d\mathbf{p}}{dt} + \mathbf{R}\begin{pmatrix} \mathbf{f}_1 \\ \mathbf{f}_2 \end{pmatrix} + \mathbf{K}\begin{pmatrix} \mathbf{q}_1 \\ \mathbf{q}_2 \end{pmatrix} = \begin{pmatrix} \mathbf{e}_1 \\ -\mathbf{e}_2 \end{pmatrix} \tag{10.37}$$

The damping matrix $\mathbf{R}$ is defined by the relation

$$\mathbf{R} = \alpha\mathbf{M} + \beta\mathbf{K} \tag{10.38}$$

where α and β are suitable constants. This type of damping is known as Rayleigh damping [2].

The equation of element motion, as given by Eqs. (10.26) and (10.37), can be readily represented by a bond graph model of the BFE component shown in Fig. 10.14.

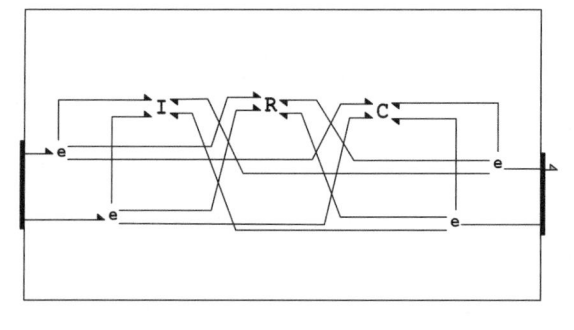

Fig. 10.14. Bond graph beam element BFE

Note that the first row of these equations corresponds to the left port variables, and the second row to those of the right port. The variable at one effort junction at the left component port is the linear velocity; at the other, it is the angular velocity. A similar situation holds for the right effort junctions. The inertia of the beam element is represented by a four-port inertial element I, corresponding to Eq. (10.26) and the first term in Eq. (10.37). Elasticity of the element is represented by a four-port capacitive element C, corresponding to the third term of Eq. (10.37) and Eq. (10.13). Finally, a four-port resistive element R describes Rayleigh type damping in the beam, as given by the second terms of Eq. (10.37) and of Eq. (10.38).

To complete the model, it is necessary to define model parameters and element constitutive relations. Model parameters can be defined conveniently at level of beam element document. We define the physical parameters first. These include the element length, mass, modulus of elasticity. We then define expressions for mass, stiffness, and damping matrix elements. The constitutive relations of I, C, and R elements are defined at corresponding ports as given by Eqs. (10.26) and (10.37), respectively.

The model of a beam is created using component models that correspond to the elements into which the beam is divided. To simplify generating models that consist of a large number of elements, a technique similar to that used in Sect. 10.3 can be applied. Thus, we define a component consisting of five element component models; then another consisting of the five previously defined components; and so forth. This way, using hierarchical decomposition of the beam, it is relatively easy to generate a beam model consisting of a large number of finite element component models by copying and inserting. We apply this technique in the next sections.

10.5 A Packaging System Analysis

10.5.1 Description of the Problem

A product-package system typically consists of an outer container, the cushion, the product, and a critical element. The critical element is the most fragile component of the product (e.g. electric boards). It is the part that is most easily damaged by a mechanical shock or by vibrations. The goal in distribution packaging (transport of packaging) is to provide a correct design for packaging so that its contents arrive safely at its destination. Vibration is associated with all transportation modes, although each mode has its own characteristic frequencies and amplitudes. The most troublesome frequencies are below 30 Hz because they are most prevalent in vehicles and it is difficult to isolate products from them [11].

In a recent thesis [11], a simple packaging system is designed consisting of container body with a cantilever beam as the critical element carrying a device at its end (Fig. 10.15). The package container plate was fixed on the table of a testing machine used for vibration testing of the package. One of the thesis goals was to predict the system's behaviour under vibration testing by simulation based on a model of the package system. The model of the package was developed using the bond graph methodology described in this book together with BondSim.

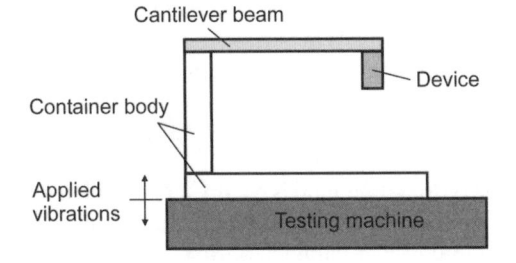

Fig. 10.15. The packaging system subjected to vibration testing

This section is partially based on this work. The goal is to show how the bond graph model of typical vibration testing set-up can be developed in a systematic way within the BondSim programming environment. Using basic data from [11], a frequency response of the system will be generated to simulate a typical laboratory vibration testing procedure.

10.5.2 Bond Graph Model Development

The package system model will be developed using an approach similar to the floating frame approach of [9]. The coordinate frames used for description of system motion are shown in Fig. 10.16. The critical part of the system is a beam clamped to a container body that is parallel to the container base plate.

The motion of the complete system is described with respect to the reference coordinate frame $O_r x_r z_r$ fixed to the ground (the testing machine body). Next, the container body frame $O_b x_b z_b$ is defined with axes parallel to that of the base frame, and that translates vertically with respect to the former.

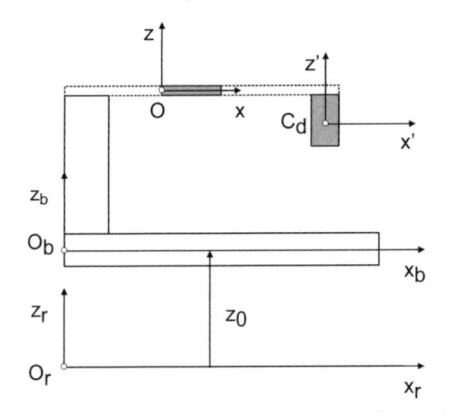

Fig. 10.16. Co-ordinate systems of the package

The beam is modelled as an assemblage of the Euler-Bernoulli beam elements of Sect. 10.4. Fig. 10.16 shows a typical beam element with the attached element frame. In the initial (undeformed) position, the element frames are parallel to the container body frame. There is also a device body frame $C_d x' z'$ attached to the body of the device at its mass centre and which moves with it.

The motion of the body frame is described by coordinate z_0, given as a function of time $z_0 = z_0(t)$. This describes motion of the package during vibration testing. We later will specialise it to two common forms, impulse and sinusoidal. Other forms could be used, as well.

Model of Beam Elements

Beam elements will be represented using the component model of Sec. 3.4. In this problem the elements frames are not fixed with respect to a global reference frame, but translate jointly with the container body frame. Thus, in addition to the generalised displacements of the element with respect to the container frame, the matrix of generalised displacements also will contain the z_0 co-ordinate of the container body frame. The generalised displacement matrix of a typical element thus reads

$$\mathbf{q} = \begin{pmatrix} z_0 \\ \mathbf{q}_1 \\ \mathbf{q}_2 \end{pmatrix} \tag{10.39}$$

where $\mathbf{q}_1$ and $\mathbf{q}_2$ are defined by Eqs. (10.10) and (10.11), respectively.

Transverse displacement of the beam axis with respect to the base frame is given by

$$z = z_0 + \mathbf{S} \begin{pmatrix} \mathbf{q}_1 \\ \mathbf{q}_2 \end{pmatrix} \tag{10.40}$$

where $\mathbf{S}$ is the shaping function matrix of Eqs. (10.15) and (10.16). Corresponding velocities are

$$v = v_0 + \mathbf{S}\mathbf{f} \tag{10.41}$$

where $\mathbf{f}$ is vector of flows (see Eq. (10.13)) and

$$v_0 = \frac{dz_0}{dt} \tag{10.42}$$

To develop the equation for beam element motion, Lagrange's equation are used, as in Sect. 10.4. These read

$$\frac{d}{dt}\left(\frac{\partial T}{\partial \dot{\mathbf{q}}}\right) - \left(\frac{\partial T}{\partial \mathbf{q}}\right) = \mathbf{Q} \tag{10.43}$$

where T is the kinetic energy of the element and $\mathbf{Q}$ is vector of generalised forces.

The generalised forces now read (see Eqs. (10.33) and (10.35))

$$\mathbf{Q} = - \begin{pmatrix} 0 & 0 & 0 \\ 0 & \mathbf{K}_{11} & \mathbf{K}_{12} \\ 0 & \mathbf{K}_{21} & \mathbf{K}_{22} \end{pmatrix} \begin{pmatrix} z_0 \\ \mathbf{q}_1 \\ \mathbf{q}_2 \end{pmatrix} + \begin{pmatrix} F_0 \\ e_1 \\ -e_2 \end{pmatrix} \tag{10.44}$$

where $\mathbf{K}$ is the element stiffness matrix and F_0 is a force corresponding to the effect of container body translation on the element.

The kinetic energy of a beam element is given by

$$T = \frac{1}{2} \int_0^L \rho A v^2 dx \tag{10.45}$$

where ρ is mass density of the beam element, and A is element's cross-sectional area. Using Eq. (10.41) we get

$$v^2 = (v_0 + \mathbf{f}^T\mathbf{S}^T)(v_0 + \mathbf{S}\mathbf{f}) = v_0^2 + 2v_0\mathbf{S}\mathbf{f} + \mathbf{f}^T\mathbf{S}^T\mathbf{S}\mathbf{f} \qquad (10.46)$$

Where superscript T denotes the transposition operation. By substituting Eq. (10.46) into Eq. (10.45), the kinetic energy expression can be written as

$$T = \frac{1}{2}mv_0^2 + v_0\mathbf{N}\begin{pmatrix}\mathbf{f}_1\\\mathbf{f}_2\end{pmatrix} + \frac{1}{2}\begin{pmatrix}\mathbf{f}_1^T & \mathbf{f}_2^T\end{pmatrix}\mathbf{M}\begin{pmatrix}\mathbf{f}_1\\\mathbf{f}_2\end{pmatrix} \qquad (10.47)$$

Where m is the element mass, $\mathbf{M}$ is element consistent mass matrix given by Eq. (10.22), and

$$\mathbf{N} = m\int_0^1 \mathbf{S}d\xi \qquad (10.48)$$

Using the shape functions of Eqs. (10.15) and (10.16), we get

$$\mathbf{N} = \frac{m}{12}(6 \quad L \quad 6 \quad -L) \qquad (10.49)$$

This matrix can be written as 1×2 block matrix

$$\mathbf{N} = (\mathbf{N}_1 \ \mathbf{N}_2) \qquad (10.50)$$

Next, we define momentum of the beam element by

$$\mathbf{p} = \partial T / \partial \dot{\mathbf{q}} \qquad (10.51)$$

From Eq. (10.47) we get

$$\mathbf{p} = \begin{pmatrix} m & \mathbf{N}_1 & \mathbf{N}_2 \\ \mathbf{N}_1^T & \mathbf{M}_{11} & \mathbf{M}_{12} \\ \mathbf{N}_2^T & \mathbf{M}_{21} & \mathbf{M}_{22} \end{pmatrix}\begin{pmatrix} v_0 \\ \mathbf{f}_1 \\ \mathbf{f}_2 \end{pmatrix} \qquad (10.52)$$

Thus, by substituting from Eqs. (10.51), and (10.44) into Eq. (10.43), the equation of beam element motion can be written as

$$\frac{d\mathbf{p}}{dt} + \begin{pmatrix} 0 & 0 & 0 \\ 0 & \mathbf{K}_{11} & \mathbf{K}_{12} \\ 0 & \mathbf{K}_{21} & \mathbf{K}_{22} \end{pmatrix}\begin{pmatrix} z_0 \\ q_1 \\ q_2 \end{pmatrix} = \begin{pmatrix} F_0 \\ e_1 \\ -e_2 \end{pmatrix} \qquad (10.53)$$

We modify this equation further by introducing beam damping

$$\frac{d\mathbf{p}}{dt} + \begin{pmatrix} 0 & 0 & 0 \\ 0 & \mathbf{R}_{11} & \mathbf{R}_{12} \\ 0 & \mathbf{R}_{21} & \mathbf{R}_{22} \end{pmatrix}\begin{pmatrix} v_0 \\ \mathbf{f}_1 \\ \mathbf{f}_2 \end{pmatrix} + \begin{pmatrix} 0 & 0 & 0 \\ 0 & \mathbf{K}_{11} & \mathbf{K}_{12} \\ 0 & \mathbf{K}_{21} & \mathbf{K}_{22} \end{pmatrix}\begin{pmatrix} z_0 \\ q_1 \\ q_2 \end{pmatrix} = \begin{pmatrix} F_0 \\ e_1 \\ -e_2 \end{pmatrix} \qquad (10.54)$$

where $\mathbf{R}$ is defined by Eq. (10.38)

Comparing Eqs. (10.54) and (10.52) to Eqs. (10.37) and (10.26), we conclude that the same component model as in Fig. 10.14 can be used with the added port corresponding to the body frame translation effect (Fig. 10.17). However, the constitutive relation at the inertial elements ports should be changed to the form given by Eq. (10.52). This additional port shows that there is transfer of power, because of the motion of the container body frame.

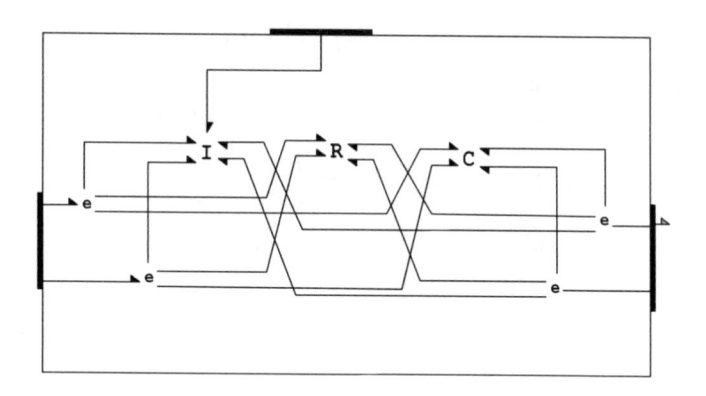

Fig. 10.17. Model of the package beam element

Motion of The Device

To model motion of the device, the rigid body model of Sect. 9.2 is used. There is some difference, however. Motion of the device is referred to the container body frame, not to the base (inertial) frame. Because the container body translates with respect to the base, there is an additional (inertial) force acting at the device mass centre, as was found above for the beam elements. Thus, in addition to the port corresponding to joining the end of the cantilever beam, there is a port that corresponds to the effects of translation of the container body frame (Fig. 10.18). The effort and flow of the container port are inertial force—owing to the container frame acceleration—and velocity of the container frame translation, respectively. The efforts and flows at the beam ports are of the form given by Eqs. (10.9) and (10.7), respectively. It was assumed that the beam axis undergoes z-axis transverse displacement only, ignoring the beam x-axis strains. There is also an output port to supply information on the relative transverse displacement of the device mass centre with respect to the container body frame.

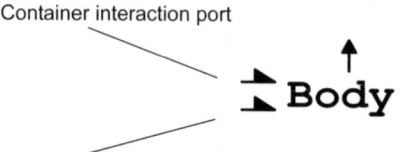

Container interaction port

Beam interaction port

Fig. 10. 18. Representation of device body in a translating frame

The model of the device body created using the plane body model of Fig. 9.3 is shown in Fig. 10.19. The effort junction at the top-left of the document window corresponds to the transverse velocity component of the device mass centre with respect to the base frame.

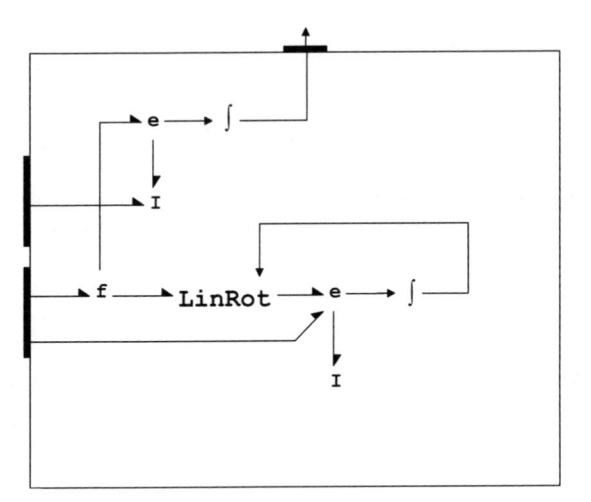

Fig. 10.19. Component model of the device body

This velocity can be expressed as

$$v_d = v_0 + v_d^b \tag{10.55}$$

where the term with the superscript b denotes the velocity component with respect to the container body frame, and v_0 is its translation velocity. Thus, the kinetic energy of the device translation is given by

$$T_{Cd} = m_d(v_0 + v_d^b)^2 / 2 \tag{10.56}$$

Analogous to the approach used in the beam element model, we define the flow vector corresponding to the mass centre motion as

$$\mathbf{f}_d = \begin{pmatrix} v_0 \\ v_d^b \end{pmatrix} \tag{10.57}$$

The translation momentum of the device, defined by

$$\mathbf{p}_d = \partial T_{Cd} / \partial \mathbf{f}_d \tag{10.58}$$

now reads

$$\mathbf{p}_d = \begin{pmatrix} m_d & m_d \\ m_d & m_d \end{pmatrix} \mathbf{f}_d \tag{10.59}$$

This way, the device translation dynamics is given by

$$\dot{\mathbf{p}}_d = \mathbf{e}_d \tag{10.60}$$

where effort $\mathbf{e}_d$ consists of forces at the component ports

$$\mathbf{e_d} = \begin{pmatrix} F_0 \\ F_d \end{pmatrix} \tag{10.61}$$

Following Eqs. (10.59) and (10.60), the device translation dynamic is represented in Fig. 10.19 with a two-port inertial element, where one port corresponds to interaction with the moving container body and the other to that with the beam end to which it is connected. Because the device is clamped to the cantilever, its angular velocity is equal to the angular velocity of beam cross section.

Model of the system

We now give a short description of the model of package system vibration testing. The complete model can be accessed in the BondSim program library under the project name Package Vibration Testing. The system level model is given in Fig. 10.20.

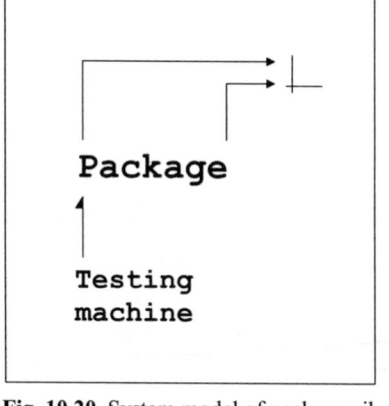

Fig. 10.20. System model of package vibration testing

It consists of two main components: Package and Testing machine. There is also an output component used to display time and frequency plots of the package container and device body positions. The Testing machine is represented simply by a flow source generating vertical motion of the machine table. We will use two types of test inputs. Similarly, as in Sect. 6.2.3, to generate a frequency response plot of device motion, we apply to the package a velocity pulse of short duration. The form of the pulse is shown in Fig. 10.21a. The corresponding package container position is shown in Fig. 10.21b.

Another type input is a sinusoidal function of frequency f

$$v_0 = v_{p0} \sin(2\pi f t) \tag{10.62}$$

Amplitude v_{p0} is chosen such that the peak acceleration is constant, i.e.

$$v_{p0} = \frac{a_{max}}{2\pi f} \tag{10.63}$$

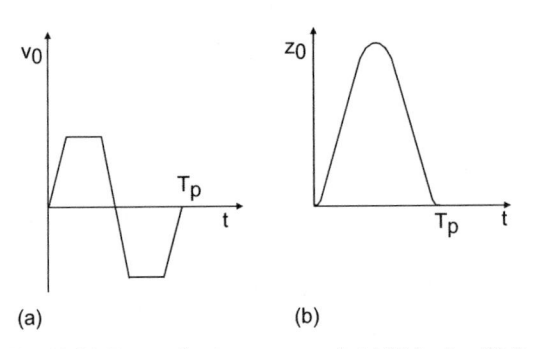

Fig. 10.21. Form of pulses generated. (a) Velocity, (b) Position

The package consists of container wall and base plate represented by the **Wall** component (Fig. 10.22), fixed to the vibration machine table and to which is connected a **Cantilever** component that carries component **Device**.

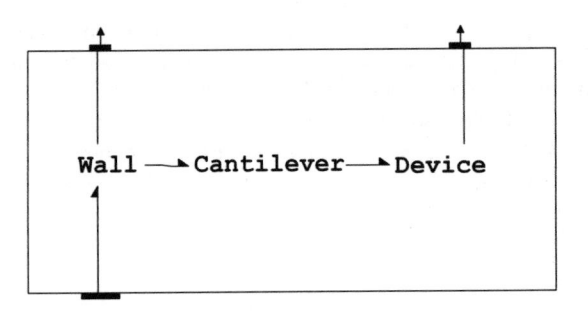

Fig. 10.22. The basic structure of the package system

Interconnection of the cantilever beam to the container is shown in Fig. 10.23. The effort junction is the transverse displacement node of the container. The junction port, together with the **Clamp** port, is connected to the **Wall** output port. The port—a compounded one—has efforts and flows at the container wall represented by

$$\mathbf{e}_{con} = \begin{pmatrix} F_0 \\ \mathbf{e}_1 \end{pmatrix} \tag{10.64}$$

and

$$\mathbf{f}_{con} = \begin{pmatrix} v_0 \\ \mathbf{f}_1 \end{pmatrix} \tag{10.65}$$

Here, v_0 and F_0 are, respectively, its velocity and the total force that the testing machine supplies to the container. Effort component $\mathbf{e}_1$ consists of the transverse force and moment at the clamping of the cantilever beam to the container. The component in the beam axis direction was neglected. Similarly, $\mathbf{f}_1$ consists of the linear velocity and the angular velocity of the beam end with respect to the container.

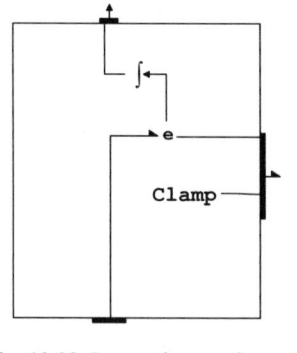

Fig. 10.23. Interactions at the container wall

The end of the cantilever is clamped to the container wall, which is represented by Clamp component. This component consists of two zero-flow sources that force the linear velocity and the angular velocity of the beam end to zero.

The cantilever beam is divided into four subsections (Fig. 10.24), each of which is represented by a SectB component consisting of five BFE components (Fig. 10.25). The bond graph representation of this finite-element beam discretization is depicted in Fig. 10.17. Overall, the cantilever is divided into 20 finite element components.

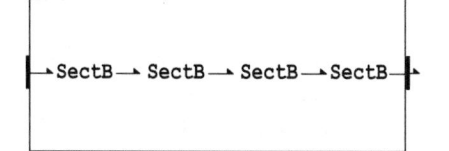

Fig. 10.24. Cantilever beam represented by four beam sections

The effort junctions appearing above each BFE component are the finite-element's translation velocity nodes. The effort variables at the ports connected to the BFE components are forces corresponding to the distributed inertia of the element translation. All such forces, including that of the device, give the total force generated by the Test machine. This way, the mechanical effect of the machine on the beam elements is clearly visible

Finally, returning to Fig. 10.22, there is a component Device that models the device at the cantilever end. This consists of the Body component of Figs. 10.18 and 10.19. It was not possible to use this model directly, owing to the way in which the ports were compounded.

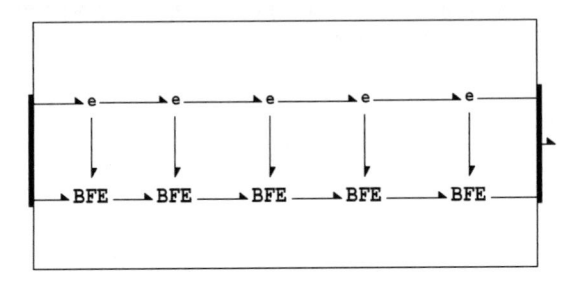

Fig. 10.25. Beam section as a super element of five BFEs

10.5.3 Evaluation of Vibration Test Characteristics

Now we can simulate the vibration characteristics of the package. The parameters used for simulation are given in Table 10.1. The damping parameters α and β of the beam model (see Eq. (10.38)) were selected such that they correspond approximately to 5% of the critical damping ratio in the range of natural frequencies of interest 20–50 Hz, e.g. $\alpha = 8$ and $\beta = 4 \cdot 10^{-4}$.

Table 10.1. Parameters of the package

Property		Value
Cantilever	Length	0.100 m
	Width	0.0258 m
	Height	0.0055 m
Material		Acrylic (PMMA)
Modulus of elasticity		3.1 GPa
Density		1200 kg/m3
Device block	Mass	0.0362 kg
	Moment of inertia	$5.168 \cdot 10^{-6}$ kg·m2
	x	- 0.0128 m
	z	0.0126 m

Note: The device body data include the part of the beam clamped to it. x and y are co-
       ordinates of the mid-point of the cantilever beam end section with respect to the de-
       vice centroidal body frame.

We first generate a frequency response diagram for the device body transverse motion. Hence, the package is subjected to a velocity pulse generated by the Test machine component according to Fig. 10.21. The strength of the pulse is 0.5 m/s and its duration is Tp = 0.001 s. The system settle-time is about 0.5 s. As noted in Sect. 6.2.3, we must append zeros to the response to create a frequency response. Thus, a simulation interval of 1 s is taken along with a fairly small output interval of 0.0001s. The rather tight error tolerances of 10^{-8} were selected. The transient of the device body position (with respect to the container) is shown in Fig. 10.26. At the beginning, the device experiences a relatively large initial jump in the negative direction, and then settles down to zero after about 0.3 s.

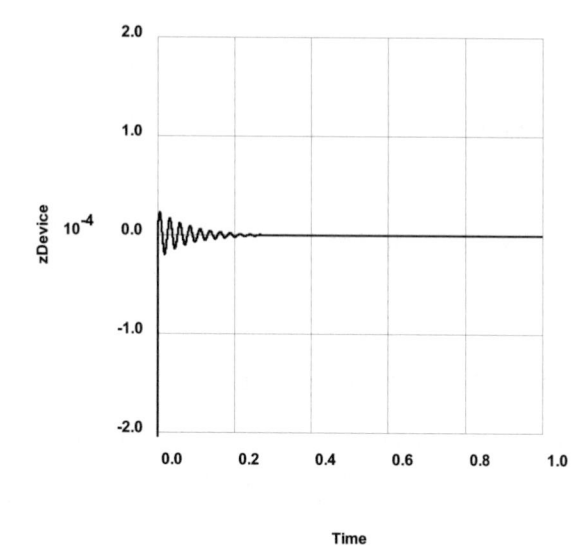

Fig. 10.26. The response to the velocity pulse input

The data gathered during the simulation are used to generate a frequency response plot by applying the Fast Fourier Transform (FFT). As explained in Sect. 6.2.3, this is done by right-clicking the time plot and selecting *Frequency Response* from the drop-down menu. We enlarge the part of the plot in the range of 0 to 100 Hz by right-clicking on the frequency plot, selecting the *Expand* command, and setting maximum frequency to 100 Hz. The resulting diagram is shown in Fig. 10.27.

The diagram resembles the familiar response diagrams of single-degree-of-freedom system vibrations induced by the motion of the base [12]. The maximum amplitude occurs in vicinity of 39.0 Hz. In a similar way, we can find the FFT of the input (Fig. 10.28). The amplitude of container displacement is practically constant over the frequency range of interest at about $1.0 \cdot 10^{-7}$ m and, hence, it approximates the ideal impulse fairly well. Comparing this with Fig. 10.27, it can be seen that the maximum displacement transmissibility of the device to the container ratio positions is a little above 10.

We next apply a sinusoidal vibration to the container according to Eqs. (10.62) and (10.63) with a frequency of 39.0 Hz and a peak acceleration of $a_{max} = 0.5$ g. This requires changing the source effort relation in the Testing machine (Fig. 10.20) component, then re-building the model. The response curve is shown in Fig. 10.29. It was generated with same simulation parameters as in the previous run. The amplitudes at beginning steadily grow until they reach a steady value of 0.0008375 m. The amplitude of the container vibration is $8.170 \cdot 10^{-5}$ m; thus, we have the transmissibility ratio of 10.25, as found previously.

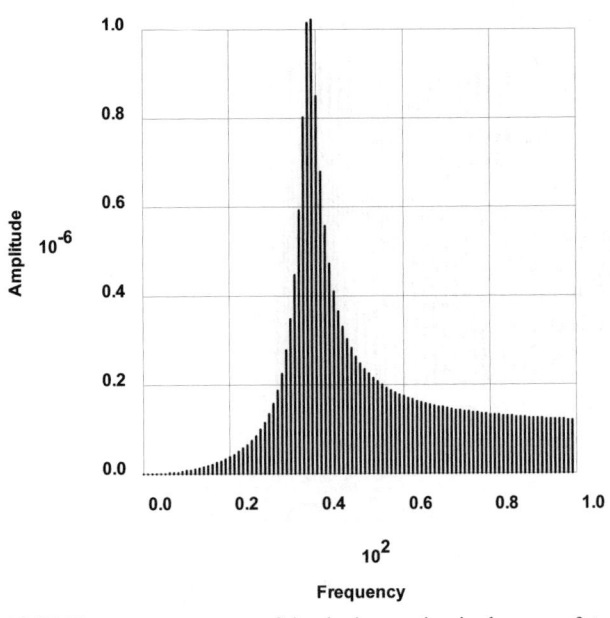

Fig. 10.27. Frequency response of the device motion in the range 0 to 100 Hz

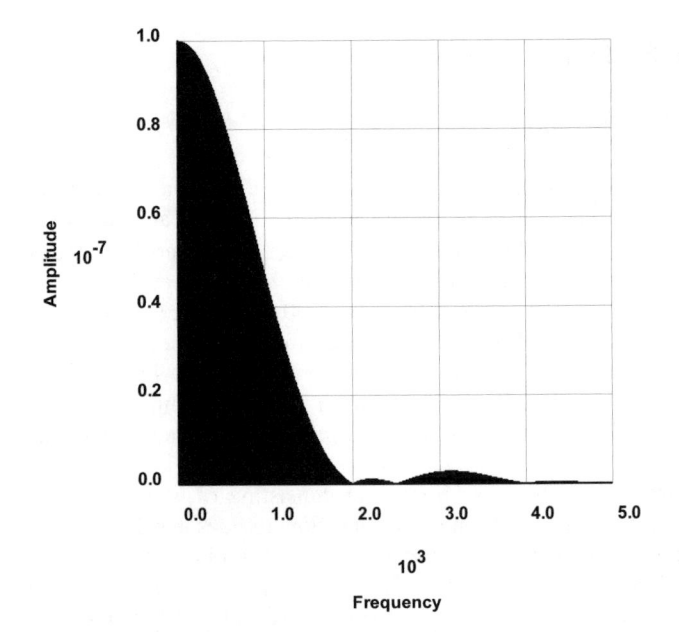

Fig. 10.28. Amplitude-frequency diagram of impulsive displacement of the container

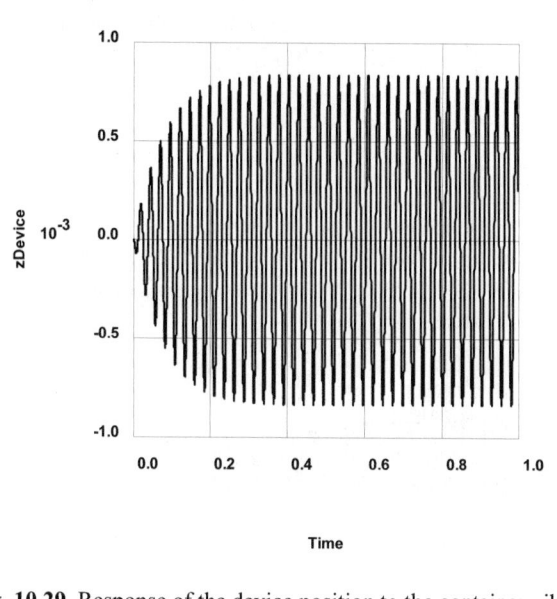

Fig. 10.29. Response of the device position to the container vibration at 39 Hz

10.6 Coriolis Mass Flowmeter

10.6.1 Principle of Operation

The Coriolis mass flowmeter (CMF) is widely used for direct mass flowmetering in the petrol, food, pharmaceutical, and chemical industries. The accuracy of typical commercial CMFs is ±0.20% of the flow rate, practically over the range from zero and up [13]. The range of fluid densities that can be measured varies widely, running from about 100 kg/m^3 or less to over 3000 kg/m^3; certain flowmeters even can be used for gases. CMFs are highly insensitive to temperature and pressure variations. They are very accurate and reliable instruments. The principle of their operation is rather simple. We will explain it with an example of a straight parallel tube CMF. There are other configurations as well, but the basic principles are the same.

Fig. 10.30 shows a Coriolis mass flowmeter consisting of two parallel tubes hydraulically connected in parallel. The incoming flow divides and flows into two separate and practically identical tubes; the flows again combine at their exits. The tubes' ends are clamped to the meter body. They are excited by a drive—located at their middle—which vibrates in their first vibration mode. On the left and the right of the driver are two sensors that detect displacements between the tubes. They are used to measure the difference between times when the tubes cross their middle positions (time delay). As a rule, CMFs use two tubes, the better to minimise the effects of vibration associated with the environment in which the meter is installed for operation.

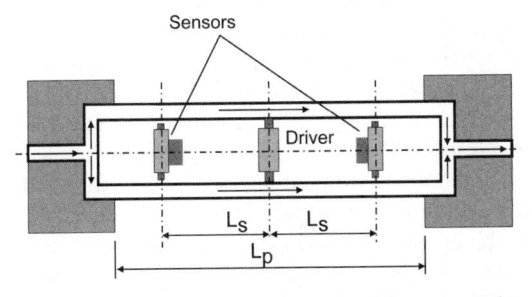

Fig. 10.30. Coriolis mass flowmeter with two straight parallel tubes

To explain CMF operation, Fig. 10.31 shows the exaggerated displacement of one of the tubes moving up with fluid flowing through it. Two small elements of the tube are shown at symmetrical positions to the left and the right of the driver.

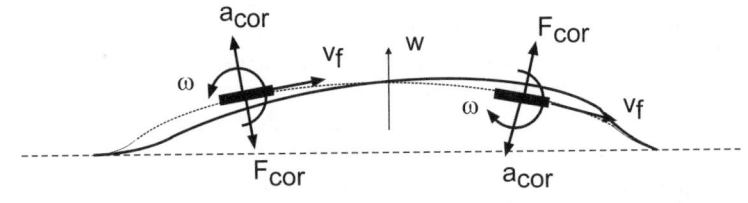

Fig. 10.31. Principle of flowmeter operation

During vibration, tube elements rotate with angular velocity ω and, because of the relative velocity v_f of the fluid flowing through the tube, there is a Coriolis acceleration given by $a_{cor} = 2v_f\omega$. As a result, there is an apparent (inertial) force on the tube element of the opposite sense. That is, Coriolis force $F_{cor} = dm \cdot a_{cor}$, where dm is the mass of the fluid contained in a tube element. If dL is the length of the tube element, the fluid mass contained in the element is equal $dm = \rho A_s dL$, where ρ is fluid density and A_s is the area of the wetted section of the tube. Thus, the Coriolis force on the element is $F_{cor} = 2\rho A_s v_f \omega dL$. Hence, because the mass flow rate through the tube is $Q_m = \rho A_s v_f$, the Coriolis force is $F_{cor} = 2 Q_m \omega dL$. That is, the Coriolis force is proportional to the mass flow rate. Becasuse it appears as a load distributed along the tube, the tube elastically deforms as shown in Fig. 10.31. Note that because the Coriolis forces on the left and right parts of the tube are of opposite directions, the left and the right tube parts deform differently. Thus, tube sections in the planes of the sensors don't cross the zero positions at the same instants of time; rather, there is a time delay. As the analysis of [14] shows, the time delay is proportional to the mass flow rate. Hence, by evaluating the time delay, accurate information on the mass flow rate can be generated. This is not a technically easy problem: The time delays are of the order of microseconds, and their accurate measurement is critical. CMF manufacturers are careful not to expose many of the details of how this is really done.

Mathematical modelling of Coriolis mass flowmeters has attracted the attention of both academic and industry researchers. The common objective is to explain more precisely the operation of the meters, the better to improve their design, testing, and application. The main focus has been evaluation of CMF sensitivity factors and the effects of meter geometry and environment conditions on performance. Different approaches have been used. Thus, in [15], rather simple lumped-parameter models were used. In [14], the tubes are modelled as Euler – Bernoulli beams and problems were solved by perturbation theory. Timoshenko beam theory was applied in [16], and the problem was analysed using a finite-element program. CMF analysis based on curved beam theory was used in [17]. In this section we show that problems can be readily solved using the bond graph component method. It is based on the Euler-Bernoulli beam model of Sect. 10.4. The exposition is based on [18].

10.6.2 Bond Graph Model of the Meter

A crucial part of the CMF is the tube where the interaction with the fluid takes place. We divide the tube into a number of finite tube elements. The elastic behaviour of the tube is described using Bernoulli-Euler beam theory. Following [14], fluid flowing through the tube is treated as a string of fluid particles. Based on such a modelling, we develop a bond graph model of a tube finite element. This leads to a relatively simple model. It is also possible, however, to develop a more detailed model of fluid motion and to include it in the overall model. The model allows analysis of the effects of pulsations in the flow [19], too. On the other hand, multiphase flow requires different models of fluid flow. Companies that produce CMFs, however, usually limit the allowable content of the gas phase [13].

We start by analysing a tube element of Fig. 10.31. Following [18], the differential equation of transverse displacement of a tube element with the flow reads

$$(m_t + m_f)\frac{\partial^2 w}{\partial t^2} + 2m_f v_f \frac{\partial^2 w}{\partial t \partial x} + m_f v_f^2 \frac{\partial^2 w}{\partial x^2} +$$
$$m_f \frac{\partial v_f}{\partial t}\frac{\partial w}{\partial x} + EI\frac{\partial^4 w}{\partial x^4} = 0 \qquad (10.66)$$

m_t and m_f are masses of the tube element and its contained fluid, respectively, per unit length. E is the modulus of elasticity of the tube, I is the second moment of the tube section, v_f is mean fluid velocity, and x denotes the coordinate along the tube axis. Fluid velocity is assumed to be constant along the tube, but can change with time.

The boundary condition at the tube ends read

$$F_1 = \left. EI\frac{\partial^3 w}{\partial x^3}(0) \right\}$$

$$M_1 = \left. -EI\frac{\partial^2 w}{\partial x^2}(0) \right\}$$

(10.67)

and

$$F_2 = \left. EI\frac{\partial^3 w}{\partial x^3}(L) \right\}$$

$$M_2 = \left. -EI\frac{\partial^2 w}{\partial x^2}(L) \right\}$$

(10.68)

Here, F_1 and M_1 are a force and moment acting at left element end, respectively. Similarly, in accordance with the convention used in Sect. 10.4, the reaction force $-F_2$ and the moment $-M_2$ act at the right end.

To discretisize Eq. (10.66), we represent transverse displacement by Eq. (10.14) using the shape functions of Eq. (10.15) and (10.16) and apply Galerkin's method [2]. Applying standard techniques of integration by parts, we arrive at

$$\mathbf{M}\frac{d^2\mathbf{q}}{dt^2} + 2\mathbf{R}v_f\frac{d\mathbf{q}}{dt} + (\mathbf{K} + \mathbf{R}\frac{\partial v_f}{\partial t} + \mathbf{C}v_f^2)\mathbf{q} = \mathbf{F}$$

(10.69)

where $\mathbf{q}$ is a vector of the element end section displacements and rotations given by Eqs. (10.10)–(10.12), and the vector of the boundary forces and moments $\mathbf{F} = (F_1\ M_1 - F_2 - M_2)^T$. $\mathbf{M}$ is the consistent mass matrix given by Eqs. (10.22) and (10.23), with the mass per unit length of the element replaced by $m_t + m_f$. $\mathbf{K}$ is the stiffness matrix of Eqs. (10.31) and (10.32). There are, however, additional matrices in this equation. The matrix $\mathbf{R}$ takes into account the Coriolis forces and is given by

$$\mathbf{R} = m_f \int_0^1 \mathbf{S}^T \frac{d\mathbf{S}}{d\xi} d\xi$$

(10.70)

After evaluation using the shape functions of Eqs. (10.15) and (10.16), we find

$$\mathbf{R} = \frac{m_f}{2}\begin{pmatrix} -1 & L/5 & 1 & -L/5 \\ -L/5 & 0 & L/5 & -L^2/30 \\ -1 & -L/5 & 1 & L/5 \\ L/5 & L^2/30 & -L/5 & 0 \end{pmatrix}$$

(10.71)

which also can be written as 2×2 block matrices

$$\mathbf{R} = \begin{pmatrix} \mathbf{R}_{11} & \mathbf{R}_{12} \\ \mathbf{R}_{21} & \mathbf{R}_{22} \end{pmatrix}$$

(10.72)

It is interesting that the same matrix appears in the tangential inertial force, the second term in parenthesis of Eq. (10.69). The other matrix, $\mathbf{C}$, takes into account the normal inertial forces given by the third term in parenthesis of Eq. (10.69). The matrix is defined by

$$C = \frac{m_f}{L} \int_0^1 S^T \frac{d^2S}{d\xi^2} d\xi \tag{10.73}$$

And, after evaluating, reads

$$C = \frac{m_f}{L} \begin{pmatrix} -6/5 & -11L/10 & 6/5 & -L/10 \\ -L/10 & -2L^2/15 & L/10 & L^2/30 \\ 6/5 & L/10 & -6/5 & 11L/10 \\ -L/10 & L^2/30 & L/10 & -2L^2/15 \end{pmatrix} \tag{10.74}$$

The matrix also can be written using 2×2 block matrices as

$$C = \begin{pmatrix} C_{11} & C_{12} \\ C_{21} & C_{22} \end{pmatrix} \tag{10.75}$$

Matrix R is not strictly skew-symmetric. The nonzero elements -1 and 1 on the main diagonal correspond to transverse displacements of two connected elements. Thus, if the matrices corresponding to all connected elements are assembled, these terms cancel out. The assembled matrix of all tube elements is strictly skew-symmetric. Hence, the overall effect of the Coriolis forces is energetically neutral, i.e. there is neither dissipation nor generation of mechanical power. Similar conclusions can be drawn regarding symmetry of matrix C. Thus, normal inertial forces of fluid particles, in effect, reduce the overall stiffness of the tube-fluid system and, hence, the natural frequency of tube vibrations. We thus expect that the frequency of CMF vibrations decrease with fluid flow velocity. This effect is analysed also in [14, 17].

We transform Eq. (10.69) to a form that can be directly represented by bond graph components in a similar way as we did in Sect. 9.4. We define flows and efforts at component ports by Eq. (10.6) – (10.9). The momentum of the tube element filled with the fluid is given by Eq. (10.26). This way, the governing equation of transversal vibration of the tube with fluid flow now reads

$$\frac{d\mathbf{p}}{dt} + 2v_f R \begin{pmatrix} f_1 \\ f_2 \end{pmatrix} + (K + R\frac{\partial v_f}{\partial t} + Cv_f^2)\begin{pmatrix} q_1 \\ q_2 \end{pmatrix} = \begin{pmatrix} e_1 \\ -e_2 \end{pmatrix} \tag{10.76}$$

We further modify this equation by adding Rayleigh damping (Sect. 10.4), i.e.

$$\frac{d\mathbf{p}}{dt} + (2v_f R + \alpha M + \beta K)\begin{pmatrix} f_1 \\ f_2 \end{pmatrix} + (K + R\frac{\partial v_f}{\partial t} + Cv_f^2)\begin{pmatrix} q_1 \\ q_2 \end{pmatrix} = \begin{pmatrix} e_1 \\ -e_2 \end{pmatrix} \tag{10.77}$$

where α and β are suitable constants.

Thus, we use the same component BFE of Sec. 9.4, but with an added control port to transfer information on fluid velocity v_f and its time rate of change $\partial v_f/\partial t$. The corresponding model is shown in Fig. 10.32.

The constitutive relations of the inertial element are the same as those used previously, but the mass of the element includes the fluid mass, as already stated. The constitutive relations of the resistive element ports have to include the Coriolis term $2v_f Rf$, where the matrix R is given by Eqs. (10.71) and (10.72), and the Rayleigh damping terms. Finally, the constitutive relations of the capacitive element ports are modified by addition of tangential and normal inertial force terms. The coefficient matrix of the first is the same matrix as in the Coriolis term, and matrix **C** of the other term is given by Eqs. (10.74) and (10.75).

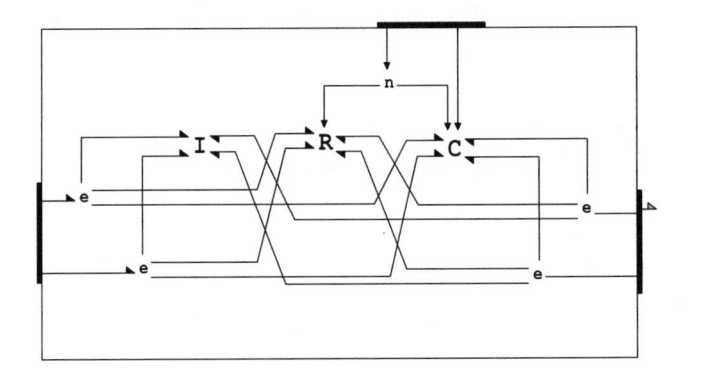

Fig. 10.32. Model of BFE component with fluid flow

After the tube element with fluid flow is developed, we start model development by defining a project called Coriolis Mass Flowmeter. The system-level model is shown in Fig. 10.33. It mimics the structure of the Coriolis flowmeter with two straight tubes in Fig. 10.30. The model contains two Measuring Tube components clamped to the Wall at both ends. This means that velocities (transversal and angular) at the tube ends are zero, a condition that is enforced by suitable source-flow components in the Wall. In the middle of the tube components are power ports to which the tubes driver is attached. The Driver is connected to the tubes by an effort node f component consisting of two flow junctions. These describe relative velocities at the tube-driver connections. The force developed by the driver is applied to both tubes simultaneously. The Driver component model consists of two source efforts. They generate a transversal force and zero torque, respectively, applied to the tubes.

Fig. 10.33 shows many signals flowing through the system. Inflow defines the velocity of the fluid flowing through the meter and its time rate of change. This component consists of two input components, one for the fluid velocity and the other for its time rate of change. They are connected to the same document port internally; thus, the output is a vector signal consisting of fluid velocity and its time rate of change. There are two output ports for the flows that branch to the two tubes. These ports are connected to the flow input ports of the Measuring Tube components. Likewise, at other end of the tubes, there is an Outflow component that calculates the total mass flow rate and the time rate of mass flow rate through the meter. The information generated is fed as separate signals to the display component at top right of the Coriolis Mass Flowmeter window.

Near the ends of the tubes are output ports that retrieve information on the tubes' positions at the sensors. Because the tubes vibrate in opposition, the position signals are subtracted by summator components denoted as diff. The outputs of diff components, hence, simulate the output of two sensors of the Coriolis flowmeter. These signals are further on summed up and subtracted. As shown in [18], summing these signals effectively eliminates the Coriolis effect on the tubes; this gives twice the mean value of displacement between the tubes. Similarly, subtraction effectively extracts the relative displacement of the tubes owing to the Coriolis forces. These signals are used to evaluate meter sensitivity, as described later (Sect. 10.6.3).

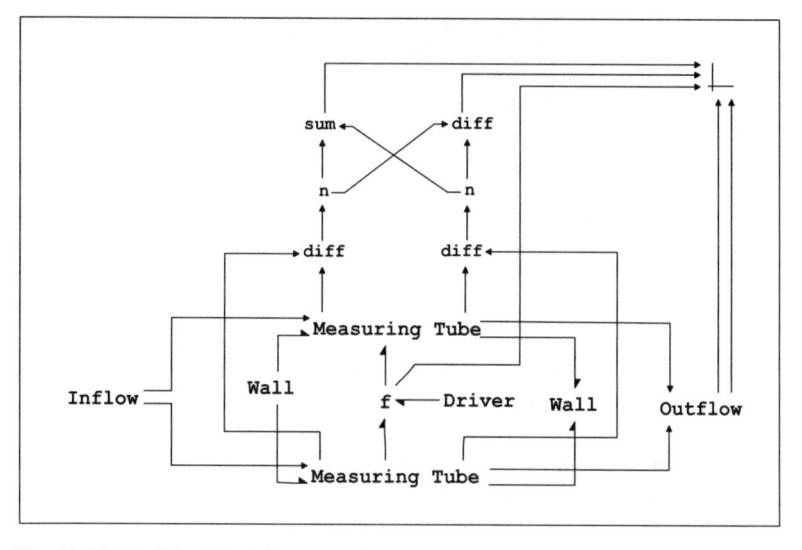

Fig. 10.33. Model of Coriolis Mass Flowmeter

The **Measuring Tube** components are modelled by four **SectBC** (section bond graph) components that represent the tube parts between the left and right tube ends and the sensors, respectively; and between the sensors and the driver. Fig. 10.34 shows the structure of the upper tube component. The lower component differs only by the position of the driver port and the direction of the power flow.

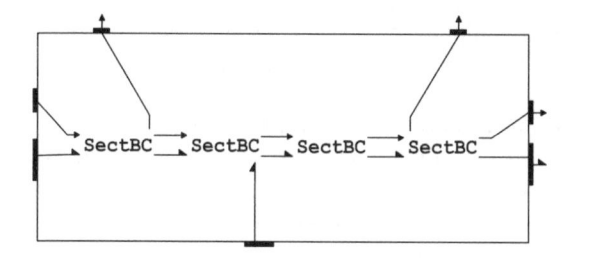

Fig. 10.34. Structure of the Measuring Tube component (upper)

SectBC components consist of five BFE components (Fig. 10.35). Signals of the fluid flow velocity and its time rate of change flows throughout the tube component and branch to individual BFE components. There are, overall, 20 elements per tube. Signals of the tube displacements are picked off the capacitive elements of 5th and 16th elements. Similarly, bonds from the driver junction in Fig. 10.33 are connected to the right effort junctions of the 10th element of each tube.

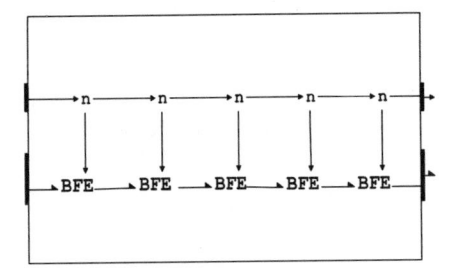

Fig. 10.35. SectBC component consisting of five BFE components

10.6.3 Evaluation of the Meter Sensitivity Factor

In commercial Coriolis mass flowmeters there are quite elaborate techniques for processing sensor signals to evaluate time delays, frequency of vibrations, the meter sensitivity factor, and the current mass-flow rate. We will not attempt to develop a model of signal processing, for the information needed to perform this task generally is not available either from the manufacturer or in published papers. It *is* possible, however, to extract the necessary information using frequency analysis of the signals generated by the system. We will follow the approach of [18], but a similar technique is also reported in [20].

Following [18], the time delay between sensor signals is given by the simple formula

$$\Delta t = \frac{a_{diff}}{\pi a_{sum} f} \qquad (10.78)$$

where a_{sum} and a_{diff} are amplitudes of the sum and difference sensors' signals at frequency f, which corresponds to the first vibration mode. The sensitivity coefficient of the flowmeter is found by dividing the time delay and mass flow rate, and is expressed in s/(kg/s). The sensitivity coefficient evaluated by simulation corresponds to testing the meter with water.

Parameters used in the model are take from [18] and are based on a commercial Coriolis flowmeter (Table 10.2). It should be noted that the length used is less than the overall length of the tubes taken from the drawing of the meter. In the real design, there are two braces joining the tubes near their ends, because of which vibration nodes appear. The exact positions of these braces are not known in advance. They can be used for fine-tuning the first mode of the natural vibration frequency. It is assumed that the tubes are clamped at both ends. Their length is calculated according to an experimental first-mode vibration frequency of 118.7 Hz. A similar approach is used in [17, 20]. Calculation of the vibration frequency is based on the classical formula for a beam with both ends clamped [12], modified by adding the mass of the fluid contained in the tube. The formula reads

$$f = \frac{K_L^2}{2\pi L_t^2} \sqrt{\frac{EI}{\rho_t A_t + \rho_f A_f}}$$
(10.79)

where A_t and I are area and second moment of tube cross section, respectively, A_f is the area of the internal section the tube, and $K_L = 4.730041$ is the coefficient of the first vibration mode. For the tube length given in Table 10.2, the calculated frequency of vibration is 118.483 Hz.

Table 10.2. Parameters of the model

Property	Value
Tube outside diameter Dout	0.0395 m
Tube inside diameter Din	0.0375
Tube length Lt	1.175 m
Modulus of elasticity E	195 GPa
Tube density ρ_t	8000 kg/m^3
Fluid density ρ_f	1000 kg/m$_3$

We further assume that tube displacement sensors are at the middle of the left and right parts of the tubes. The tube finite elements in all four components SecBC of Fig. 10.32 are set to the same length: 0.05875 m. Finally, the damping parameters are set to α = 15 and β = 2.69·10^{-5}. This corresponds to a damping coefficient for the first mode of $\zeta \approx 0.02$.

To evaluate the frequency of the first mode vibration by the model, a half-sine pulse is applied to the tubes. This is defined as

$$F_{driver} = t < Tp\,?\,F_{pulse}\,\sin(\pi * t / Tp):0$$
(10.80)

Such a pulse excites the first mode of the beam [21]. We use the pulse strength value F_{pulse} = 450 N and a duration of Tp = 1/f = 0.00844003 s. Hence, we must modify the constitutive relation of the effort source that generates the driver force in Driver (the other generates the torque, which is zero). Flow velocity was set initially to zero. Thus, there will be no time difference between the signals.

After the model is built, we start the simulation by choosing a simulation interval of 0.1 s, an output interval of 5·10^{-5} s, and error tolerances (both) of 1·10^{-8}. The results for the sum of sensor signals are shown in Fig. 10.36.

Data collected by the display window are used to generate a frequency spectrum of the signal. Owing to practical considerations, a relatively short simulation interval of 0.1 s was used. We thus expect resolution of the spectrum of 10 Hz only. On other side, after the pulse finishes, the signal behaves as a lightly damped vibration. Hence, we use as a time-window a portion of the response between the first and last zero crossings (corresponding to displacements going from positive to negative values). An integral number of full periods is included in the window. The resulting diagram, expanded to 0–500 Hz, is shown in Fig. 10.37.

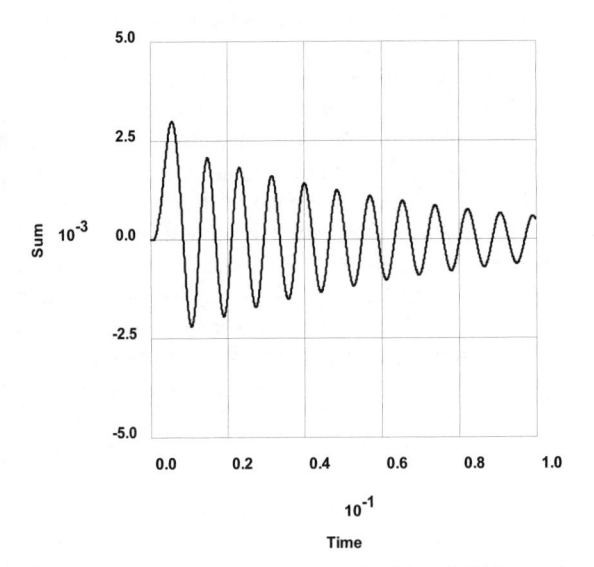

Fig. 10.36. Sum of signals response to the driver half-sine pulse (no flow)

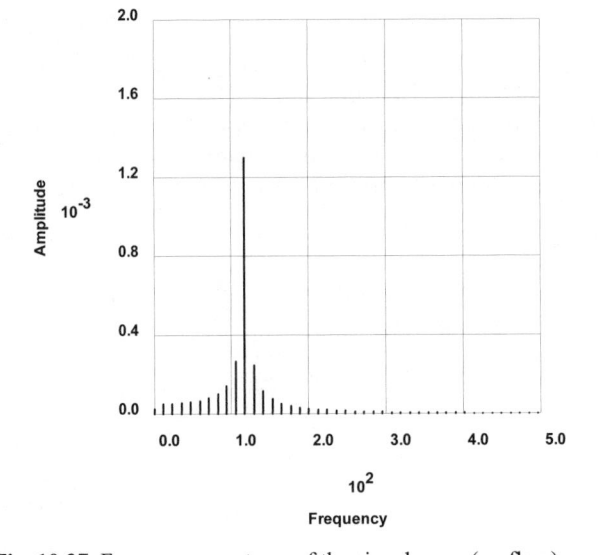

Fig. 10.37. Frequency spectrum of the signals sum (no flow)

From the diagram we find a of frequency f = 118.343 Hz and an amplitude a_s = 0.00130075 m. This is close to the analytical value obtained previously. Next we replace the pulse and drive the meter at this frequency by sinusoidal force F = $F_0 \sin(2\pi ft)$, where the amplitude F_0 = 40 N was selected low enough to limit the strains in the tube to within the elastic region. The resulting model consists of 906 equations, and its partial derivative matrix (Jacobian) consists of 4,058 nonzero elements.

We run several simulations corresponding to mass flow rates of 0, 5, 10, 15, and 20 kg/s. The simulation interval is selected to be 1 s to insure that the initial transient dies out and the vibration amplitude settles to a steady value. The output interval and error tolerances are selected as in the previous run. Figs. 10.38 and 10.39 show time histories of the sum and the difference of the sensor signals, respectively, at a mass flow rate of 10 kg/s. The complete simulation lasted 622 s. During this time 21,325 steps were made, 45,271 functions were evaluated, and 23,950 Jacobian matrix were LU decomposed. From these figures we see that signal amplitudes settle down after about 0.3 s. To be sure, we use a part of the diagram from 0.4 to 1 s to generate the frequency spectrums. We select minimum and maximum values of the time-window based on zero crossing values, as in the former case.

The frequency spectrums of the sum and difference of the signals are given in Figs. 10.40 and 10.41, expanded again to the 0 – 500 Hz range. Time delays and, hence, the sensitivity coefficient of the meter are calculated using Eq. (10.78). Thus, for the mass flow rate Qm = 10 kg/s, we find the values of the frequency and the amplitudes from frequency plots:

f = 118.3331925 Hz
a_{sum} = 0.004042818478 m
a_{diff} = 1.901149327·10^{-5} m

From Eq. (10.78) we find the delay Δt = 12.6496 μs and, hence, the sensitivity coefficient K_{meter} = 1.26496 μs/(kg/s). The complete results are given in Table 10.3. We see that the sensitivity coefficient is fairly constant over the range of mass flow rates 0 – 20 kg/s and has a value of 1.265 μs/(kg/s). This value agrees closely with the reported mean value of 1.20 μs/(kg/s) coming from water tests [18].

Table 10.3. Results of simulation of Coriolis Mass Flowmeter

Mass flow rate kg/s	Time delay μs	Sensitivity coefficient μs/(kg/s)
0	0	-
5	6.32431	1.26486
10	12.6496	1.26496
15	18.9767	1.26511
20	25.3067	1.26534

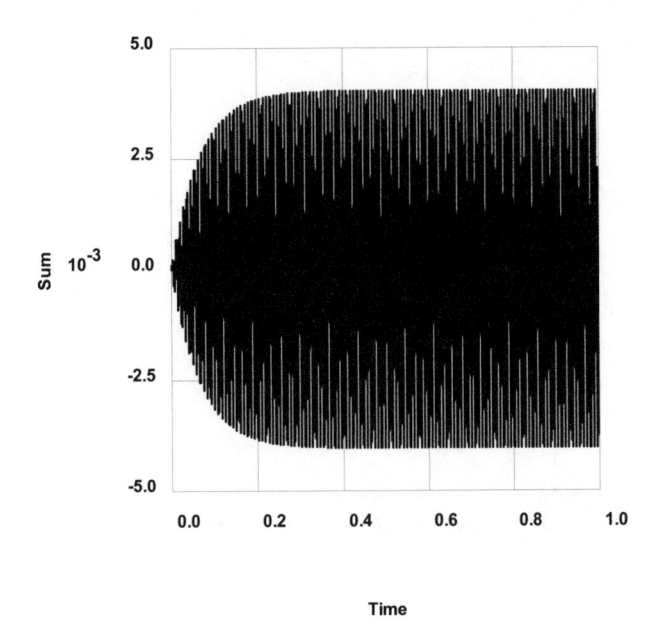

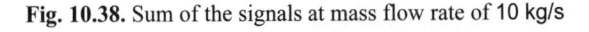

Fig. 10.38. Sum of the signals at mass flow rate of 10 kg/s

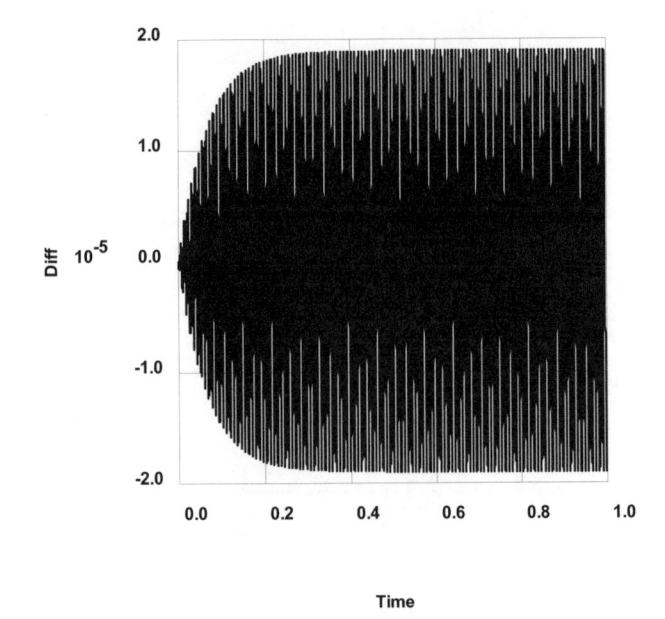

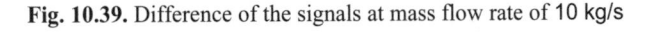

Fig. 10.39. Difference of the signals at mass flow rate of 10 kg/s

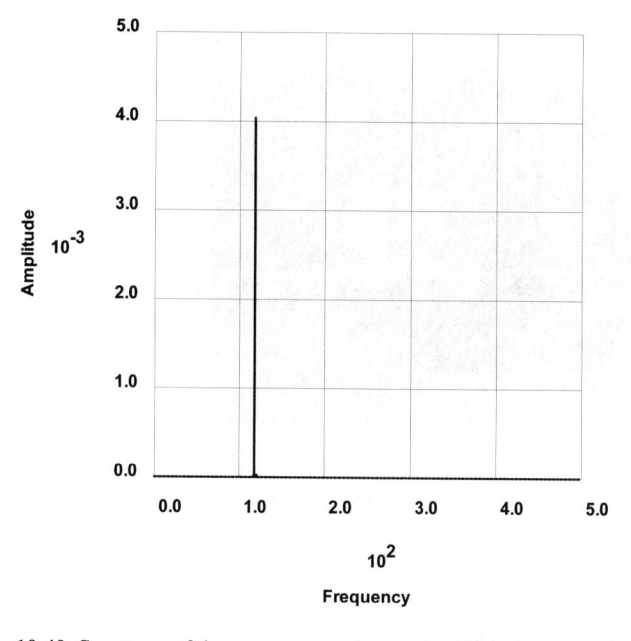

Fig. 10.40. Spectrum of the sum at mass flow rate of 10 kg/s

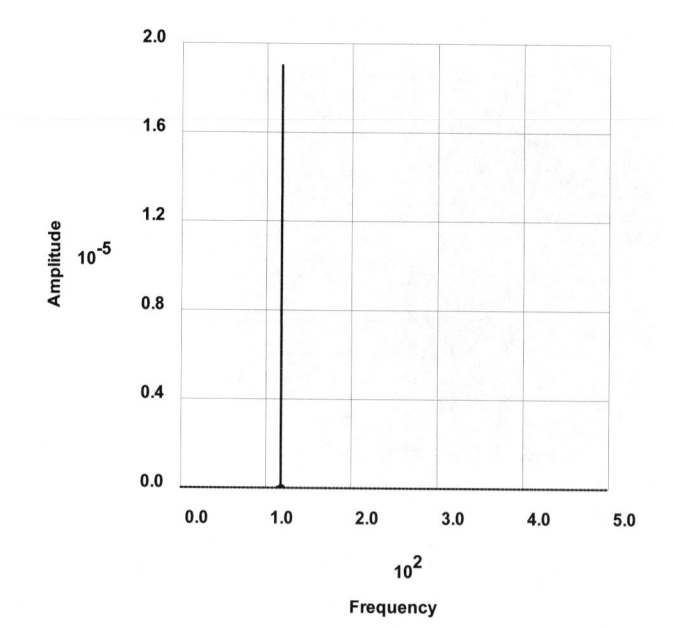

Fig. 10.41. Spectrum of the difference at mass flow rate of 10 kg/s

References

1. AR Mitchell and DF Griffiths (1980) The Finite Difference Methods in Partial Differential equations. John Wiley and Sons, Chichester
2. RD Cook, DS Malkus and ME Plesha (1989) Concept and Applications of Finite Element Analysis, 3rd edn. John Wiley and Sons, New York
3. Fahrenthold EP and Wargo JD (1994) Lagrangian Bond Graphs for Solid Continuum Dynamics Modeling. ASME J. of Dynamic Systems, Measurement and Control 116: 178 – 192
4. EP Fahrenthold and M Venkataraman (1996) Eulerian Bond Graphs for Fluid Continuum Dynamics Modeling. ASME J. of Dynamic Systems, Measurement and Control 118: 48 – 57
5. Dean C Karnopp, Donal L Margolis and Ronald C Rosenberg (2000) System Dynamics: Modeling and Simulation of Mechatronic Systems, 3rd edn. John Wiley, New York
6. WE Schiesser (1991) The Numerical Method of Lines. Academic Press, San Diego
7. J Johansson and U Lundgren (1997) EMC of Telecommunication Lines, A Master Thesis from the Fieldbusters, http://jota.sm.luth.se/~d92-uln/master/Theory/4.html
8. J Keown (2001) OrCAD PSpice and Circuit Analysis, 4th edn. Prentice Hall, Upper Saddle River
9. AA Shabana (1998) Dynamics of Multibody Systems, 2nd edn. Cambridge University Press, Cambridge
10. M Borri, L Trainelli and CL Bottasso (2000) On representation and Parameterizations of Motions. Multibody Systems Dynamics 4: 129-193
11. V Jaram (2001) Evaluation of Bond Graph Based Object Oriented Approach to Determination of Natural Frequencies of Packaging System Elements, MS Thesis, Department of Packaging Science of Rochester Institute of Technology, Rochester, New York
12. SS Rao (1995) Mechanical Vibrations, 3rd edn. Addison-Wesley, Reading
13. P Kesic, V Damic and AM Ljustina (2000) The Coriolis Flowmeter for Measurement of Petroleum and its Products. Nafta, Zagreb, 51: 103-111
14. H Raszillier and F Durst (1991) Coriolis-Effect in Mass Flow metering. Archive of Applied Mechanics, 61: 192-214
15. KO Plache (1979) Coriolis Gyroscopic Flow Meter. Mechanical Engineering, March 1979, 36-41
16. CP Stack, RB Garnett and GE Pawlas (1993) A Finite Element for the Vibration Analysis of a Fluid-Conveying Timoshenko Beam. American Institute of Aeronautics and Astronautics, AIAA-93-1552-CP
17. G Sultan and J Hemp (1989) Modelling of a Coriolis Mass Flowmeter. Journal of Sound and Vibration, 132: 473-489
18. P Kesic, V Damic and AM Ljustina (2000) Modelling of the Coriolis Mass Flowmeter with Straight Parallel Tubes. Nafta, Zagreb, 51: 91-98
19. V Damic and P Kesic (2000) Bond Graph Modelling of Flow Pulsation in the Coriolis Mass Flowmeter with Parallel Straight Tubes. In P Marovic ed. Proc. of the 3rd International Congress of Croatian Society of Mechanics, Dubrovnik, Croatia, 483 – 490
20. NM Keita (2000) Ab Initio Simulation of Coriolis Mass Flowmeter. In AP Pereira ed. Collection of Abstracts with Conference Programme & CD Rom of Proceedings of FLOMEKO 2000, Salvador, Brasil

21. IM Smith and DV Griffiths (1998) Programming the Finite Element Method, 3rd edn. John Wiley and Sons, Chichester

Appendix

Installing BondSim

The book contains the program *BondSim Research Pack* on a separate CD ROM. The program can be run on PC computers with the following minimal requirements:

Processor	Pentium class
Memory	64 MB
Disc space	20 MB
Operating systems	Windows 2000, Windows NT 4.0
	Windows XP, Windows 98, Windows 95

Recommended the screen area setting is 1024 by 768 pixels or higher.

To install the program, put the CD ROM in corresponding drive and close the door. The installation process starts automatically. You simply need to follow the procedure by answering questions, or making suitable selections, in the dialogues. On computers with Windows 2000, you will typically need administrator privilege.

If you are using some other platform than Windows 2000—such as Windows NT 4.0, Windows 95 or Windows 98—you may also need to install Window Installer. The setup program on the CD ROM (setup.exe) first checks if the Windows Installer is already installed on your system. If it is not, or if a more recent version is needed, the setup program will install the Window Installer. The system then may need to be restarted before you can proceed with *BondSim* installation.

At anytime you may uninstall *BondSim* by double-clicking *Add/Remove Programs* icon in *Control Panel*, selecting *BondSimRP*, then clicking the *Remove* button. (The *Control Panel* is accessed from the *Start* button in the left-bottom corner, choosing Settings, and then *Control Panel*.)

Launching and Using BondSim

BondSim can be launched as any other Windows application, e.g. by double-clicking the *BondSimRP* icon on the desktop, or by using *Start*, *Programs*, and then choosing *BondSimRP*. A splash-screen appears first, which after few seconds disappears, and the main *BondSim* program window appears.

The book explains how to use the program. Chapt. 4, for example, describes the main program commands. Modelling and simulation of mechatronic systems using *BondSim* is described in the application part of the book, starting with Chapt. 6.

The first section of these chapters describes the procedure on relatively simple problems in a step-by-step manner. The later sections treat more demanding problems.

BondSim RP bundled with the book is a restricted version of the full program. It is designed for use with the book. All projects developed in the book are stored in the project library, from which they can be copied to the program workspace by using command *Get From* in the *Project* menu. The reader can develop his or her own models, as well. These are, however, more restricted. In particular, they are limited to two-level structures: The models can be developed as a system level models that may consist of simple component only, i.e. those described exclusively in terms of elementary bond graph components. We hope that these will allow the reader to follow explanations given in the book.

More advanced operations are not included with this version of the program, such as multilevel modelling. There is no online Help; neither is the collaboration support described in the book.

Interested readers can order a full version of the program from the first author, at the address given bellow. We would also be very happy for any feedback, criticism, advice or support for further work. Suggestions on other mechatronic or other relevant problems are truly welcome.

Contact Addresses

The authors can be reached at

vdamic@vdu.hr
haedickemontgome@compuserve.de

The readers are welcome to visit our web page
http://www.vdu.hr/~vdamic

Index

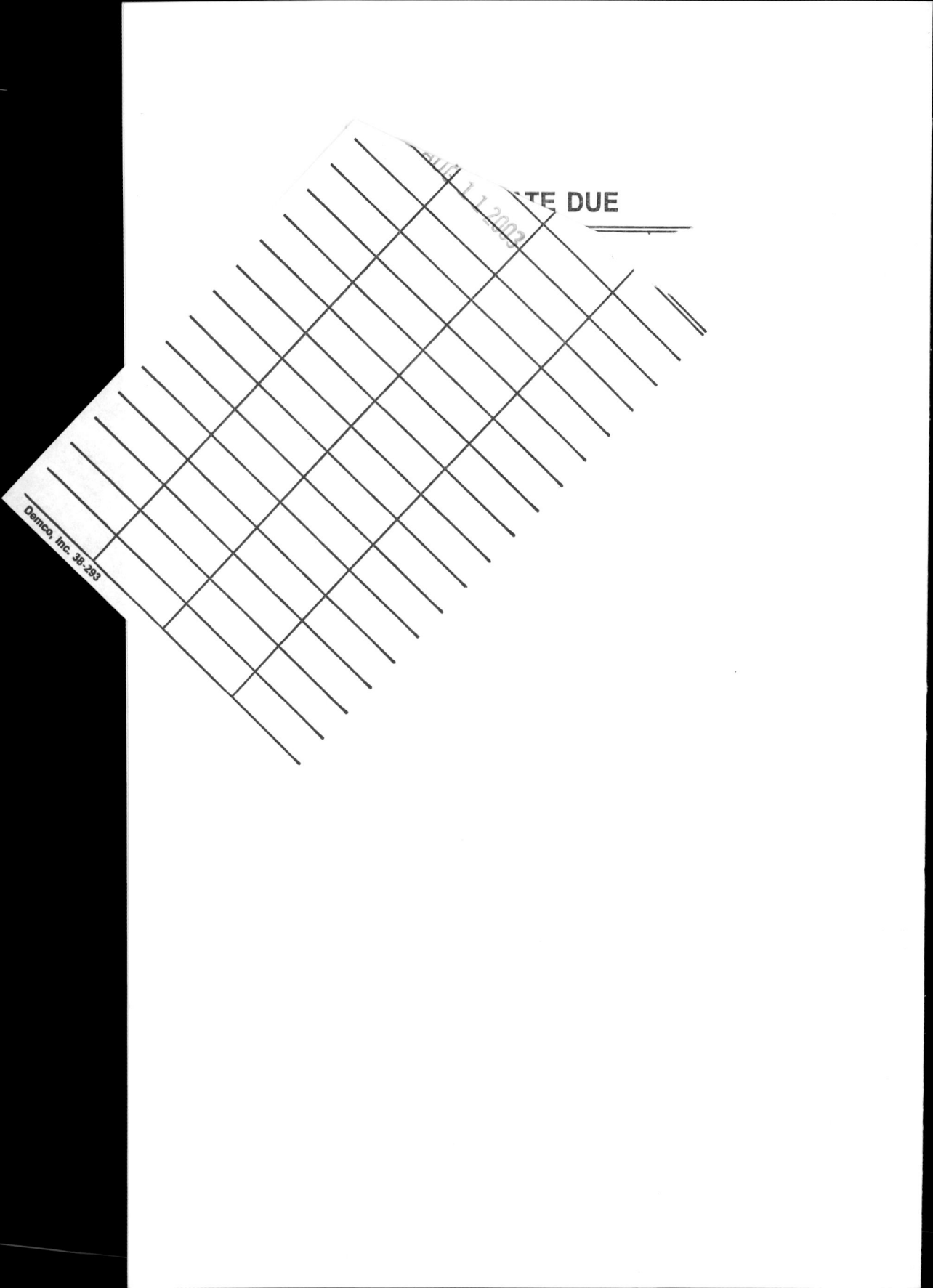